ECOSYSTEM HEALTH

ECOSYSTEM HEALTH

Edited by:

David Rapport, PhD, FLS
Professor, Faculty of Environmental Sciences, University of Guelph;
Professor, Department of Pharmacology and Toxicology, Faculty of
Medicine, University of Western Ontario, London, Ontario, Canada

Robert Costanza, PhD
Professor, Department of Biological Sciences; Director, University of
Maryland Institute for Ecological Economics, Center for Environmental
Science, University of Maryland, Solomons, Maryland

Paul R. Epstein, MD, MPH
Associate Director, Center for Health and the Global Environment, Harvard
Medical School, Boston, Massachusetts

Connie Gaudet, PhD
Environment Canada, Ottawa, Ontario, Canada

Richard Levins, PhD
John Rock Professor of Population Science, Harvard School of Public
Health, Boston, Massachusetts

**Blackwell
Science**

© 1998 by Blackwell Science, Inc.
Editorial Offices:
350 Main Street, Malden, MA 02148-5018, USA
Osney Mead, Oxford OX2 0EL, England
25 John Street, London WC1N 2BL, England
23 Ainslie Place, Edinburgh EH3 6AJ, Scotland
54 University Street, Carlton, Victoria 3053,
    Australia

Other Editorial Offices:
Blackwell Wissenschafts-Verlag GmbH
Kurfürstendamm 57
10707 Berlin, Germany

Blackwell Science KK
MG Kodenmacho Building
7–10 Kodenmacho Nihombashi
Chuo-ku, Tokyo 104, Japan

DISTRIBUTORS
USA
    Blackwell Science, Inc.
    Commerce Place
    350 Main Street
    Malden, Massachusetts 02148
    (Telephone orders: 800-215-1000 or 781-388-
        8250; fax orders: 781-388-8270)

Canada
    Login Brothers Book Company
    324 Saulteaux Crescent
    Winnipeg, Manitoba R3J 3T2
    Canada
    (Telephone orders: 204-224-4068)

Australia
    Blackwell Science Pty, Ltd.
    54 University Street
    Carlton, Victoria 3053
    (Telephone orders: 03-9347-0300;
        fax orders: 03-9349-3016)

Outside North America and Australia
    Blackwell Science, Ltd.
    c/o Marston Book Services, Ltd.
    P.O. Box 269
    Abingdon
    Oxon OX14 4YN
    England
    (Telephone orders: 44-01235-465500;
        fax orders: 44-01235-465555)

First published 1998

Acquisitions: Nancy Hill-Whilton
Production: Kevin Sullivan
Manufacturing: Lisa Flanagan
Typeset by Best-set Typesetter Ltd., Hong Kong
Printed and bound by Edwards Brothers
Printed in the United States of America
98 99 00 01 5 4 3 2 1

The Blackwell Science logo is a trade mark of
Blackwell Science Ltd, registered at the United
Kingdom Trade Marks Registry

Library of Congress Cataloging-in-Publication
Data
Ecosystem health
    David Rapport . . . [et al.].
        p.   cm.
    Includes bibliographical references and index.
    ISBN 0-632-04368-7
    1. Ecological assessment (Biology)
2. Environmental health.   3. Ecosystem
management.   I. Rapport, David.
QH541.15.E22E36   1998
577—dc21   97-27825
                                         CIP

CONTENTS

PART III. ECOSYSTEM HEALTH AND SUSTAINABILITY

PART IV. CASE STUDIES

CONTRIBUTING AUTHORS

Juan Almendares, MD
Tegucigalpa, Honduras

Pamela K. Anderson, PhD
Virology Unit, CIAT
Cali, Colombia

David Barkin, PhD
Professor de Economia
Universidad Autonoma Metropolitana/Unidad Xochimilco
Xochimilco, Mexico

John Cairns, Jr., PhD
University Distinguished Professor of Environmental Biology Emeritus
Virginia Polytechnic Institute and State University
Blacksburg, Virginia

René Capote, PhD
Centro Nacional de Biodiversidad
Instituto de Ecología y Sistemática
Ministerio de Ciencia, Technología y Medio Ambiente (CITMA)
Boyeros, Ciudad Havana, Cuba

Brian F. Cumming, PhD
Assistant Professor, Department of Biology
Paleoecological Environmental Assessment and Research Laboratory
    (PEARL)
Queen's University
Kingston, Ontario, Canada

Geoff Dates, PhD
Science Coordinator
River Watch Network
Montpelier, Vermont

Jack Greer, PhD
Sea Grant College Program
College Park, Maryland

Yrjö Haila, PhD
Professor of Environmental Policy
Department of Regional Studies and Environmental Policy
University of Tampere
Tampere, Finland

Mikael Hildén, PhD
Division Manager, Environmental Policy Instruments Division
Finnish Environment Institute
Helsinki, Finland

Patricia A. Lane, PhD
Professor of Biology
Dalhousie University
Halifax, Nova Scotia, Canada;
Consultant, Department of Population and International Health
Harvard School of Public Health
Harvard University School of Medicine
Boston, Massachusetts

Cynthia Lopez, MPIA, PhD
Assistant Professor, Department of Family and Community Medicine
University of New Mexico School of Medicine
Albuquerque, New Mexico;
Visiting Scientist, Department of Population and International Health
Harvard University School of Public Health
Boston, Massachusetts

Michael Mageau, AbD
University of Maryland Institute for Ecological Economics
Center for Environmental Science
University of Maryland
Solomons, Maryland

Leda Menéndez, PhD
Centro Nacional de Biodiversidad
Instituto de Ecología y Sistemática
Ministerio de Ciencia, Technología y Medio Ambiente (CITMA)
Boyeros, Ciudad Havana, Cuba

Bryan Norton, PhD
School of Public Policy
Georgia Institute of Technology
Atlanta, Georgia

Bernard C. Patten, PhD
Institute of Ecology
University of Georgia
Athens, Georgia

Manuel Sierra, MD
Harvard School of Public Health
Boston, Massachusetts

John P. Smol, PhD, FRSC
Professor of Biology
Paleoecological Environmental Assessment and Research Laboratory
   (PEARL)
Queen's University
Kingston, Ontario, Canada

Deysi Vilamajó, PhD
Centro Nacional de Biodiversidad
Instituto de Ecología y Sistemática
Ministerio de Ciencia, Technología y Medio Ambiente (CITMA)
Boyeros, Ciudad Havana, Cuba

Walter G. Whitford, PhD
Senior Research Ecologist (ST)
National Exposure Laboratory
Environmental Sciences Division
U.S. Environmental Protection Agency
Las Vegas, Nevada;
Collaborating Scientist
USDA-ARS Jornada Experimental Range
Las Cruces, New Mexico

PREFACE

Ecosystem health is an integrative field exploring the interrelations between human activity, social organization, natural systems, and human health. Its focus is the human condition at all levels, from the community, to the regional, to the global. The interface of disciplines involved in ecosystem health forms a rich "ecotone" in which one might expect the diversity of ideas and productivity to be unusually high. This rapidly evolving field more than meets that expectation.

Ecosystem health embraces a large territory. Its concerns are the functions of the social, ecological, health, political, economic, and legal dimensions of systems on local and regional scales. The environmental problems at local, regional, and global levels are reflections of a global ecosystem distress syndrome. A few decades ago, one might have said that this syndrome posed potential threats to the future of humankind. Now, sadly, many of the potential threats have been realized in the failure of ecosystems to sustain "Nature's services" and in the human health consequences of ecological change.

How is humankind adapting and coping with the many environmental issues that will command center stage as we move into the twenty-first century? Approaches on an issue-by-issue basis abound, but despite some local successes (e.g., partial rehabilitation of some degraded bays and harbors in the Great Lakes, improvements in air quality with respect to particulates in certain urban centers), the global environment continues to deteriorate, registering unprecedented losses in biodiversity, placing whole ecosystems on the "endangered list," and posing a host of human health risks, manifest in the resurgence and spread of old epidemics and the emergence of new ones. These are but a few of the immense problems that will continue to challenge us in the future.

Reviewing recent environmental history, one might say that we have done well with tactics and technical solutions to the most pressing and highly focused local issues, but poorly in devising strategies for dealing with the host of interrelated environmental issues. We have to deal with the paradox that while our insights and methods for analyzing the parts of the complex web of relations have become increasingly sophisticated and our tools for measurement incredibly precise, our conceptual framework for integrating the growing mass of information is still lagging behind. This has occurred because of both institutional barriers and intellectual traditions. Therefore, as we confront the substantive issues of ecosystems as wholes we also have to explore the methodological problems of this emergent field and examine ourselves as participant-observers, questioning our received ways of doing things and the obstacles presented by the way intellectual life is structured.

In this field, the result of a collaboration that spans the natural, health, and social sciences, a holistic approach to environmental management is emerging—one that recognizes the intimate linkages between human activity, social organization, health, and the natural environment. Ecosystem health is about preserving and enhancing the capacity of regional environments to be resilient and maintain inherent productivity and a degree of organization that preserves Nature's services. It is not about maintaining or "fixing" the status quo. One must acknowledge the accelerated rates of change inherent in the earth's ecosystems. The trick, as Aldo Leopold stated long ago, is how to humanely colonize the earth without rendering it dysfunctional.

In *Ecosystem Health: Principles and Practice*, some of the principles and case studies that strive to accomplish this are set forth. This book is not about failings but rather the opportunities for our future, discovered by reexamining the relationship between our species and the rest of nature and our relations with each other around this enterprise on the broadest scale, and applying this perspective to our dealings with the environment on all scales. It is about the rationale for management as well as the methods for a more holistic approach to environmental management.

This work was supported by a large group of institutions and agencies, too numerous to mention in detail. However, we are particularly grateful to the Tri-Council Eco-Research Chair Program in Ecosystem Health at the University of Guelph for its support in enabling the development of the book. The Tri-Council comprises the Medical Research Council, the Social Sciences and Humanities Research Council, and the Natural Sciences and Engineering Research Council of Canada. We would also like to acknowledge the support of their partner agencies, including Statistics Canada, the Canadian Forest Service, the Ontario Ministry of Agriculture, Food and Rural Affairs, and Environment Canada.

We also thank Catherine Jefferson of Environment Canada for volunteering her time to assist in the editing of the text, and Ellen Woodley of the University of Guelph for her assistance in the final stages of manuscript preparation.

PART I

Ecosystem Health:
An Integrative Science

"States shall cooperate in a spirit of global partnership to conserve, protect and restore the health and integrity of the Earth's ecosystem. In view of the different contributions to global environmental degradation, States have common but differentiated responsibilities . . ."

Rio Declaration on Environment and Development, 1992, Principle 7

If the Earth and its inhabitants are to have a viable future, we urgently need to reexamine the consequences of present patterns of human activity, with a view toward seeking those that are compatible with sustaining the health and integrity of the world's ecosystems. With a greater awareness of how human activity affects the environment and how human health depends on ecosystem health comes opportunities for environmental management in a new, more integrated mode.

In the first four chapters of this book, we will examine ecosystem health. This study brings together socioeconomic, biophysical, and human health dimensions of environmental change. It draws upon our long experience with the health sciences related to the recognition of ills, diagnostic protocols, and health promotion.

In chapter 1, the phenomena of environmental deterioration at local, regional, and global scales are reviewed. Mounting evidence for deterioration indicates the need for more effective means to cope with environmental change and the limitations of discipline-based approaches to environmental management. These limitations lead us to consider the prospect of a more integrative approach.

In chapter 2, we explore the concept of health as a metaphor. Its advantage lies primarily in facilitating communication with the non-scientist about environmental change in terms of health descriptors—for example, by using concepts such as signs, syndromes, diagnostics, and dysfunction. Use of such a metaphor also stimulates questions about the relationship of human health to ecosystem health. Ecosystem health is more than a metaphor, however. Thus, this chapter also introduces the concept of ecosystem health as an emerging integrative science with novel methods for evaluating environmental change.

Chapter 3 examines the many dimensions of ecosystem health as a transdisciplinary science. Chapter 4 responds to the critics who disparage the idea of ecosystem health. Many critiques rest on the notion of ecosystem as organism—a concept we recognize as a "red herring." Nevertheless, some critics claim that unless one accepts the notion of ecosystem as organism, the concept of ecosystem health itself has little validity. This chapter emphasizes the need to move beyond such a limited view and to adopt the view of ecosystem health as an integrative science that is emerging at the interface of societal values and scientific understanding. Finally, in answering the critics, important distinctions between the overlapping concepts of ecological integrity, sustainable development, and ecosystem health are made. This chapter concludes that ecosystem health is a template for binding together the social, natural, and health science dimensions of our changing environment.

CHAPTER 1

Need for a New Paradigm

David Rapport

THE present environmental crisis is on a scale and magnitude of daunting proportions. Human activities have resulted in adverse environmental changes at scales ranging from local levels to the biosphere. Ozone depletion, acidification of lakes, rivers, and forests, desertification of arid drylands, eutrophication of marine coastal systems, and bioaccumulation of toxic substances are all examples of how stress affects the Earth's ecosystems (Tolba et al 1992; Arrow et al 1995). One should not underestimate the gravity of the present situation. The World Scientists' Warning to Humanity, a document that speaks of the clear and present dangers from this source, was signed by over 100 Nobel laureates and more than 1600 leading scientists, many of them members of the U.S. National Academy of Sciences and its counterpart organizations in other countries (Union of Concerned Citizens 1992).

When *Time* magazine featured "The State of the Planet" as its cover story, it chose the subtitle of "gloomy forecast," emphasizing that continued degradation of the planet's ecosystems portends dire consequences for human futures. The evidence of ecosystem breakdown, even in the most remote areas (Linden 1995) with its attendant social, economic, and human health costs, provides grounds for pessimistic concern.

At the same time, hope remains. Public awareness of how ecosystem breakdown may diminish economic opportunity and pose risks to human health is prompting concerted efforts in areas of environmental protection and conservation. Efforts to date have been fragmented, however. Economic policy seldom takes environmental consequences into account. Likewise, health issues are treated independently from ecological concerns. What is needed are integrated approaches to environmental management.

One such approach is to treat the planet as patient (Somerville 1995), differentiating between a well-functioning ecosystem and a dysfunctional one. Focusing on the ecosystem as patient goes far beyond analyzing the system for its own sake. The very concept of "health" is a value-driven, mission-oriented notion. The integration of the ecosystem and health science brings to the fore both analysis and values. Analysis enables one to gain an understanding of the complex behavior of ecosystems under various stress pressures. Societal values determine what one does with this new knowledge. A coupling of analysis with societal values permits an assessment of ecosystem health, a diagnosis of probable consequences of current behavior, and options for changing course. Health becomes a goal, and the concept begs for the indicators and techniques by which to achieve that goal.

PRESSURES ON THE ENVIRONMENT

Rachel Carson, the celebrated author of *Silent Spring* (1962), first sounded the alarm as to the potentially devastating effects of synthetic chemicals—particularly herbicides and pesticides—on all life systems. Today, case studies abound on the myriad impacts that these substances have on life systems, ranging from impacts on single organisms to effects on populations and entire ecosystems (Colburn 1996; Woodwell 1994). While the use of a handful of highly toxic substances has been curtailed or banned in some regions of the world, the proliferation of species of synthetic chemicals has steadily increased. Worldwide industry now discharges some 70,000 species of chemicals, which go largely untested for their environmental effects (Myers 1991). In the United States alone, approximately 1.2 million tons of chemicals are emitted annually. Continued reliance on synthetic materials such as plastics, for example, entails the release of highly toxic polychlorinated biphenyls (PCBs) to the environment.

The cumulative effects of the release of synthetic chemicals (including toxic substances) from human activities have resulted in serious deleterious impacts on our life systems (Woodwell 1994; Colburn 1996). These harmful effects include deformities in living organisms, reproductive failures, local extinction of species (Minns et al 1990) and, indeed, destruction of entire landscapes (Gunn 1995). The risks of global atmospheric changes to human health are considerable (McMichael and Martens 1995).

While pollution of the environment continues to represent a major concern, environmental protection and conservation involves far more than restrictions on the use of synthetic chemicals. The following sections describe some other significant pressures on the environment.

Overharvesting

Overharvesting has had severe impacts on both terrestrial and aquatic eco-systems. In the former case, overgrazing by livestock has resulted in the trans-formation of large areas of semi-arid grasslands into virtual deserts (Whitford 1995); overharvesting of forests has depleted many of the world's forested areas and is a primary cause of extinction of species in tropical regions. Over-harvesting of marine and freshwater fisheries has significantly depleted fish stocks and dramatically reduced yields in some regions. In the Yellow Sea ecosystems, for example, overharvesting resulted in declines in the demersal fish stocks from 180,000 tonnes to less than 10,000 tonnes (Sherman and Busch 1995). In the Laurentian Great Lakes, overharvesting was among the major stresses that led to the extinction of high-valued commercial fish stocks (Regier and Hartmann 1973; Regier and Baskerville 1986).

Physical Restructuring

Physical restructuring is always purposeful. One type, however, involves the outright obliteration of one kind of system in favor of another—for example, from forest to field, or from field to urban. Another type involves restructur-ing existing systems, though the interventions often lead to at least partial dysfunction and degradation.

The transformation of the central Rio Grande Valley, from a natural river system with riparian vegetation characterized by impressive cottonwood "gallery forests" to the largest intensively engineered floodplain landscape in North America, is a prime example of the first type of physical restructuring. The river channel, now confined between levees, runs through the center of the valley. Farm fields have been laser-leveled for maximum efficiency of flood irrigation. In turn, flood irrigation water is delivered via concrete-lined canals to laterals that feed water to fields. Excess water is returned to the river by irrigation return drains, a feature essential to avoid salt accumulation in fields and orchards.

To the engineer, this sort of transformation would hardly be classified as ecosystem degradation. After all, it was an engineered transformation with unquestioned economic benefits. The valley now supports year-round pro-duction of a variety of specialty crops that are among the most valuable in the North American economy, including lettuce, onions, alfalfa, spinach, cabbage, walnuts, pecans, pistachios, and cotton. From an ecosystem health perspec-tive, however, the situation appears quite differently. The economic benefits are tempered by the considerable reductions in ecological services and new risks to human health (Whitford et al 1996). For example, the transformation has been linked to reductions in biodiversity, lower water quality, increased circulation of toxic substances, and new human health risks from untreated sewage and toxic chemicals that enter the groundwater and drainage.

The other sort of physical restructuring is intended not to obliterate an ecosystem, but rather to provide additional services (for example, transportation or hydroelectric power). While the change is intended to enhance the net flow of benefits, unforeseen results may seriously disrupt natural system function. The construction of marinas, for example, has led to damage of natural habitats, elimination of valued wetlands, modification of hydrology, and detrimental impacts on fish communities. Extensive transportation corridors and pipelines common in all parts of the world are also a source of stress, particularly in northern ecosystems where migration patterns of caribou have been disrupted. These cases are but a few examples of unplanned "side effects" of physical restructuring.

Introduction of Exotic Species

Whether intended or accidental, introductions of non-native fauna and flora have often led to extinction of native species and degradation of the ecosystem. For example, the purposeful introduction of Leymanns lovegrass (*Eragrostis lehmanniana*) to stem erosion in the semi-arid grasslands of New Mexico and Arizona was all too successful. Not only was erosion reduced, but the hardy, drought-tolerant species also significantly outcompeted native grasses, reducing biodiversity in the area. In the Upper Laurentian Great Lakes, the accidental introduction of the sea lamprey through the Welland Canal had deleterious effects on commercially desirable fish stocks such as lake trout, which were heavily preyed upon by the sea lamprey (Regier and Hartmann 1973).

Modification of Natural Perturbations

Until quite recently, efforts to smooth out the natural fluctuations in ecosystems would hardly be viewed as a source of stress. Rather, actions such as fire suppression in forested and grassland systems, modification of stream flow, and controls over natural pest outbreaks were viewed as positive actions that moderated fluctuations in ecosystems. Today, such actions are seen as major sources of stress, often leading to severe degradation of the systems they target (Holling 1995; Covington and Moore 1994). Attempts to control ecosystems have, in effect, backfired largely because most ecosystems are dependent on perturbations from natural disturbance (Vogl 1980). Counteracting natural fluctuations inevitably leads to a weakening of ecosystem resilience and ultimately results in degradation (Holling 1995).

The collective impact of the various stress pressures on ecosystems, as well as extreme natural events (such as volcanic activity), is to modify, transform, and often degrade ecosystems (Sherman and Busch 1995). In the following section, we review the consequences of single and multiple stresses for the Earth's ecosystems.

ECOSYSTEM PATHOLOGY

As several of the case studies presented in Part 4 of this book illustrate, the cumulative impacts of stress from a variety of human activities have resulted in ecosystem degradation (Hildén and Rapport 1993). In many cases, the synergistic impact of stresses overrides the natural resilience of ecosystems. Figure 1.1 illustrates how the cumulative effects of fire suppression and grazing pressure have acted on the structure and functions of the Ponderosa pine forests in the southwestern United States. This example is but one of many (Tolba et al 1992) showing a significant reduction in the quality and quantity of ecological services of ecosystems under stress relative to those found in an unstressed condition.

As the Earth's ecosystems come under greater pressure from population growth and rising consumption levels in both developing and developed countries, pressures on ecosystems cannot help but increase. Whether this

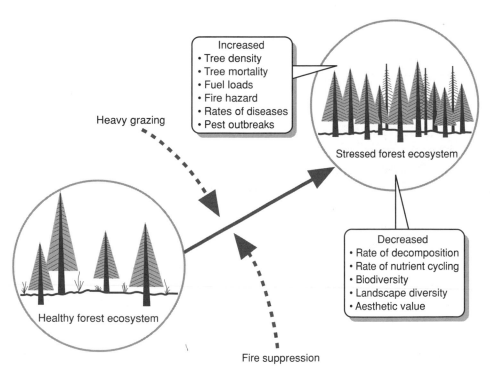

FIGURE 1.1. *The transformation of a healthy presettlement Ponderosa pine forest ecosystem into a stressed ecosystem through heavy livestock grazing and suppression of fires (adapted from Yazvenko and Rapport 1996).*

increase in pressure must necessarily degrade ecosystems to the point that human populations can no longer be sustained is the critical question. In the views of some, this outcome is the most likely in the absence of a radical change of course in environmental management. A recent United Nations Environmental Program assessment of the state of the global environment pointed to numerous examples of global environmental deterioration (UNEP 1995). According to the UNEP assessment, between 150 and 200 species of life become extinct per day—a catastrophic extinction rate fueled by unsustainable production and consumption activity. Other examples of deterioration include freshwater shortages in approximately 80 countries that support 40% of the world's population; unacceptable air quality levels in more than half of the 54 cities monitored by UNEP, including Paris, Madrid, Rio de Janeiro, Bangkok, and Tehran; tropical forests disappearing more than 50% more quickly than a decade ago; and more than 40 key marine fish species overexploited or depleted.

Nevertheless, the UNEP report gives a guardedly hopeful outlook. If the public support and political will existed, many of the pressures on the world's ecosystems could be relieved. Ecosystem health has a vital role to play in this scenario, for its focus is the planet as patient. By carefully monitoring the signs of ecosystem health, and pointing to those systems where early warning indicators suggest risks of deterioration, it should be possible to mobilize public interest in preventing the kinds of ecosystem degradation that have already occurred (Bloom 1995) and in becoming more sensitive to those human activities directly linked to risks of ecosystem pathology. Changing course is the only option if humankind is to survive the next century. This development will require a radical shift, away from the almost exclusive focus on more efficient technologies and toward a consideration of the functionality of the ecosystem.

As as we have previously stated, the issue is one of focusing on the ecosystem, the landscape, and the biosphere as patient. Adopting a stance of "ecosystem as patient" focuses concern on what types of dysfunctional systems exist, how they entered that state, and what can be done to prevent more ecosystem casualties. It is well to recognize the costs after the occurrence, which can include environmental damage resulting in losses in forestry, fisheries, grazing lands, productive agricultural lands, potable drinking water, and the like. There are also the accompanying losses of economic opportunity (such as unemployment resulting from the closure of mills and the loss of fisheries), human health risks (such as increased risks of skin cancers, eye disorders, and possible immune suppression due to increased exposure to ultraviolet radiation, and a host of indirect consequences related to the loss of cereal crop production [McMichael 1993; Parry and Rosenzweig 1993]), and social disruption resulting from the flood of environmental refugees as land becomes less sustaining for rural communities because of environmental degradation (Jacobson 1988). It is far better, however, to design new approaches to environmental management that focus on ecosystem health and to implement ways and means of ensuring that ecosystem health is not further compromised by human activity. In other words, the action must shift to prevention.

SHORTCOMINGS OF PRESENT APPROACHES TO ENVIRONMENTAL MANAGEMENT

There are two reasons to doubt the efficacy of present approaches to environmental management. First, the health of ecosystems continues to deteriorate on local, regional, national, and biosphere scales (UNEP 1995). This trend offers a prima facie case for the insufficiency of existing approaches to identify risks at an early stage and to lessen or reverse damage once it has occurred. While failings here might, of course, be attributable to the politics rather than the present paradigms governing environmental management, obviously these paradigms have failed to convince the body politic that a pressing need for changing course exists.

Second, present approaches to environmental management are largely driven by disciplinary models from economics, engineering, human health, and ecology, with little attention being paid to the pressing need for integration across these fields. This lack of integration has inevitably led to simplistic solutions that are generally based on the belief that all powerful technologies can deal with the problems after they have occurred. Experience now speaks to the contrary (Woodwell 1994; UNEP 1995).

The dominant approaches to environmental management stem from economic, ecological, and engineering-based models. Economic approaches generally rely on cost/benefit analysis, contingency valuation, and similar strategies. In contrast, ecological approaches are based on systems models that often remain confined to the biophysical side of the equation and leave out the human activities and conditions that influence ecosystem function. Engineering-based models tend to focus on technical solutions to pollution abatement and environmental restoration. Each of these approaches suffers from major limitations. Furthermore, their deficiencies tend to exacerbate environmental problems rather than contribute to their solution. In the following sections, we will review the limitations of each of these approaches to environmental assessment and management.

Limitations of Economic Approaches

Nearly all aspects of human life are dominated by economic considerations. Unfortunately, the environment proves no exception, even though not all of its aspects can be managed in strictly economic terms. Many economic models are also seriously flawed in that they have generally ignored the critical role of natural capital in the production process; they tend to assume that no environmental limits govern growth (Common 1995; Beckerman 1992). From a strictly economic perspective, nature—or "natural capital" as it is euphemistically known—is assumed to be a "free good," or one that is renewable and undamaged by human activity.

Of course, the assumption that natural capital is unaffected by human activity does not stand up to even cursory scrutiny, as the degradation of tropical forests, world fisheries, and arid grasslands well illustrates. Perhaps, with some qualifications, this assumption was more realistic in the age of Adam Smith (*Inquiry into the Nature and Causes of the Wealth of Nations,* 1776) and David Ricardo (*Principles of Political Economy and Taxation,* 1817), both of whom were pioneers in economic theory. After all, the Industrial Revolution was just getting under way at that time and was then largely confined to Great Britain. Pre-industrial technologies, coupled with the existence of a global population only one-fifth of that found today, would have had relatively little impact on the environment on a regional scale.

Myths in economics continue to persist, however. Even today, with the massive evidence gathered about wholesale transformations of ecosystems and unprecedented losses in biodiversity (UNEP 1995), economists tend to ignore environmental issues. They cling to the belief that natural resources, while not free, have a price, like any other good; they also assume that when resources became scarce, the price system will curtail demand and restore supply. This concept ignores biophysical realities. Extinct species do not magically reappear. Numerous documented cases of ecosystem collapse, directly linked to pressures from human activity, argue strongly that the economic paradigm of environment is completely wrong. The classic case is the recent collapse of the cod fishery of the northeast coast of North America. No self-correcting mechanisms snapped into place to stem the overharvesting of cod until the fishery was decimated—necessitating a complete and indefinite closure of this fishing area. The same fate has befallen many other substocks, such as herring, ocean perch, salmon, and lake trout (Ludwig et al 1993).

Traditional economics tends to ignore the complex interplay of human activities in radically restructuring and transforming environmental systems. No ecosystem has proved itself immune to the pressures from human activity. Deforestation, desertification, eutrophication, and salinization of agricultural lands all attest to this fact. Economics also leaves out important considerations related to future generations. Market prices reflect today's scarcities, and no allowances are made for the consequences for future generations. Rather, the economic calculus severely discounts the future—so much so that long-term investments in environment or anything else are seldom viewed as profitable, as such distant returns have negligible present value when discounted to the present.

In short, human activity is intimately associated with the environment, though this relationship has long been neglected in formal economic models. The assumption of environmental immunity from the impacts of the human enterprise is incorrect and short-sighted. Furthermore, market prices reflect only present scarcities. No technological substitutions exist for ecosystems that have been degraded beyond repair and losses in biodiversity cannot be compensated.

Economists have, of course, long recognized "externalities"—that is, costs or benefits associated with production and consumption activity that accrue to society as a whole but are not taken into account in market incentives to engage in production or consumption. In principle, market prices can be adjusted to reflect externalities. Pursuing this line of thought leads to the assumption of a potential for pricing ecosystem services—not in the market-place, for real markets seldom exist for such externalities, but according to the benefits these services confer to society (Holmes and Kramert 1996; Freeman 1993). If ecosystem services could be properly priced and traded, would this exchange provide a solution to environmental degradation?

Proposals along these lines have been made by some observers (for example, by Mitchell and Carson 1989; Costanza et al 1989). While they have merit relative to existing practices, which largely ignore environmental exter-nalities, an insurmountable difficulty remains—namely, that many ecosystem services cannot be valued in strictly monetary terms. The question of what Nature's services are worth has never been fully resolved (Westman 1990).

Another difficulty also limits the value of the economic paradigm. Not only does the marketplace fail to capture all of the value of ecosystem services, but markets also reflect present values and cannot account for future social costs of environmental degradation. In practice, stress pressures can and have intensified without corrective market signals. As the history of desertification in the southwest U.S. semi-arid grasslands illustrates, by the time the problem becomes obvious, any possible market responses may come far too late to restore ecosystem health (Whitford 1995).

Limitations of Ecological Approaches

If economics encounters difficulty in properly valuing the role of the natural system in production and consumption, would ecology provide a better base for environmental management? After all, ecologists focus on the myriad interactions among and between species and their environment. Finding the pattern that governs those interactions, describing their dynamics, and eluci-dating mechanisms whereby ecosystems evolve and change over time is the mainstay of this discipline.

Ecology, of course, has much of relevance to say about the environment. The rise of "Green" parties and their entry into mainstream politics in parts of North America and Western Europe demonstrate that ecological considerations have much merit in guiding the political process. The key strength of ecology lies in its elucidation of an interactive model, where all elements of the system are tied together by the flows of energy and nutrients.

Ironically, just as economic models have generally ignored nature, ecological approaches have tended to leave out human societies and economic activity. The focus of mainstream ecology is nature—without humans. Humans are viewed as an "outside" force that serves to modify ecological systems, but not as part and parcel of such systems. Because of this slant, ecology has proved far less relevant to environmental problem solving than it might otherwise be. It needs to integrate the economy with ecological process to deal effectively with questions of resource stewardship. Ludwig et al (1993) have pointed to the inadequacies of traditional ecological approaches to resource management, noting that they fail to integrate social process within the ecosystem. Reflecting on the lessons learned from the history of resource overexploitation, they suggest that it is more likely that resources manage humans, rather than the reverse.

Limitations of Engineering Approaches

If economic approaches to environmental problem solving fail to consider the functioning of natural systems and ecological approaches ignore the economy, would the engineering approach provide a more appropriate balance? After all, "command and control"—the heart of engineering—could in principle incorporate and integrate both ecological and economic considerations. Unfortunately, this approach suffers from yet another flaw: control of natural systems has proved deleterious in many cases. It is not control that is needed, but rather an understanding of these complex systems' natural dynamics and a strategy that works with, rather than against, these dynamics.

In many cases, attempts to regulate or control natural fluctuations that characterize ecological systems have increased the likelihood of serious ecosystem degradation. Holling (1995) provides several examples in which modern engineering, technological, economic, and administrative experience succeeded in stabilizing fluctuations in a targeted component of the ecosystem for a time. In each instance, the goal was to reduce the fluctuations in a part of the system that had caused problems. For example, insecticides were used to control spruce budworm outbreaks; increased fire detection and suppression techniques were applied to control the outbreak of fire; and modern rangeland management practice was implemented in the savannas of South Africa to reduce fluctuations in cattle grazing. Control of the target caused the system to become more homogenized, however, which rendered it more vulnerable to the very stress that was presumably thwarted. In the case of forests, for example, the exclusion of fire and the minimization of defoliating insects changed the forest architecture, producing more even-aged stands; this development, in turn, made the forest more vulnerable to an outbreak of defoliating insects or fire. In rangeland management, introducing more productive grasses, while initially favoring grazing, made the system more vulnerable to drought as the introduced grass species displaced drought-resistant ones. These consequences well illustrate the futility often encountered when attempting to intervene in the natural dynamics of complex natural ecosystems.

THE NEED FOR INTEGRATIVE KNOWLEDGE

The obvious shortcomings with respect to environmental management inherent in each of the three paradigms lead one to consider the potential held by a more integrated approach. If each approach has a weakness that is, in effect, another approach's strength, would a hybrid prove more successful? Would a combination of approaches provide a more comprehensive framework for environmental analysis than any single approach?

In recent years, prominent scientists have emphasized the need for integrating knowledge gained from the separate disciplines. Holling (1993), for example, suggests that mainstream ecology strives for consensus on the detailed mechanisms at work, by taking a narrow focus. According to this author, this stream is appropriately conservative and unambiguous, but "it achieves that by being incomplete and fragmentary" (Holling 1993, 553). Holling points to another stream emerging from evolutionary biology that is fundamentally integrative and expands to consider inherent surprise and uncertainty within a systems model. Such an integrated approach has some affinity with social sciences that are historical, analytical, and integrative.

Levins et al (1994) make a convincing case that, to understand the resurgence of old epidemics and the emergence of new ones, it is essential to integrate knowledge from social, natural, and health sciences. In this spirit, a number of cross-disciplinary or interdisciplinary approaches to environmental management have emerged in the last 10 years. Among these are ecological economics (1989), ecological engineering (1992), and ecosystem health (1995).

Ecosystem health is perhaps the most integrative of these interfaces. Its essence is to marry knowledge of how environmental systems behave (that is, impacts of stress pressures on ecosystems) with knowledge of what is desirable and acceptable. This question necessitates considering the widest possible implications of change—change in human health, economic opportunity, cultural integrity, and aesthetics—in terms of societal values. Values relate to what is acceptable and desirable; in essence, they define what is healthy. As an integrative science, ecosystem health goes far beyond the boundaries of stress ecology, a field that concerns itself with the strictly biophysical aspects of the problem. It explicitly takes into account the wider territory of socioeconomic, human health, legal, and policy aspects (Rapport 1995). Of course, no one model can be expected to accomplish all of these goals at once, but practical evaluations of changes in ecosystem health must draw upon all of these dimensions (Rapport 1995).

In the following sections, we explore the concept of ecosystem health, its role as a metaphor to facilitate communication and understanding of environmental transformation, its influence as a societal goal in placing constraints on human activity, and its value as a holistic approach to environmental management. The ecosystem health approach draws upon and integrates natural, social, and health sciences in the recognition of ecosystem dysfunction, the diagnosis of the causes of dysfunction, and the analysis of the potential solutions both to redress environmental degradation and to take preventive actions to reduce the risks of ecosystem pathology.

The integration of social, natural, and health sciences in developing the principles and practice of ecosystem health emphasizes several features of this approach.

Ecosystem health is a goal-directed enterprise. It is not merely an effort to understand how complex ecosystems function under stress, although that element is vital to the enterprise. Its aim is to evaluate changes in ecological function with respect to societal goals. Science can elucidate how the system works from a biophysical perspective. Values inform us about what is desired. Without an explicit consideration of values, health is devoid of meaning. What is considered "health" from one perspective (conservationists) may not be so from another (farmers).

Systematic and efficient methods are needed for diagnosis of ecosystem dysfunction, and effective methods for preventive and rehabilitative actions. Although the health sciences have long supported expert diagnostic systems to aid in identifying early signs and symptoms of disease, nothing comparable exists in ecosystem sciences.

Human health and ecosystem health are intimately linked (McMichael and Martens 1995). Recent reports suggest that degradation of the Earth's ecosystems is having a negative impact on human health (Haines et al 1993; Epstein 1995). Cholera outbreaks, for example, are correlated with eutrophication of coastal waters (Epstein 1995); pulmonary disease, such as acute asthma, has been correlated with periods of high ground-level ozone; and the risk of skin cancer increases with exposure to ultraviolet radiation. These examples may actually prove to be only "the tip of the iceberg." Changes in human health—such as fetal abnormalities, resurgence of infectious diseases, and prevalence of skin cancer—may provide key indicators of environmental change.

REFERENCES

Arrow K, Bollin B, Costanza R, Dasgupta P, Folke C, Holling CS, Jansson B-O, Levin S, Maler K-G, Perrings C, Pimentel D. Economic growth, carrying capacity, and the environment. Science 1995;268:520–521.

Beckerman W. Economic growth and the environment; whose growth? whose environment? World Dev 1992;20:481–496.

Bloom DE. International public opinion on the environment. Science 1995; 269:354–358.

Carson R. Silent spring. Boston: Houghton-Mifflin, 1962.

Colburn T, Dumanoski D, Myers JP. Our stolen future. Toronto: Penguin Books, 1996.

Common M. Economists don't read *Science*. Ecol Econ 1995;15:101–103.

Costanza R, Farber SC, Maxwell J. The valuation and management of wetland ecosystems. Ecol Econ 1989;1:335–361.

Covington WW, Moore MM. Southwestern Ponderosa forest structure: changes since Euro-American settlement. J Forestry 1994;92:39–47.

Epstein PR. Emerging diseases and ecosystem instability: new threats to public health. Am J Public Health 1995;85:168–172.

Freeman AM III. The measurement of environmental and resource values. Washington, DC: Resources for the Future, 1993.

Gunn JM, ed. Restoration and recovery of an industrial region: progress in restoring the smelter-damaged landscape near Sudbury, Canada. New York: Springer-Verlag, 1995.

Haines A, Epstein PR, McMichael AJ. Global health watch: monitoring impacts of environmental change. Lancet 1993;342:1464–1469.

Hildén M, Rapport DJ. Four centuries of cumulative impacts on a Finnish river and its estuary: an ecosystem health-approach. J Aquatic Ecosystem Health 1993; 2:261–275.

Holling CS. Sustainability: the cross-scale dimension. In: Munasinghe M, Shearer W, eds. Defining and measuring sustainability: the biogeophysical foundations. Washington, DC: World Bank, 1995.

———. Investing in research for sustainability. Ecological Applications 1993;3:552–555.

Holmes TP, Kramet RA. Contingent valuation of ecosystem health. Ecosystem Health 1996;2:56–60.

Jacobson J. Environmental refugees: a yardstick of habitability. Worldwatch paper no. 86. Washington DC: Worldwatch Institute, 1988.

Levins R, Awebuch T, Brinkmann U, et al. The emergence of new diseases. Am Scientist 1994;82:52–60.

Linden E. The tortured land. Time Sept. 4, 1995:36–47.

Ludwig D, Hillborn R, Walters C. Uncertainty, resource exploitation, and conservation: lessons from history. Science 1993;260:17–36.

McMichael AJ. Planetary overload, global environmental change and the health of the human species. Cambridge, UK: Cambridge University Press, 1993.

McMichael AJ, Martens WJM. The health impacts of global climate change: grappling with scenarios, predictive models, and multiple uncertainties. Ecosystem Health 1995;1:23–34.

Minns CK, Moore JE, Schindler DW, Jones NL. Assessing the potential extent of damage to inland lakes in Eastern Canada due to acidic deposition IV. Predicted impacts on species richness in seven groups of aquatic biota. Can J Fisheries Aquatic Sci 1990;47:821–830.

Mitchell RC, Carson RT. Using surveys to value public goods: the contingent valuation method. Washington, DC: Resources for the Future, 1989.

Myers M. The Gaia atlas of future worlds: challenge and opportunity in an age of change. New York: Doubleday, 1991.

Parry ML, Rosenzweig C. Food supply and the risk of hunger. Lancet 1993; 342:1345–1347.

Rapport DJ. Ecosystem health: more than a metaphor? Environ Values 1995; 4:287–309.

Regier HA, Baskerville GL. Sustainable redevelopment of regional ecosystems degraded by exploitive development. In: Clark WC, Munn RE, eds. Sustainable development of the biosphere. London: Cambridge University Press, 1986:75–103.

Regier HA, Hartmann WL. Lake Erie's fish community: 150 years of cultural stresses. Science 1973;180:1248–1255.

Sherman K, Busch DA. Assessment and monitoring of large marine ecosystems. In: Rapport DJ, Gaudet CL, Calow P, eds. Evaluating and monitoring the health of large-scale ecosystems. New York: Springer, 1995.

Somerville MA. Planet as patient. Ecosystem Health 1995;1:61–71.

Tolba MK, El-Kholy OA, El-hinnawi E, Holdgate MW, McMichael DF, Munn RE. The world environment. London: Chapman & Hall, 1992.

UNEP (United Nations Environmental Programme). Global biodiversity assessment. Nairobi, Kenya: UNEP, 1995.

Union of Concerned Citizens. World scientists' warning to humanity. Nucleus 1992;14:1–3.

Vogl RJ. The ecological factors that produce perturbation-dependent ecosystems. In: Cairns J, ed. The recovery process in damaged ecosystems. Ann Arbor, MI: Ann Arbor Science, 1980:63–64.

Westman WE. Evaluating the benefits of ecosystem integrity. In: Edwards CJ, Regier HA, eds. An ecosystem approach to the integrity of the Great Lakes in turbulent times. Great Lakes Fish Comm Spec Pub 90-4, 1990:91–103.

Whitford WG. Desertification: implications and limitations of the ecosystem health metaphor. In: Rapport DJ, Gaudet D, Calow P, eds. Evaluating and monitoring the health of large-scale ecosystems. Heidelberg: Springer-Verlag, 1995:273–294.

Whitford WG, Rapport DJ, Goothousen RM. The central Rio Grande Valley—organizing and interpreting ecosystem health assessment data. GIS World 1996; 9:60–62.

Woodwell GM. Ecology. The restoration. Restoration Ecol 1994;2:1–3.

Yazvenko SB, Rapport DJ. A framework for assessing forest ecosystem health. Ecosystem Health 1996;2:41–55.

CHAPTER 2

Defining Ecosystem Health

David Rapport

ECOSYSTEM HEALTH AS A METAPHOR

Metaphors are fundamental to the way we think and are often used in problem solving. Their use is clearly not restricted to poetry, but has a legitimate place in science as well. In both areas, metaphors stimulate associations, thereby bringing into juxtaposition phenomena that might initially appear to have little in common (Rapoport 1983). In the realm of science, their use often points to phenomena in very different spheres that bear some similarity, making it appropriate to transfer concepts and models across disciplines (Rapport and Turner 1977). The use of metaphor also serves as a powerful communication device. In ecosystem assessment, the health metaphor provides a language—that is, signs, diagnostic indicators, dysfunction, and ecosystem ills—all terms with which the public is familiar.

Under closer scrutiny, the use of metaphor in this and many other applications may be a double-edged sword—to make use of yet another metaphor—that simply drives home the point that we can't seem to manage without these comparisons. On the one hand, the health metaphor extends the notion of health from the level of the individual (clinical medicine, veterinary medicine) or population (public health) to the ecosystem (many species in interaction with their environment). On the other hand, its use begs a question: Are we resurrecting the much discredited analogy of "ecosystem as organism" (Rapport et al 1985; Rapport 1995)? This outcome is hardly our intent, so the question becomes one of whether it is possible to derive the advantages of this strategy without incurring undesirable costs. That is, can one accept that the health concept is relevant to the ecosystem level while simultaneously rejecting the notion of ecosystem as organism?

Let us examine this question by analyzing its two components. First, rejecting the notion of ecosystem as organism is rather straightforward. The suggestion that ecosystems—or the whole Earth, for that matter—can be viewed as a superorganism first appeared in the writings of James Hutton, a Scottish physician and geologist. Hutton delivered a paper to the Royal Society of Edinburgh in 1788 on a theory of the Earth as a superorganism capable of self-maintenance (Hutton 1788). In recent times, the notion of ecosystem as organism was associated with the theory of Clements (1916), an ecologist who described ecological succession in terms of a life cycle. Nevertheless, it is clear that organisms, unlike ecosystems, have clearly defined boundaries, reproduce, and are subject to genetic selection and evolution. Ecosystems, while consisting of many species of organisms, possess none of these features. At best, ecosystems are loosely or feebly organized and bear no resemblance to the intricate physiological feedback controls that characterize organisms. In addition, one can categorically say that the behavior of many organisms is characterized by "goal seeking"—a concept that would be absurd if applied to the behavior of ecosystems in the absence of human intervention and management.

Once the concept of ecosystem as organism is discarded, we turn to the second part of the question: In what ways might the metaphor of ecosystem health remain valid and useful? A lot, one might argue! Writing in 1941, the famed naturalist Aldo Leopold advanced the notion of "land health" (Leopold 1941). It is clear from Leopold's various writings that he meant "land" to be the entire ecosystem—that is, the web of relationships between organisms with each other and their environment. Substituting ecosystem for land—a substitution wholly justified by Leopold's comprehensive view of natural systems—we find that Leopold set as the goal of the enterprise of ecosystem health to "determine the ecological parameters within which land may be humanly occupied without making it dysfunctional."

The idea that ecosystems can and do become dysfunctional lies at the core of environmental concerns. The dysfunction in ecosystems can be measured in a variety of ways (see Part 2 of this book). The most dramatic indicator, ecosystem services, relates to properties of the ecosystem that offer direct benefits for the human community (for example, as a sink for waste disposal and detoxification of chemicals, as a source of food and fiber, or as a provider of potable water) (Cairns and Pratt 1995). In many instances, these services are sharply curtailed when ecosystems come under stress. As a consequence, clean air, clean water, and renewable resources such as fish and timber can no longer be taken for granted.

In most cases, declines in ecological services are permanent, and efforts to restore such services have met with meager results. It appears to be the exception rather than the rule when apparent damage to ecosystems proves to be temporary and the system "bounces back" when the stress disappears. One classic case of such a "miraculous" recovery involved Lake Washington, where diversion of a sewage outfall largely reversed the process of eutrophication (Edmondson 1968). In most cases, however, transformations of ecosystems under stress result in irreversible damage, where even heroic efforts are unlikely to succeed in re-establishing ecosystem services (Whitford 1995). In general, once degradation has proceeded to moderate levels, efforts to rehabilitate ecosystems achieve at best only partial success and at very high cost (Maini 1992).

It is precisely in these cases that use of the health metaphor is so apt. While entire ecosystems function very differently than biotic components (that is, the resident organisms), and are not nearly as tightly or centrally organized, a host of similar problems and approaches to the diagnosis, prognosis, and restoration of system function exist. In both cases, one needs to assess the loss of function, to diagnose probable causes (bearing in mind that misdiagnosis is as characteristic of ecosystem ills as it is of human ills), and to determine the appropriate interventions (Rapport et al 1981). Likewise, in both cases, these steps must rely on a suite of indicators to confirm dysfunction and to establish probable causes. Once the causes are identified, one needs to know what treatment (intervention) is feasible and to what extent functions can be restored. Here the metaphor represents a powerful tool for suggesting methodologies already in place in the health sciences that have applications to evaluating dysfunction in ecosystems.

The health metaphor also is proving to be a powerful communication device. The Canadian Centre for Inland Waters has a banner in its lobby proclaiming that "Ecosystem Health is Everyone's Business." This concept is not difficult to grasp. Individuals know, from personal experience, what it means to be dysfunctional. They are already educated about the process of general screening, diagnosis, monitoring of signs, and other phases of health care. That a very similar goal-oriented process is applicable to evaluating ecosystem condition is readily understandable. With the elucidation of direct health risks to humans from degradation of ecosystems, the case becomes even more convincing.

What Is Implied by the Health Metaphor at the Ecosystem Level

The use of the health metaphor at the ecosystem level has a number of implications (after Rapport 1995).

Ecosystem health can be defined in an operational manner, and assessments of health status can, at least in part, be based on objective criteria (see, for example, Mageau et al 1995). Viewing the ecosystem as a patient enables its functions to be evaluated in terms of objective standards that relate to the system's capacity for organization, vigor, and resilience.

TABLE 2.1   Evidence for ecosystem distress syndrome (EDS) in three ecosystems under a suite of stresses

Stress Factors	Lower Laurentian Great Lakes	Kyrönjoki River and Estuary	Jornada Rangelands
System Property			
Primary productivity[a]	+	+	0/−
Horizontal nutrient transport	+	+	+
Species diversity[b]	−	−	+/−
Disease prevalence[c]	+	+	+
Population regulation[d]	−	−	−
Reversal of succession	+	+	+
Metastability[e]	−	−	−
Community Structure			
r-Selected species[f]	+	+	+
Short-lived species	+	+	+
Smaller biota	+	+	?
Exotic species	+	+	+
Mutualistic interactions between species	−	−	−
Boundary linearity	+	+	+
Extinction of habitat specialists	?	+	+

[a] With respect to Jornada Rangelands, the productivity goes from 0 to − as stress intensifies; species diversity in birds and mammals goes from + to − as stress increases.

[b] Species diversity in the Jornada Rangelands increases in avian fauna, smaller biota, and small mammals because of the habitat provided by shrub layers; it decreases in grasses and annual plants. In the Great Lakes, species diversity declines in fish and aquatic vegetation. In the Kyrönjoki, diversity declines in fish and aquatic vegetation.

[c] Disease prevalence can be monitored in mistletoe (Jornada), fish tumors and fish parasites (Great Lakes), and crayfish (Kyrönjoki).

[d] Population regulation declines in fishes for the aquatic systems and grasses in the Jornada.

[e] Metastability refers to local stability and resilience of dominant biotic communities.

[f] r-Selected species dominate in disturbed ecosystems, particularly within the grasses of the Jornada. This measurement refers to annuals in the Jornada and fish communities of aquatic systems.

+ = increase; − = decrease; 0 = no change.

tious disease. Similarly, in ecosystems, while a primary cause of degradation might exist, its effects are most potent when the source works in conjunction with other stress factors. For example, pollution has been implicated in a number of current environmental problems, including the decline of amphibians, bleaching of coral reefs, appearance of phytoplankton bloom,

Systematic diagnosis of ecosystem condition is possible. When a p
arises, it usually indicates the presence of a dysfunction in the system. S
atic ways to identify sources of stress and relate these sources to si
ecosystem dysfunction should be possible. Case studies detailing the et.
of ecosystem dysfunction provide this required understanding o
ecosystem-by-ecosystem basis (Schindler 1990). Comparative studies 1
many common features (Rapport and Whitford 1997).

Careful study of the etiology of ecosystems under stress can yield
warning indicators of ecosystem degradation. Patterns of ecosystem breakd(
can be identified, and ecosystems under specified stress pressures can
shown to exhibit a characteristic sequence of signs of dysfunction. As sho
in Table 2.1, a comparative study of the behavior of the Baltic, Laurent
Great Lakes, and Jornada Rangelands (Chihuahuan Desert, New Mexic
under various stress pressures suggests a number of common properties
ecosystem response, despite the large differences in the ecosystems und(
study (Rapport and Whitford 1997).

Ecosystem health practice requires not only diagnostic and curative capabili
ties, but also, most importantly, preventive measures. The need for and value o1
preventive measures are now obvious for both human health and ecosystem
health. Once a complex system becomes severely damaged, whether at the
level of the organism or the ecosystem, the costs of rehabilitation greatly
increase and the probability of success declines sharply. For both human and
ecosystem health, in many cases, it is far more cost-effective to implement
preventive measures than to attempt cures after system damage occurs.

What Is Not Implied by the Health Metaphor at the Ecosystem Level

The use of the health metaphor at the ecosystem level does not imply that
every application of the health concept for organisms has an exact counterpart in ecosystems. In other words, metaphors are based on analogy rather
than homology. Thus, the health metaphor at the ecosystem level does not
imply that ecosystems are organisms or superorganisms. Further, it does not
imply that these systems are organized in a similar manner to organisms, nor
does it indicate that all aspects of health care should or ought to be incorporated into ecosystem management.

In particular, the concept of health in relation to ecosystems can be
defined by what it is not.

The health metaphor is not an endorsement of the conventional biomedical
model. Conventional biomedical models tend to be overly simplistic, relating
ills to proximate causes and often ignoring the complexity of many contributing and enabling factors (Ehrenfeld 1995; Levins et al 1994). Such simplification is appropriate for neither medicine nor ecology (Holling 1995; Ehrenfeld
1995). In medicine, many codeterminants may affect susceptibility to infec-

mass mortality among seals and dolphins, and cancer epizootics in fish (Myers 1995). Pollution seems to cause the most harm, however, when it occurs in conjunction with sources of stress such as habitat disruption and eutrophication (Myers 1995; Hildén and Rapport 1993).

The health metaphor is not an endorsement of the prevailing medical practice whereby profits derive primarily from illness rather than health. The practice of medicine emphasizes treatment rather than prevention. Smoking, for example, is known to be a primary cause of lung cancer. Yet almost all resources to cope with this problem are allocated to research on treating cancer, rather than research on lifestyle habits and prevention. This approach is used with many human health problems—efforts are made to "fix" the problem once it occurs, rather than to reduce the probability of occurrence in the first place.

The health metaphor is not an endorsement of the common medical practice in which responsibilities for health assessments and treatment reside almost entirely with medical professionals. Lack of empowerment of individuals is often detrimental to recovery. Instead, decisions on options for ecosystem health need to empower the community as a whole. Scientists and resource managers ought to play the role of "independent counsel"—that is, clarifying available options (Lackey 1996). Stakeholders, as community representatives, ought to have the final word on decisions affecting their environments.

ECOSYSTEM HEALTH AS A SOCIETAL GOAL

Given the marked deterioration in major ecosystems of the world, both quantitatively and qualitatively (World Watch Institute 1994; Tolba et al 1992), enhancing ecosystem health represents a goal with critical importance to our future. Without ecosystem health, the very foundations of our social and economic systems are undermined. Economic activity, social organization, and human health maintenance are all tied closely to the viability and health of ecosystems. Their continued degradation threatens the future of the entire human community (UNEP 1995; Union of Concerned Citizens 1992; Tolba et al 1992; Myers 1995). Linkages between threats to ecosystem health—for example, from eutrophication of coastal marine systems or climate change—and threats to human health resulting from the resurgence of old epidemics and the emergence of new ones are linked (Patz 1996; Levins et al 1994; Epstein 1995). These relationships only underscore the critical importance to society of fostering ecosystem health.

The question is not whether ecosystem health ought to be a super societal goal—of course, the answer is a resounding and unqualified "yes." The real question is what priority society will give to achieving that goal. In a period of faltering economies, will the seemingly more tangible goals of providing jobs win out over the longer-term goals of ensuring ecosystem health? Will the political system, as it has so often in the past, opt for short-term gain at the cost of long-term pain?

Reaching consensus on the answers to such questions has proved difficult. Official proclamations tend to move in the direction of more focus and attention to the goals of ecosystem health and integrity, such as through those issued at conventions, and treaties concluded at the United Nations Conference on Environment and Development in Rio (1992) offer hope for broad cooperation among nations in conserving biological diversity and forest resources and in reducing greenhouse gas emissions. Such proclamations are not limited only to government bodies. In 1990, the Business Council for Sustainable Development was established for the purpose of taking action on the environment (Schmidheiny 1992). The Declaration of the Business Council for Sustainable Development, signed by some 50 business leaders representing major corporations worldwide, commits its members to the Brundtland Agenda, namely "to meeting the needs of the present without compromising the welfare of future generations." The declaration goes on to acknowledge "that economic growth and environmental protection are inextricably linked, and that the quality of present and future life rests on meeting basic human needs without destroying the environment on which all life depends." Concerted action plans have also been proposed, such as those implemented to restore the health of the Areas of Concern in the Laurentian Great Lakes. Some 41 areas were identified, and for each of these a remedial action plan has been drafted; indeed, significant actions have already been taken for many areas (Hartig and Zarull 1992).

Against these positive gains, one must consider recent reports (UNEP 1995; Tolba et al 1992) pointing to markedly deteriorating environmental conditions on a worldwide scale. In addition, a recent tendency has been to weaken or otherwise dismantle long-standing environmental protection legislation in the United States and within some provinces of Canada.

Thus, we may conclude that our human future continues to hang in the balance between the concerns for tomorrow and the immediate gratification of today. To tip that balance in favor of the future of humankind, we argue, requires an integrative perspective that relates the health of our species to the health of our planet.

ECOSYSTEM HEALTH AS A TRANSDISCIPLINARY SCIENCE

In the preceding sections, we considered ecosystem health as a metaphor and as a societal goal. In this section, we examine it as a transdisciplinary science, integrating social, natural, and health sciences as they pertain to environmental change. Transdisciplinary (sometimes referred to as "interdisciplinary" or "cross-disciplinary") study involves at least a partial integration of two or more fields. In the process, transcending concepts are identified, and new approaches to problem identification and resolution may emerge. Transdisciplinary science thus contrasts with multidisciplinary science, which refers to activities enlisting contributions from different disciplines without necessarily attempting to integrate them in any fundamental way. The hallmark of transdisciplinary science is the integration of knowledge and the transcendence of disciplinary boundaries (Rapport 1995).

The motivation for transdisciplinarity lies in both the development of novel ideas and the efficiency of problem solving. Real-world problems are rarely confined to a domain covered by a single discipline. For example, consider the present phenomena in which epidemics of infections once thought eradicated (such as tuberculosis, cholera, and smallpox) have re-emerged. It might be thought that understanding this resurgence lies entirely within the purview of epidemiology. Taking such an approach, however, would miss critical factors of an economic and ecological nature that are critical to an understanding of this phenomena (Levins et al 1994). As the need for integration across the sciences becomes more widely recognized, particularly as it relates to finding solutions to environmental problems (Holling 1995), it is encouraging that national and international agencies are restructuring their research by moving away from traditional discipline-based missions and toward integrated-theme or problem-solving missions. For example, the Canadian International Development Research Council has identified ecosystem health as one of its foci.

The question of what constitutes ecosystem health remains somewhat perplexing and controversial. Paradoxically, ecosystem breakdown has always been easier to define. Aldo Leopold, the great U.S. naturalist and conservationist, clearly recognized this asymmetry more than half a century ago (Leopold 1941). He was readily able to identify signs of "land sickness" such as erosion, loss of fertility, hydrological abnormalities, occasional irruption of certain species, mysterious local extinction of others, and qualitative deterioration in farm and forest products. When it came to identifying features of land health, however, Leopold drew attention to the fact that while land doctoring is an ancient art "practiced with vigor . . . land health is a job for the future" (Callicott 1992).

Leopold was correct in surmising that it is typically easier to detect signs of "sickness" than to establish criteria for health, whether working at the level of the organism, the population, or the ecosystem. Obviously, one can more readily identify signs of "something gone wrong" than one can specify necessary and sufficient criteria for "something gone right." Further, the concept of "health" implies value judgments—that is, a determination of what constitutes a desirable condition. Value judgments, of course, open the door to differing opinions.

The World Health Organization (WHO) has grappled with definitions of human health for some time. Its official definitions have been changed by degrees to incorporate more subjective elements, such as mental well-being, psychological well-being, and satisfaction of individual life goals and aspirations (Porn 1984; Last 1993). No agency has yet taken on the responsibility for defining ecosystem health—although in view of the recent United Nations declaration obligating states to safeguard "the health and integrity" of the Earth's ecosystems, one might expect an official definition of ecosystem health to be forthcoming. For now, however, the literature remains replete with a plethora of definitions that reflect the views of researchers (Rapport 1995) and environmental/resource interest groups. For example, the Montreal Process has developed explicit criteria for forest ecosystem health (Yazvenko and Rapport 1996). A number of efforts to rehabilitate damaged environments have explicit goals that essentially include a definition of a desirable or healthy state for the particular ecosystem (Hartig and Zarull 1992).

Efforts to define the healthy or desirable state for any complex system suffer from the "perennial dilemma," as succinctly stated by Amos Hawley, the father of human ecology. According to Hawley (1950, 7), how the effectiveness of a system is to be measured—"as durability, productivity or efficiency, or all together"—remains open to debate. All three criteria have in various forms been advanced as measures of ecosystem health. For example, Mageau et al (1995) suggest that ecosystem health be assessed in terms of "vigor," "resilience," and "organization." Vigor may be operationally measured in terms of productivity, or throughput of material or energy in the system. Resilience may be calculated in terms of a system's ability to maintain its structure and pattern of behavior in the presence of stress (Holling 1986). Organization may be assessed in terms of both the diversity of components and their degree of mutual dependence—that is, by studying the exchange pathways between them. In Table 2.2, ecosystem health criteria proposed by various authors are related to the primary characteristics suggested by Mageau et al (1995). These comparisons suggest that Leopold's 1941 prediction that "the science of land health is a task for the future" has now come to pass.

Ideally, indicators of ecosystem health should reflect both biophysical conditions and societal values. Consider, for example, the task of assessing the health of a wetland. One might first base the assessment on strictly biophysical criteria: Is the wetland self-sustaining? This determination would necessarily take into account the following factors: aspects of hydrology, particularly the maintenance of fluctuating water levels upon which many of the plant species depend; water quality, particularly questions of nutrient concen-

TABLE 2.2 Specific ecosystem health criteria related to primary components of health

Primary Components*	Ecosystem Health Criteria
Vigor	Primary productivity; nutrient cycling
Organization	Ratio of r-selected species to K-selected species; ratio of short-lived species to long-lived species; ratio of exotic to endemic species; degree of mutualism; extinction of habitat specialists
Resilience	Recovery rates from natural perturbations; resistance to natural perturbations

*As suggested by Mageau et al (1995).

trations and contaminant levels; and habitat diversity, which governs the species associations and species richness of the wetland. Yet such an assessment would still be incomplete because it ignores the question of societal values. This assessment entrains a second set of queries: What are the desired properties of the wetland—water regulation, sequestering of contaminants, waterfowl production, recreational opportunities, or all of these? Depending on the answers to this second set of questions, one can render a final judgment on the direction in which the health of the system is changing.

It might first appear from this example that ecosystem health depends on whimsy—that is, on human likes and dislikes. Do any absolute criteria exist for ecosystem health? Can the equation eliminate value judgments? We would claim not. Health here is viewed from a psychological perspective, involving human perceptions and values. Without such judgments, criteria for what constitutes health would be lacking. We do not mean to suggest that ecosystem changes do not affect the well-being of component species. Of course they do. Rather, we suggest that the assertion that one state is healthy and another less so embodies a value judgment. Humans often assign the highest value to the "natural" environment, but at other times they may favor the human-constructed culturally productive environment (for example, Dubos 1968). As we have emphasized, however, health is more than survival—it is survival in a condition and manner that satisfies human goals.

CRITERIA FOR ECOSYSTEM HEALTH

Previously, we identified three overarching attributes of ecosystem health: vigor, resilience, and organization. These attributes, which are derived from ecological perspectives, give rise to specific criteria for evaluating ecosystem health. These criteria may then be applied to the biophysical, socioeconomic, and human health dimensions of the ecosystem. In addition, several other emerging attributes address broader aspects of ecosystem health, such as the ability of a system to sustain management options. The eight criteria described in the following sections form a checklist that might be used in evaluating ecosystem health. Each of these criteria, in principle, may be specified in terms of operational measures, as has already been done for the first three criteria (Mageau et al 1995).

Vigor

Vigor refers to energy or activity. In an ecosystem context, it refers to a throughput of energy that can be measured in terms of nutrient cycling and productivity. It is not true, however, that the higher the throughput, the healthier the system. Far from it! Some major problems, particularly within aquatic ecosystems, are caused by excessive throughput that may be stimulated by increased nutrients from land disturbance and land run-off.

We have now reached a critical point to which we will return many times—namely, that it is necessary to evaluate the health of an ecosystem against norms that are characteristic of the specific type of ecosystem. Just as one would not evaluate the health of an adult male against norms derived from an infant female, one could not assess the functioning of a temperate forest by comparing it with the norms for a tropical lake. Obviously, norms and expected temporal dynamics are ecosystem-dependent. For example, high-latitude systems are characterized by low average energy throughput and seasonal "bursts" of activity (Rapport et al 1996). As a consequence, these ecosystems have characteristic behavior that differs considerably from tropical systems, which are distinguished by high overall throughput and less dramatic seasonal fluctuations. Nevertheless, one might hypothesize, at least for most ecosystems, that more stress is associated with less vigor in terms of productivity and throughput.

Resilience

Resilience refers to the capacity of a system to cope with stress and to bounce back when the stress diminishes (Holling 1986; Pimm 1984). This capacity, elsewhere referred to as "counteractive capacity" (Stebbing 1981; Rapport 1989), is measured by the system's capacity to return after a perturbation. In this instance, one might conceive of an ecosystem as playing a role equivalent to the stress tolerance tests commonly seen in medicine. Tests for cardiovascular and respiratory functions, for example, often involve monitoring system recovery after stress is introduced. Ecosystem resilience could be tested in a similar manner.

emerging attributes address broader aspects of ecosystem health, such as the ability of a system to sustain management options. The eight criteria described in the following sections form a checklist that might be used in evaluating ecosystem health. Each of these criteria, in principle, may be specified in terms of operational measures, as has already been done for the first three criteria (Mageau et al 1995).

Vigor

Vigor refers to energy or activity. In an ecosystem context, it refers to a throughput of energy that can be measured in terms of nutrient cycling and productivity. It is not true, however, that the higher the throughput, the healthier the system. Far from it! Some major problems, particularly within aquatic ecosystems, are caused by excessive throughput that may be stimulated by increased nutrients from land disturbance and land run-off.

We have now reached a critical point to which we will return many times—namely, that it is necessary to evaluate the health of an ecosystem against norms that are characteristic of the specific type of ecosystem. Just as one would not evaluate the health of an adult male against norms derived from an infant female, one could not assess the functioning of a temperate forest by comparing it with the norms for a tropical lake. Obviously, norms and expected temporal dynamics are ecosystem-dependent. For example, high-latitude systems are characterized by low average energy throughput and seasonal "bursts" of activity (Rapport et al 1996). As a consequence, these ecosystems have characteristic behavior that differs considerably from tropical systems, which are distinguished by high overall throughput and less dramatic seasonal fluctuations. Nevertheless, one might hypothesize, at least for most ecosystems, that more stress is associated with less vigor in terms of productivity and throughput.

Resilience

Resilience refers to the capacity of a system to cope with stress and to bounce back when the stress diminishes (Holling 1986; Pimm 1984). This capacity, elsewhere referred to as "counteractive capacity" (Stebbing 1981; Rapport 1989), is measured by the system's capacity to return after a perturbation. In this instance, one might conceive of an ecosystem as playing a role equivalent to the stress tolerance tests commonly seen in medicine. Tests for cardiovascular and respiratory functions, for example, often involve monitoring system recovery after stress is introduced. Ecosystem resilience could be tested in a similar manner.

TABLE 2.2   Specific ecosystem health criteria related to primary components of health

Primary Components*	Ecosystem Health Criteria
Vigor	Primary productivity; nutrient cycling
Organization	Ratio of *r*-selected species to *K*-selected species; ratio of short-lived species to long-lived species; ratio of exotic to endemic species; degree of mutualism; extinction of habitat specialists
Resilience	Recovery rates from natural perturbations; resistance to natural perturbations

* As suggested by Mageau et al (1995).

trations and contaminant levels; and habitat diversity, which governs the species associations and species richness of the wetland. Yet such an assessment would still be incomplete because it ignores the question of societal values. This assessment entrains a second set of queries: What are the desired properties of the wetland—water regulation, sequestering of contaminants, waterfowl production, recreational opportunities, or all of these? Depending on the answers to this second set of questions, one can render a final judgment on the direction in which the health of the system is changing.

It might first appear from this example that ecosystem health depends on whimsy—that is, on human likes and dislikes. Do any absolute criteria exist for ecosystem health? Can the equation eliminate value judgments? We would claim not. Health here is viewed from a psychological perspective, involving human perceptions and values. Without such judgments, criteria for what constitutes health would be lacking. We do not mean to suggest that ecosystem changes do not affect the well-being of component species. Of course they do. Rather, we suggest that the assertion that one state is healthy and another less so embodies a value judgment. Humans often assign the highest value to the "natural" environment, but at other times they may favor the human-constructed culturally productive environment (for example, Dubos 1968). As we have emphasized, however, health is more than survival—it is survival in a condition and manner that satisfies human goals.

CRITERIA FOR ECOSYSTEM HEALTH

Previously, we identified three overarching attributes of ecosystem health: vigor, resilience, and organization. These attributes, which are derived from ecological perspectives, give rise to specific criteria for evaluating ecosystem health. These criteria may then be applied to the biophysical, socioeconomic, and human health dimensions of the ecosystem. In addition, several other

tutes a desirable condition. Value judgments, of course, open the door to differing opinions.

The World Health Organization (WHO) has grappled with definitions of human health for some time. Its official definitions have been changed by degrees to incorporate more subjective elements, such as mental well-being, psychological well-being, and satisfaction of individual life goals and aspirations (Porn 1984; Last 1993). No agency has yet taken on the responsibility for defining ecosystem health—although in view of the recent United Nations declaration obligating states to safeguard "the health and integrity" of the Earth's ecosystems, one might expect an official definition of ecosystem health to be forthcoming. For now, however, the literature remains replete with a plethora of definitions that reflect the views of researchers (Rapport 1995) and environmental/resource interest groups. For example, the Montreal Process has developed explicit criteria for forest ecosystem health (Yazvenko and Rapport 1996). A number of efforts to rehabilitate damaged environments have explicit goals that essentially include a definition of a desirable or healthy state for the particular ecosystem (Hartig and Zarull 1992).

Efforts to define the healthy or desirable state for any complex system suffer from the "perennial dilemma," as succinctly stated by Amos Hawley, the father of human ecology. According to Hawley (1950, 7), how the effectiveness of a system is to be measured—"as durability, productivity or efficiency, or all together"—remains open to debate. All three criteria have in various forms been advanced as measures of ecosystem health. For example, Mageau et al (1995) suggest that ecosystem health be assessed in terms of "vigor," "resilience," and "organization." Vigor may be operationally measured in terms of productivity, or throughput of material or energy in the system. Resilience may be calculated in terms of a system's ability to maintain its structure and pattern of behavior in the presence of stress (Holling 1986). Organization may be assessed in terms of both the diversity of components and their degree of mutual dependence—that is, by studying the exchange pathways between them. In Table 2.2, ecosystem health criteria proposed by various authors are related to the primary characteristics suggested by Mageau et al (1995). These comparisons suggest that Leopold's 1941 prediction that "the science of land health is a task for the future" has now come to pass.

Ideally, indicators of ecosystem health should reflect both biophysical conditions and societal values. Consider, for example, the task of assessing the health of a wetland. One might first base the assessment on strictly biophysical criteria: Is the wetland self-sustaining? This determination would necessarily take into account the following factors: aspects of hydrology, particularly the maintenance of fluctuating water levels upon which many of the plant species depend; water quality, particularly questions of nutrient concen-

tal change. Transdisciplinary (sometimes referred to as "interdisciplinary" or "cross-disciplinary") study involves at least a partial integration of two or more fields. In the process, transcending concepts are identified, and new approaches to problem identification and resolution may emerge. Transdisciplinary science thus contrasts with multidisciplinary science, which refers to activities enlisting contributions from different disciplines without necessarily attempting to integrate them in any fundamental way. The hallmark of transdisciplinary science is the integration of knowledge and the transcendence of disciplinary boundaries (Rapport 1995).

The motivation for transdisciplinarity lies in both the development of novel ideas and the efficiency of problem solving. Real-world problems are rarely confined to a domain covered by a single discipline. For example, consider the present phenomena in which epidemics of infections once thought eradicated (such as tuberculosis, cholera, and smallpox) have re-emerged. It might be thought that understanding this resurgence lies entirely within the purview of epidemiology. Taking such an approach, however, would miss critical factors of an economic and ecological nature that are critical to an understanding of this phenomena (Levins et al 1994). As the need for integration across the sciences becomes more widely recognized, particularly as it relates to finding solutions to environmental problems (Holling 1995), it is encouraging that national and international agencies are restructuring their research by moving away from traditional discipline-based missions and toward integrated-theme or problem-solving missions. For example, the Canadian International Development Research Council has identified ecosystem health as one of its foci.

The question of what constitutes ecosystem health remains somewhat perplexing and controversial. Paradoxically, ecosystem breakdown has always been easier to define. Aldo Leopold, the great U.S. naturalist and conservationist, clearly recognized this asymmetry more than half a century ago (Leopold 1941). He was readily able to identify signs of "land sickness" such as erosion, loss of fertility, hydrological abnormalities, occasional irruption of certain species, mysterious local extinction of others, and qualitative deterioration in farm and forest products. When it came to identifying features of land health, however, Leopold drew attention to the fact that while land doctoring is an ancient art "practiced with vigor . . . land health is a job for the future" (Callicott 1992).

Leopold was correct in surmising that it is typically easier to detect signs of "sickness" than to establish criteria for health, whether working at the level of the organism, the population, or the ecosystem. Obviously, one can more readily identify signs of "something gone wrong" than one can specify necessary and sufficient criteria for "something gone right." Further, the concept of "health" implies value judgments—that is, a determination of what consti-

win out over the longer-term goals of ensuring ecosystem health? Will the political system, as it has so often in the past, opt for short-term gain at the cost of long-term pain?

Reaching consensus on the answers to such questions has proved difficult. Official proclamations tend to move in the direction of more focus and attention to the goals of ecosystem health and integrity, such as through those issued at conventions, and treaties concluded at the United Nations Conference on Environment and Development in Rio (1992) offer hope for broad cooperation among nations in conserving biological diversity and forest resources and in reducing greenhouse gas emissions. Such proclamations are not limited only to government bodies. In 1990, the Business Council for Sustainable Development was established for the purpose of taking action on the environment (Schmidheiny 1992). The Declaration of the Business Council for Sustainable Development, signed by some 50 business leaders representing major corporations worldwide, commits its members to the Brundtland Agenda, namely "to meeting the needs of the present without compromising the welfare of future generations." The declaration goes on to acknowledge "that economic growth and environmental protection arc inextricably linked, and that the quality of present and future life rests on meeting basic human needs without destroying the environment on which all life depends." Concerted action plans have also been proposed, such as those implemented to restore the health of the Areas of Concern in the Laurentian Great Lakes. Some 41 areas were identified, and for each of these a remedial action plan has been drafted; indeed, significant actions have already been taken for many areas (Hartig and Zarull 1992).

Against these positive gains, one must consider recent reports (UNEP 1995; Tolba et al 1992) pointing to markedly deteriorating environmental conditions on a worldwide scale. In addition, a recent tendency has been to weaken or otherwise dismantle long-standing environmental protection legislation in the United States and within some provinces of Canada.

Thus, we may conclude that our human future continues to hang in the balance between the concerns for tomorrow and the immediate gratification of today. To tip that balance in favor of the future of humankind, we argue, requires an integrative perspective that relates the health of our species to the health of our planet.

ECOSYSTEM HEALTH AS A TRANSDISCIPLINARY SCIENCE

In the preceding sections, we considered ecosystem health as a metaphor and as a societal goal. In this section, we examine it as a transdisciplinary science, integrating social, natural, and health sciences as they pertain to environmen-

mass mortality among seals and dolphins, and cancer epizootics in fish (Myers 1995). Pollution seems to cause the most harm, however, when it occurs in conjunction with sources of stress such as habitat disruption and eutrophication (Myers 1995; Hildén and Rapport 1993).

The health metaphor is not an endorsement of the prevailing medical practice whereby profits derive primarily from illness rather than health. The practice of medicine emphasizes treatment rather than prevention. Smoking, for example, is known to be a primary cause of lung cancer. Yet almost all resources to cope with this problem are allocated to research on treating cancer, rather than research on lifestyle habits and prevention. This approach is used with many human health problems—efforts are made to "fix" the problem once it occurs, rather than to reduce the probability of occurrence in the first place.

The health metaphor is not an endorsement of the common medical practice in which responsibilities for health assessments and treatment reside almost entirely with medical professionals. Lack of empowerment of individuals is often detrimental to recovery. Instead, decisions on options for ecosystem health need to empower the community as a whole. Scientists and resource managers ought to play the role of "independent counsel"—that is, clarifying available options (Lackey 1996). Stakeholders, as community representatives, ought to have the final word on decisions affecting their environments.

ECOSYSTEM HEALTH AS A SOCIETAL GOAL

Given the marked deterioration in major ecosystems of the world, both quantitatively and qualitatively (World Watch Institute 1994; Tolba et al 1992), enhancing ecosystem health represents a goal with critical importance to our future. Without ecosystem health, the very foundations of our social and economic systems are undermined. Economic activity, social organization, and human health maintenance are all tied closely to the viability and health of ecosystems. Their continued degradation threatens the future of the entire human community (UNEP 1995; Union of Concerned Citizens 1992; Tolba et al 1992; Myers 1995). Linkages between threats to ecosystem health—for example, from eutrophication of coastal marine systems or climate change—and threats to human health resulting from the resurgence of old epidemics and the emergence of new ones are linked (Patz 1996; Levins et al 1994; Epstein 1995). These relationships only underscore the critical importance to society of fostering ecosystem health.

The question is not whether ecosystem health ought to be a super societal goal—of course, the answer is a resounding and unqualified "yes." The real question is what priority society will give to achieving that goal. In a period of faltering economies, will the seemingly more tangible goals of providing jobs

Systematic diagnosis of ecosystem condition is possible. When a problem arises, it usually indicates the presence of a dysfunction in the system. Systematic ways to identify sources of stress and relate these sources to signs of ecosystem dysfunction should be possible. Case studies detailing the etiology of ecosystem dysfunction provide this required understanding on an ecosystem-by-ecosystem basis (Schindler 1990). Comparative studies reveal many common features (Rapport and Whitford 1997).

Careful study of the etiology of ecosystems under stress can yield early warning indicators of ecosystem degradation. Patterns of ecosystem breakdown can be identified, and ecosystems under specified stress pressures can be shown to exhibit a characteristic sequence of signs of dysfunction. As shown in Table 2.1, a comparative study of the behavior of the Baltic, Laurentian Great Lakes, and Jornada Rangelands (Chihuahuan Desert, New Mexico) under various stress pressures suggests a number of common properties in ecosystem response, despite the large differences in the ecosystems under study (Rapport and Whitford 1997).

Ecosystem health practice requires not only diagnostic and curative capabilities, but also, most importantly, preventive measures. The need for and value of preventive measures are now obvious for both human health and ecosystem health. Once a complex system becomes severely damaged, whether at the level of the organism or the ecosystem, the costs of rehabilitation greatly increase and the probability of success declines sharply. For both human and ecosystem health, in many cases, it is far more cost-effective to implement preventive measures than to attempt cures after system damage occurs.

What Is Not Implied by the Health Metaphor at the Ecosystem Level

The use of the health metaphor at the ecosystem level does not imply that every application of the health concept for organisms has an exact counterpart in ecosystems. In other words, metaphors are based on analogy rather than homology. Thus, the health metaphor at the ecosystem level does not imply that ecosystems are organisms or superorganisms. Further, it does not imply that these systems are organized in a similar manner to organisms, nor does it indicate that all aspects of health care should or ought to be incorporated into ecosystem management.

In particular, the concept of health in relation to ecosystems can be defined by what it is not.

The health metaphor is not an endorsement of the conventional biomedical model. Conventional biomedical models tend to be overly simplistic, relating ills to proximate causes and often ignoring the complexity of many contributing and enabling factors (Ehrenfeld 1995; Levins et al 1994). Such simplification is appropriate for neither medicine nor ecology (Holling 1995; Ehrenfeld 1995). In medicine, many codeterminants may affect susceptibility to infec-

TABLE 2.1   Evidence for ecosystem distress syndrome (EDS) in three ecosystems under a suite of stresses

Stress Factors	Lower Laurentian Great Lakes	Kyrönjoki River and Estuary	Jornada Rangelands
System Property			
Primary productivity[a]	+	+	0/−
Horizontal nutrient transport	+	+	+
Species diversity[b]	−	−	+/−
Disease prevalence[c]	+	+	+
Population regulation[d]	−	−	−
Reversal of succession	+	+	+
Metastability[e]	−	−	−
Community Structure			
r-Selected species[f]	+	+	+
Short-lived species	+	+	+
Smaller biota	+	+	?
Exotic species	+	+	+
Mutualistic interactions between species	−	−	−
Boundary linearity	+	+	+
Extinction of habitat specialists	?	+	+

[a] With respect to Jornada Rangelands, the productivity goes from 0 to − as stress intensifies; species diversity in birds and mammals goes from + to − as stress increases.
[b] Species diversity in the Jornada Rangelands increases in avian fauna, smaller biota, and small mammals because of the habitat provided by shrub layers; it decreases in grasses and annual plants. In the Great Lakes, species diversity declines in fish and aquatic vegetation. In the Kyrönjoki, diversity declines in fish and aquatic vegetation.
[c] Disease prevalence can be monitored in mistletoe (Jornada), fish tumors and fish parasites (Great Lakes), and crayfish (Kyrönjoki).
[d] Population regulation declines in fishes for the aquatic systems and grasses in the Jornada.
[e] Metastability refers to local stability and resilience of dominant biotic communities.
[f] r-Selected species dominate in disturbed ecosystems, particularly within the grasses of the Jornada. This measurement refers to annuals in the Jornada and fish communities of aquatic systems.
+ = increase; − = decrease; 0 = no change.

tious disease. Similarly, in ecosystems, while a primary cause of degradation might exist, its effects are most potent when the source works in conjunction with other stress factors. For example, pollution has been implicated in a number of current environmental problems, including the decline of amphibians, bleaching of coral reefs, appearance of phytoplankton bloom,

temporary and the system "bounces back" when the stress disappears. One classic case of such a "miraculous" recovery involved Lake Washington, where diversion of a sewage outfall largely reversed the process of eutrophication (Edmondson 1968). In most cases, however, transformations of ecosystems under stress result in irreversible damage, where even heroic efforts are unlikely to succeed in re-establishing ecosystem services (Whitford 1995). In general, once degradation has proceeded to moderate levels, efforts to rehabilitate ecosystems achieve at best only partial success and at very high cost (Maini 1992).

It is precisely in these cases that use of the health metaphor is so apt. While entire ecosystems function very differently than biotic components (that is, the resident organisms), and are not nearly as tightly or centrally organized, a host of similar problems and approaches to the diagnosis, prognosis, and restoration of system function exist. In both cases, one needs to assess the loss of function, to diagnose probable causes (bearing in mind that misdiagnosis is as characteristic of ecosystem ills as it is of human ills), and to determine the appropriate interventions (Rapport et al 1981). Likewise, in both cases, these steps must rely on a suite of indicators to confirm dysfunction and to establish probable causes. Once the causes are identified, one needs to know what treatment (intervention) is feasible and to what extent functions can be restored. Here the metaphor represents a powerful tool for suggesting methodologies already in place in the health sciences that have applications to evaluating dysfunction in ecosystems.

The health metaphor also is proving to be a powerful communication device. The Canadian Centre for Inland Waters has a banner in its lobby proclaiming that "Ecosystem Health is Everyone's Business." This concept is not difficult to grasp. Individuals know, from personal experience, what it means to be dysfunctional. They are already educated about the process of general screening, diagnosis, monitoring of signs, and other phases of health care. That a very similar goal-oriented process is applicable to evaluating ecosystem condition is readily understandable. With the elucidation of direct health risks to humans from degradation of ecosystems, the case becomes even more convincing.

What Is Implied by the Health Metaphor at the Ecosystem Level

The use of the health metaphor at the ecosystem level has a number of implications (after Rapport 1995).

Ecosystem health can be defined in an operational manner, and assessments of health status can, at least in part, be based on objective criteria (see, for example, Mageau et al 1995). Viewing the ecosystem as a patient enables its functions to be evaluated in terms of objective standards that relate to the system's capacity for organization, vigor, and resilience.

Let us examine this question by analyzing its two components. First, rejecting the notion of ecosystem as organism is rather straightforward. The suggestion that ecosystems—or the whole Earth, for that matter—can be viewed as a superorganism first appeared in the writings of James Hutton, a Scottish physician and geologist. Hutton delivered a paper to the Royal Society of Edinburgh in 1788 on a theory of the Earth as a superorganism capable of self-maintenance (Hutton 1788). In recent times, the notion of ecosystem as organism was associated with the theory of Clements (1916), an ecologist who described ecological succession in terms of a life cycle. Nevertheless, it is clear that organisms, unlike ecosystems, have clearly defined boundaries, reproduce, and are subject to genetic selection and evolution. Ecosystems, while consisting of many species of organisms, possess none of these features. At best, ecosystems are loosely or feebly organized and bear no resemblance to the intricate physiological feedback controls that characterize organisms. In addition, one can categorically say that the behavior of many organisms is characterized by "goal seeking"—a concept that would be absurd if applied to the behavior of ecosystems in the absence of human intervention and management.

Once the concept of ecosystem as organism is discarded, we turn to the second part of the question: In what ways might the metaphor of ecosystem health remain valid and useful? A lot, one might argue! Writing in 1941, the famed naturalist Aldo Leopold advanced the notion of "land health" (Leopold 1941). It is clear from Leopold's various writings that he meant "land" to be the entire ecosystem—that is, the web of relationships between organisms with each other and their environment. Substituting ecosystem for land—a substitution wholly justified by Leopold's comprehensive view of natural systems—we find that Leopold set as the goal of the enterprise of ecosystem health to "determine the ecological parameters within which land may be humanly occupied without making it dysfunctional."

The idea that ecosystems can and do become dysfunctional lies at the core of environmental concerns. The dysfunction in ecosystems can be measured in a variety of ways (see Part 2 of this book). The most dramatic indicator, ecosystem services, relates to properties of the ecosystem that offer direct benefits for the human community (for example, as a sink for waste disposal and detoxification of chemicals, as a source of food and fiber, or as a provider of potable water) (Cairns and Pratt 1995). In many instances, these services are sharply curtailed when ecosystems come under stress. As a consequence, clean air, clean water, and renewable resources such as fish and timber can no longer be taken for granted.

In most cases, declines in ecological services are permanent, and efforts to restore such services have met with meager results. It appears to be the exception rather than the rule when apparent damage to ecosystems proves to be

Defining Ecosystem Health

David Rapport

ECOSYSTEM HEALTH AS A METAPHOR

Metaphors are fundamental to the way we think and are often used in problem solving. Their use is clearly not restricted to poetry, but has a legitimate place in science as well. In both areas, metaphors stimulate associations, thereby bringing into juxtaposition phenomena that might initially appear to have little in common (Rapoport 1983). In the realm of science, their use often points to phenomena in very different spheres that bear some similarity, making it appropriate to transfer concepts and models across disciplines (Rapport and Turner 1977). The use of metaphor also serves as a powerful communication device. In ecosystem assessment, the health metaphor provides a language—that is, signs, diagnostic indicators, dysfunction, and ecosystem ills—all terms with which the public is familiar.

Under closer scrutiny, the use of metaphor in this and many other applications may be a double-edged sword—to make use of yet another metaphor—that simply drives home the point that we can't seem to manage without these comparisons. On the one hand, the health metaphor extends the notion of health from the level of the individual (clinical medicine, veterinary medicine) or population (public health) to the ecosystem (many species in interaction with their environment). On the other hand, its use begs a question: Are we resurrecting the much discredited analogy of "ecosystem as organism" (Rapport et al 1985; Rapport 1995)? This outcome is hardly our intent, so the question becomes one of whether it is possible to derive the advantages of this strategy without incurring undesirable costs. That is, can one accept that the health concept is relevant to the ecosystem level while simultaneously rejecting the notion of ecosystem as organism?

Whitford WG, Rapport DJ, Goothousen RM. The central Rio Grande Valley—organizing and interpreting ecosystem health assessment data. GIS World 1996; 9:60–62.

Woodwell GM. Ecology. The restoration. Restoration Ecol 1994;2:1–3.

Yazvenko SB, Rapport DJ. A framework for assessing forest ecosystem health. Ecosystem Health 1996;2:41–55.

Ludwig D, Hillborn R, Walters C. Uncertainty, resource exploitation, and conservation: lessons from history. Science 1993;260:17–36.

McMichael AJ. Planetary overload, global environmental change and the health of the human species. Cambridge, UK: Cambridge University Press, 1993.

McMichael AJ, Martens WJM. The health impacts of global climate change: grappling with scenarios, predictive models, and multiple uncertainties. Ecosystem Health 1995;1:23–34.

Minns CK, Moore JE, Schindler DW, Jones NL. Assessing the potential extent of damage to inland lakes in Eastern Canada due to acidic deposition IV. Predicted impacts on species richness in seven groups of aquatic biota. Can J Fisheries Aquatic Sci 1990;47:821–830.

Mitchell RC, Carson RT. Using surveys to value public goods: the contingent valuation method. Washington, DC: Resources for the Future, 1989.

Myers M. The Gaia atlas of future worlds: challenge and opportunity in an age of change. New York: Doubleday, 1991.

Parry ML, Rosenzweig C. Food supply and the risk of hunger. Lancet 1993; 342:1345–1347.

Rapport DJ. Ecosystem health: more than a metaphor? Environ Values 1995; 4:287–309.

Regier HA, Baskerville GL. Sustainable redevelopment of regional ecosystems degraded by exploitive development. In: Clark WC, Munn RE, eds. Sustainable development of the biosphere. London: Cambridge University Press, 1986:75–103.

Regier HA, Hartmann WL. Lake Erie's fish community: 150 years of cultural stresses. Science 1973;180:1248–1255.

Sherman K, Busch DA. Assessment and monitoring of large marine ecosystems. In: Rapport DJ, Gaudet CL, Calow P, eds. Evaluating and monitoring the health of large-scale ecosystems. New York: Springer, 1995.

Somerville MA. Planet as patient. Ecosystem Health 1995;1:61–71.

Tolba MK, El-Kholy OA, El-hinnawi E, Holdgate MW, McMichael DF, Munn RE. The world environment. London: Chapman & Hall, 1992.

UNEP (United Nations Environmental Programme). Global biodiversity assessment. Nairobi, Kenya: UNEP, 1995.

Union of Concerned Citizens. World scientists' warning to humanity. Nucleus 1992;14:1–3.

Vogl RJ. The ecological factors that produce perturbation-dependent ecosystems. In: Cairns J, ed. The recovery process in damaged ecosystems. Ann Arbor, MI: Ann Arbor Science, 1980:63–64.

Westman WE. Evaluating the benefits of ecosystem integrity. In: Edwards CJ, Regier HA, eds. An ecosystem approach to the integrity of the Great Lakes in turbulent times. Great Lakes Fish Comm Spec Pub 90-4, 1990:91–103.

Whitford WG. Desertification: implications and limitations of the ecosystem health metaphor. In: Rapport DJ, Gaudet D, Calow P, eds. Evaluating and monitoring the health of large-scale ecosystems. Heidelberg: Springer-Verlag, 1995:273–294.

actually prove to be only "the tip of the iceberg." Changes in human health—such as fetal abnormalities, resurgence of infectious diseases, and prevalence of skin cancer—may provide key indicators of environmental change.

REFERENCES

Arrow K, Bollin B, Costanza R, Dasgupta P, Folke C, Holling CS, Jansson B-O, Levin S, Maler K-G, Perrings C, Pimentel D. Economic growth, carrying capacity, and the environment. Science 1995;268:520–521.

Beckerman W. Economic growth and the environment; whose growth? whose environment? World Dev 1992;20:481–496.

Bloom DE. International public opinion on the environment. Science 1995; 269:354–358.

Carson R. Silent spring. Boston: Houghton-Mifflin, 1962.

Colburn T, Dumanoski D, Myers JP. Our stolen future. Toronto: Penguin Books, 1996.

Common M. Economists don't read *Science*. Ecol Econ 1995;15:101–103.

Costanza R, Farber SC, Maxwell J. The valuation and management of wetland ecosystems. Ecol Econ 1989;1:335–361.

Covington WW, Moore MM. Southwestern Ponderosa forest structure: changes since Euro-American settlement. J Forestry 1994;92:39–47.

Epstein PR. Emerging diseases and ecosystem instability: new threats to public health. Am J Public Health 1995;85:168–172.

Freeman AM III. The measurement of environmental and resource values. Washington, DC: Resources for the Future, 1993.

Gunn JM, ed. Restoration and recovery of an industrial region: progress in restoring the smelter-damaged landscape near Sudbury, Canada. New York: Springer-Verlag, 1995.

Haines A, Epstein PR, McMichael AJ. Global health watch: monitoring impacts of environmental change. Lancet 1993;342:1464–1469.

Hildén M, Rapport DJ. Four centuries of cumulative impacts on a Finnish river and its estuary: an ecosystem health-approach. J Aquatic Ecosystem Health 1993; 2:261–275.

Holling CS. Sustainability: the cross-scale dimension. In: Munasinghe M, Shearer W, eds. Defining and measuring sustainability: the biogeophysical foundations. Washington, DC: World Bank, 1995.

———. Investing in research for sustainability. Ecological Applications 1993;3:552–555.

Holmes TP, Kramet RA. Contingent valuation of ecosystem health. Ecosystem Health 1996;2:56–60.

Jacobson J. Environmental refugees: a yardstick of habitability. Worldwatch paper no. 86. Washington DC: Worldwatch Institute, 1988.

Levins R, Awebuch T, Brinkmann U, et al. The emergence of new diseases. Am Scientist 1994;82:52–60.

Linden E. The tortured land. Time Sept. 4, 1995:36–47.

Similarly, recovery of ecosystems after disturbance ought to provide a measure of resilience. One might, for example, hypothesize that stressed ecosystems are less resilient than unstressed ecosystems (Rapport and Regier 1995; Rapport and Whitford 1996). We have tested such a hypothesis in the field, in the semi-arid grasslands of southwest New Mexico. In that location, a stress gradient had been established by grazing pressure and trampling by cattle in the vicinity of deep water wells (Rapport and Whitford 1997). As shown in Figure 2.1, recovery of grasses after the severe drought of 1994–1995 was correlated with distance from the well. Recovery or "bounce-back" of grasses was much greater farther from the well (in the less stressed area) than close to the well.

Organization

Organization refers to ecosystem complexity; this characteristic varies from system to system but generally tends to increase with secondary succession in terms of number of species and variety and intricacy of interactions (such as symbiosis, mutualism, and competition). Over time, interdependence increases among biotic and abiotic elements of the ecosystem. These trends are reversed by stress (Odum 1985). Stressed ecosystems generally show reduced species richness, fewer symbiotic relationships, and more opportunistic species (Rapport et al 1985).

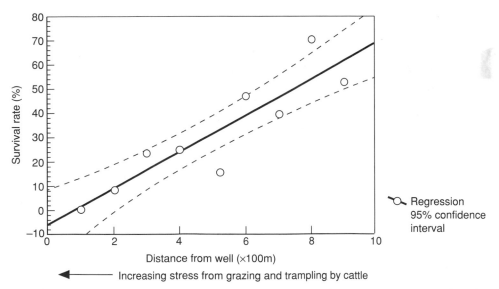

FIGURE 2.1. *Survival of* Sporobolus *sp. along a stress gradient following a severe drought. The solid line represents the statistically derived regression line. The dashed lines represent the 95% confidence interval of the regression.*

Maintenance of Ecosystem Services

Maintenance of ecosystem services is emerging as a key criteria for evaluating ecosystem health. It refers to functions that benefit the human community, such as detoxification of chemicals, water purification, production of game species, and reduced soil erosion. Stress generally reduces both the quality and quantity of such services (Cairns and Pratt 1995). In contrast, healthy ecosystems possess a full range of ecosystem services.

Management Options

Healthy ecosystems are supportive of many potential uses, such as harvesting of renewable resources, recreation, and provision of drinking water. Degraded ecosystems no longer support those options. For example, semi-arid grasslands in the U.S. Southwest (New Mexico/Arizona) once represented important grazing lands for domestic livestock (cattle and sheep). They also served as vegetative buffers regulating run-off to streams and rivers (Whitford 1995). Today, much of this landscape has been degraded to shrub land or mesquite dunes because of overgrazing. As a result, grazing at the levels of the historic stocking rates is no longer an option (Milton et al 1994).

Reduced Subsidies

All managed ecosystems depend on external inputs. These inputs constitute a subsidy—that is, they serve as additional inputs to those required by natural systems (such as the use of pesticides, herbicides, and fossil fuels in agriculture). Healthy ecosystems do not require an increasing subsidy to maintain productivity. Thus, one indicator of health is reduced subsidy in terms of required external energy and material inputs to sustain productivity. Another form of subsidy—one measured in terms of economic incentives, such as price supports, tax rebates, and the like—encourages excessive exploitation by not requiring that production cover full costs. This type of input leads, inevitably, to nonsustainable production (Schmidheiny 1992). In general, reduced dependence on external subsidies of all kinds provides an indicator of ecosystem health.

Subsidies may also take hidden forms, such as increased human health costs associated with environmental change. For example, human activity resulting in biosphere and ecosystem degradation is associated with increased risks and prevalence of human diseases (Epstein 1995; Patz 1996). These higher health costs are a form of hidden subsidy to those activities producing environmental damage, as these costs are not borne by industry but by society as a whole.

In summary, all managed ecosystems depend on some form of subsidy, whether it be human labor or fossil fuels. A healthy managed ecosystem is characterized by a diminishing (or at least not increasing) subsidy per unit output. It is also characterized by no increase in human health risks. In effect, healthy ecosystems have reduced dependence on subsidies in the biophysical, economic, and human health domains.

Damage to Neighboring Systems

In the great economic Depression (1929–1933), countries were able to stimulate their domestic economies temporarily at the expense of their neighbors by blocking imports. Using these "beggar thy neighbor" policies (that is, gains at the expense of neighboring economies), one country could temporarily gain an advantage over another. When these actions inspired retaliation from the targeted country, the net effect became economic losses all around because of the sharp reduction in world trade.

Many examples exist to show that ecosystems can thrive at the expense of others. This situation generally occurs, for instance, when waste residuals are transported to neighboring systems, as when pollution or sediments are exported. In Costa Rica, banana plantations have had severe impacts on off-shore coral reefs through increased siltation that, when transported by currents offshore, tends to smother coastal reefs. On Canada's Pacific Coast, logging practices have reduced the viability of coastal fisheries. In this case, siltation run-off degraded salmon spawning and feeding grounds in coastal estuaries and streams. Manure management and cropping practices in agriculture often damage aquatic systems. Run-off from agricultural fields, which may contain increased nutrients, toxic substances, and suspended solids, becomes a diffuse source of stress to the receiving waters and ultimately takes its toll in increased health risks to humans, reduced quality of groundwater, and loss of recreational opportunities.

Human Health Effects

Given the many pathways by which ecosystem change can adversely impact human health (see, for example, McMichael 1993; Epstein 1995; Patz 1996), human health itself might serve as a synoptic measure of ecosystem health. Healthy ecosystems are characterized by their capability to sustain healthy human populations.

REFERENCES

Cairns J Jr., Pratt JR. The relationship between ecosystem health and delivery of ecosystem services. In: Rapport DJ, Gaudet C, Calow P, eds. Evaluating and monitoring the health of large-scale ecosystems. Heidelberg: Springer-Verlag, 1995:273–294.

Callicott JB. Aldo Leopold's metaphor. In: Costanza R, Norton B, Haskell B, eds.

Ecosystem health: new goals for environmental management. Washington, DC: Island Press, 1992:42–56.

Clements FE. Plant succession; analysis of the development of vegetation. Publ. Carnegie Institute 1916;242:1–512.

Dubos R. Man, medicine and environment. New York: Praeger, 1968.

Edmondson WT. Water-quality management and lake eutrophication: The Lake Washington Case. In: Campbell TH, Sylvester RO, eds. Water resources management and public policy. Seattle: University of Washington Press, 1968:139–178.

Ehrenfeld D. The marriage of ecology and medicine: are they compatible? Ecosystem Health 1995;1:15–22.

Epstein PR. Diseases and ecosystem instability: new threats to public health. Am J Public Health 1995;85:168–172.

Hartig JH, Zarull MA. Towards defining aquatic ecosystem health for the Great Lakes. J Aquatic Ecosystem Health 1992;1:97–108.

Hawley A. Human ecology: a theory of community structure. New York: Roland Press, 1959.

Hildén M, Rapport DJ. Four centuries of cumulative impacts on a Finnish river and its estuary: an ecosystem health-approach. J Aquatic Ecosystem Health 1993; 2:261–275.

Holling CS. Sustainability: the cross-scale dimension. In: Munasinghe M, Shearer W, eds. Defining and measuring sustainability: the biogeophysical foundations. Washington: World Bank, 1995.

————. The resilience of terrestrial ecosystems: local surprise and global change. In: Clark WC, Munn RE, eds. Sustainable development of the biosphere. New York: Cambridge University Press, 1986:292–317.

Hutton J. Theory of the earth; or an investigation of laws observable in the composition, dissolution and restoration of land upon the globe. Trans Roy Soc Edinburgh 1788;1:209–304.

Lackey RT. Pacific salmon, ecological health, and public policy. Ecosystem Health 1996;2:61–68.

Last JM. Global change; ozone depletion, greenhouse warming, and public health. Ann Rev Public Health 1993;14:115–136.

Leopold A. Wilderness as a land laboratory. Living Wilderness 1941;6:3.

Levins R, Awebuch T, Brinkman U, et al. The emergence of new diseases. Am Scientist 1994;82:52–60.

Mageau MT, Costanza R, Ulanowicz RE. The development and initial testing of a quantitative assessment of ecosystem health. Ecosystem Health 1995;1:201–213.

Maini JS. Sustainable development of forests. Unasylva 1992;43:3–7.

McMichael AJ. Planetary overload, global environmental change and the health of the human species. Cambridge, UK: Cambridge University Press, 1993.

Milton SJ, Dean WRJ, duPlessis MA, Siegfried WR. A conceptual model of arid rangeland degradation. BioScience 1994;44:70–76.

Myers N. Environmental unknowns. Science 1995;269:358–360.

Odum E. Trends expected in stressed ecosystems. BioScience 1985;35:419–422.

Patz JA. Global climate change and emerging infectious diseases. JAMA 1996; 275:217–223.

Pimm SL. The complexity and stability of ecosystems. Nature 1984;307:321–326.

Porn I. An equilibrium model of health. In: Mordenfelt L, Lindahl B, eds. Health, disease and causal explanations in medicine. Dordrecht: Reidel, 1984.

Rapoport A. The metaphor in the language of science. Semiotische Berlichte 1983; 12/13:25–43.

Rapport DJ. Ecosystem health: an emerging integrative science. In: Rapport DJ, Gaudet CL, Calow P, eds. Evaluating and monitoring the health of large-Scale ecosystems. Berlin: Springer-Verlag, 1995:5–31.

———. Ecosystem health: more than a metaphor? Environ Values 1995;4:287–309.

———. What constitutes ecosystem health? Perspect Biol Med 1989;33:120–132.

Rapport DJ, Hilden JE, Roots EF. Transformation of northeastern ecosystems under stress: arctic ecological changes from the perspective of ecosystem health. Stress and Disturbance in Northern Ecosystems. Springer Verlag, 1996 (in press).

Rapport DJ, Regier HA. Disturbance and stress effects on ecological systems. In: Patten BC, Jorgenson SE, eds. Complex ecology: the part–whole in ecosystems. Englewood Cliffs, NJ: Prentice-Hall PTR, 1995.

Rapport DJ, Regier HA, Hutchinson TC. Ecosystem behaviour under stress. Am Naturalist 1985;125:617–640.

Rapport DJ, Regier HA, Thorpe C. Diagnosis, prognosis and treatment of ecosystems under stress. In: Barrett GW, Rosenberg R, eds. Stress effects on natural ecosystems. New York: Wiley, 1981:269–280.

Rapport DJ, Turner JE. Economic models in ecology. Science 1977;195:367–373.

Rapport DJ, Whitford WG. Common properties of ecosystems under stress (in preparation). 1997.

Schindler DW. Experimental perturbations of whole lakes as tests of hypotheses concerning ecosystem structure and function. Oikos 1990;57:25–41.

Schmidheiny S. Changing course: a global business perspective on development and the environment. Cambridge, MA: MIT Press, 1992.

Stebbing ARD. Stress, health and homeostasis. Mar Pollut Bull 1981;12:326–329.

Tolba MK, E1-Kholy OA, El-hinnawi E, Holdgate MW, McMichael DF, Munn RE. The world environment, 1972–1992. London: Chapman & Hall, 1992.

UNEP (United Nations Environmental Programme). Global biodiversity assessment. Nairobi, Kenya: UNEP, 1995.

Union of Concerned Citizens. World scientists' warning to humanity. Nucleus 1992;12:1–3.

Whitford WG. Desertification: implications and limitations of the ecosystem health metaphor. In: Rapport DJ, Gaudet C, Calow P, eds. Evaluating and monitoring the health of large-scale ecosystems. Heidelberg: Springer-Verlag, 1995:273–294.

World Watch Institute. State of the world 1994. London: Earthscan Publications, 1994.

Yazvenko SB, Rapport DJ. A framework for assessing forest ecosystem health. Ecosystem Health 1996;2:41–55.

CHAPTER 3

Dimensions of Ecosystem Health

David Rapport

ECOSYSTEM health is most generally viewed from a biophysical perspective—that is, in terms of structure and functions of the ecosystem. This viewpoint illuminates only one aspect of health, however. Other important dimensions concern socioeconomic features and human health itself (Rapport 1995). These dimensions emphasize the pervasive effects that changing environments have on our economic and physical health. Healthy economies and healthy humans are seen as essential to maintaining healthy ecosystems. In turn, an unhealthy economy invariably results in more stress on ecosystems. Unhealthy ecosystems may have considerable adverse consequences for the economy (such as loss of renewable resources or ecological services).

Ecosystem transformation has profound implications for human health. A clear need exists to better understand both its direct effects, such as the increased risks of certain cancers brought by the thinning of the stratospheric ozone layer, and the indirect effects, such as those resulting in poorer human nutrition and resistance to disease (McMichael 1993). Changes in all three dimensions should be taken into account to properly assess ecosystem health.

BIOPHYSICAL DIMENSION

The biophysical domain focuses on the basic properties of ecosystem function and structure, such as nutrient cycling, energy flow, biodiversity, plant and animal species dominance, cycling or sequestering of toxic substances, and habitat diversity. Such features provide a characterization of the ecosystem and are directly related to ecological services upon which the human community depends.

Of particular interest is a group of indicators characteristic of the Ecosystem Distress Syndrome (EDS) (Rapport, Regier, and Hutchinson 1985). This syndrome has been documented in a large number of terrestrial and aquatic ecosystems (Hildén and Rapport 1993; Rapport, Hildén, and Roots 1996; Birkett and Rapport 1996; Yazvenko and Rapport 1996). Signs of EDS include impaired primary productivity, reduced biodiversity, alterations in biotic structure that favor short-lived opportunistic species, and reduced population regulation, resulting in larger population oscillations and more disease outbreaks.

SOCIOECONOMIC DIMENSION

Although healthy ecosystems must be economically viable, this requirement does not imply that the viability must be self-contained. Economic viability must be seen in a global context. Obviously, in sensitive ecosystems such as the Arctic and sub-Arctic, not all requirements for human existence can be met solely from the resources of those systems. Historically, however, this case held true with less population pressure and low-impact technology (Rapport, Hildén, and Roots 1996).

Economic viability is a most acute issue in some developing countries, where it is often said that "environment is a luxury." Existing poverty encourages quick solutions, often at the expense of ecosystem health. Through overharvesting of resources, pollution, and the like, these solutions sharply reduce future options for economic viability. At the same time, unbridled consumption in the developed countries has led to impoverishment of the environment—witness the smog impacts in Los Angeles, and the pressures on tropical forests driven by consumer interests in North America and Japan. Globally, much of the immediate impact of environmental degradation falls upon the developing countries hungry for economic opportunities at almost any cost. For example, oil exploitation in Nigeria has recently taken place with the cost of destroying the regional environment and the culture of indigenous peoples—a pattern often repeated elsewhere in both developing and developed countries.

Economic viability is assessed by means of traditional indicators such as gross national product, corporate profits, income distribution, and debt burdens. Such assessments are often misleading as they fail to account for the depletion of natural capital (Jansson et al 1992; Repetto 1992). Furthermore, as advocated by the Business Council for Sustainable Development (Schmidheiny 1992), economic health must be assessed within the context of full-cost pricing—that is, in the absence of subsidies that support activities that are uneconomical and encourage environmental degradation.

As we have mentioned previously, economic viability remains very much dependent upon ecosystem health. Stressed ecosystems become increasingly vulnerable to catastrophes such as crop failures and fishery collapse, as has recently occurred along the North American East Coast with cod stocks and other ground fisheries (Ludwig et al 1993). Such developments have serious economic consequences for regional employment and incomes. Economic losses also may be triggered by increases in the frequency and severity of disturbances such as fires and floods that are, in turn, linked to changing environments at regional and global levels.

HUMAN HEALTH DIMENSION

The human health dimension includes many considerations. For example, health risks as a result of increased exposure to ultraviolet (UV) radiation (caused by a thinning of protective stratospheric ozone layer) can be measured in terms of likely increased rates of skin cancer. Threats to human health have also been related to global warming, though this link is less well established (McMichael 1993). In the summer of 1995, a record heat wave in Chicago—possibly a manifestation of climate warming—led to more than 500 deaths.

Other consequences of environmental change for human health involve increased exposure to contaminants from air, water, and food, and a potential increase in air- and water-borne infectious diseases and new diseases spread from other species to human hosts (Levins et al 1994; McMichael and Martens 1995; McMichael 1993).

Large-scale, but difficult-to-quantify indirect impacts on human health arise through the degradation of forests and agricultural systems. As these food- and fiber-producing systems become more degraded, the effects on nutrition can be severe. In turn, malnutrition leads to enhanced vulnerability to infectious diseases by lowering natural resistance (McMichael 1993; Levins et al 1994).

SPATIAL/TEMPORAL DIMENSION

Ecosystems exhibit complex spatial and temporal responses to stress, rendering analysis of cause and effect difficult. Signs of dysfunction may first appear in isolated components of the ecosystem (for example, in alterations of behavior of sensitive species) and then become more ubiquitous over time (Hildén and Rapport 1993; Schindler 1990; Smith 1981). Often gradients of stress impacts surround the point-sources (Gunn 1995), which combine with stress impacts from diffuse sources to create complex spatial/temporal patterns.

These complex patterns inherent in ecosystem response to stress also raise questions about scale dependency. Might dysfunction at a fine scale be seen as a minor disturbance at a larger scale? Suppose a small tributary is highly degraded because of a dam upstream. These impacts may have negligible consequences for conditions pertaining to the river as a whole, however. In such a case, an evaluation of health is clearly tied to the issue of scale. At the scale of the tributary, signs of distress are present and that part of the system is in poor health; at the scale of the watershed, the damage is insignificant and the watershed may be considered to be in good health.

Similarly, in an agricultural landscape context where most farms employ land stewardship practices that minimize erosion, nitrate run-off in groundwater, and other harmful effects, the refusal of a few farms to adopt these practices may have a negligible impact on the health of the agricultural landscape, but a pronounced local effect within those farming systems. Here again, ecosystem health assessments are necessarily dependent on scale. At the level of the farming system, the agro-ecosystem is unhealthy; at the landscape level, it is healthy.

STRESS AND RESPONSE

Single Stressors

Single causal agents resulting in stress on ecosystems provide the simplest cases for analysis of the relationships between stress and ecosystem response. Lake Washington offers a classic study. In this case, sewage from Seattle, Washington, represented the predominant stress. Once its role was identified, the diversion of the inflow away from the lake resulted in a rapid recovery of the ecosystem (Edmonson 1968)

Multiple Stressors

The vast majority of ecosystem transformations are characterized by multiple stress factors, such as physical restructuring, the introduction of exotic species, waste residuals, and harvesting, whose cumulative effects change the viability of the system. Consider the well-documented, large-scale ecosystem transformations such as those occurring in the Laurentian Great Lakes and the Baltic Sea (Regier and Baskerville 1986; Hartig and Zarull 1992; Harris et al 1988). In such cases, relationships between stress and ecosystem response can prove very complex.

Even in the case of only two dominant stress factors, the picture becomes highly complex. In Figure 1.1 (P. 7), we showed the transformation of the Ponderosa pine forests in the Southwest United States under the combination

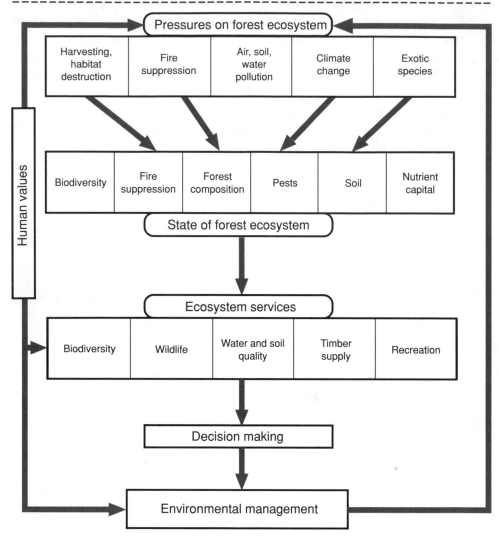

FIGURE 3.1. *Multiple stresses on forest ecosystems and their impact on ecosystem services and decision making (from Rapport and Whitford 1997).*

of two stresses: overgrazing and fire suppression. In this case, both stresses have clearly played a major role in the degradation of the pine forest, but it is not possible to apportion the total impact between the stresses.

The picture becomes even more complex when, as in many real situations, a variety of stresses simultaneously affect the ecosystem (Figure 3.1). In such cases, it can be virtually impossible to attribute cause and effect on a stress-by-stress basis for a number of reasons. First, the impact of each stress depends upon its intensity and history, as well as the resistance or buffering capacity of the receiving ecosystem and the effect of other stresses that are present. Second, the temporal dynamics of the stress are often crucial to the outcome. Acidification of estuaries initiated by spring run-off over acidic soils, for example, has a far greater impact on the fish community than would the identical stress if it occurs in late summer or early autumn, after the breeding season ends (Hildén and Rapport 1993). Third, potential synergism can arise between stresses as well as occasional antagonism (Rapport et al 1985). Fourth, it is rare to find an instantaneous, ecosystem-wide response to stress. Time lags—that is, gaps between the onset of stress and detectable change—often span decades. For example, in the demise of the Central European forest, an approximately 25-year interval separated the onset of air pollution stress and the ultimate collapse of the forest. This lag reflects the buffering capacity of the ecosystem and the time required for toxic effects to cause tree mortality. Fifth, stress may originate outside the system and have complex spatial patterns of impact. The dominant stresses are not necessarily those in closest proximity to the impacted ecosystem. For example, a brilliant piece of detective work led to the discovery that Swedish lakes and forests were threatened by pollution coming from a distant source, the United Kingdom. This hypothesis was first put forth by the Swedish soil scientist, Svante OdJn, much to the scepticism, if not outright scorn, of some of his colleagues. Detecting a precipitous decline in the pH in the soils of southern and central Swedish forests and lakes, OdJn (1968) hypothesized that the likely cause was air pollution from England and Western Europe. This hypothesis has subsequently been confirmed (OdJn 1977).

REFERENCES

Birkett S, Rapport DJ. Marine ecosystem health: a comparative study of the Baltic Sea and the Gulf of Mexico. Ecosystem Health 1996;2 (in press).

Edmonson WT. Water-quality management and lake eutrophication: the Lake Washington case. In: Campbell TH, Sylvester RO, eds. Water resources management and public policy. Seattle: University of Washington Press, 1968:139–178.

Gunn JM, ed. Restoration and recovery of an industrial region: progress in restoring the smelter-damaged landscape near Sudbury, Canada. New York: Springer-Verlag, 1995.

Harris HJ, Harris VA, Regier HA, Rapport DJ. Importance of the nearshore area for sustainable redevelopment in the Great Lakes with observations on the Baltic Sea. Ambio 1988;17:112–120.

Hartig JH, Zarull MA. Towards defining aquatic ecosystem health for the Great Lakes. J Aquatic Ecosystem Health 1992;1:97–108.

Hildén M, Rapport DJ. Four centuries of cumulative impacts on a Finnish river and its estuary: an ecosystem health-approach. J Aquatic Ecosystem Health 1993; 2:261–275.

Jansson A-M, Hammer M, Folke C, Costanza R. Investing in natural capital. Washington, DC: Island Press, 1992.

Levins R, Awebuch T, Brinkmann U, et al. The emergence of new diseases. Am Scientist 1994;82:52–60.

Ludwig D, Hilborn R, Walters C. Uncertainty, resource exploitation, and conservation: lessons from history. Science 1993;260:17–36.

McMichael AJ. Planetary overload, global environmental change and the health of the human species. Cambridge, UK: Cambridge University Press, 1993.

McMichael AJ, Martens WJM. The health impacts of global climate change: grappling with scenarios, predictive models, and multiple uncertainties. Ecosystem Health 1995;1:23–34.

OdJn D. The acidification of air and precipitation and its consequences on the natural environment. Swedish National Science Resource Council. Ecology Committee Bulletin 1968;1:68 (in Swedish).

OdJn S. The OECD programme on long range transport of air pollutants: measurements and findings. Paris: Organization for Economic Cooperation and Development, 1977.

Rapport DJ. Ecosystem health: an emerging integrative science. In: Rapport DJ, Gaudet CL, Calow P, eds. Evaluating and monitoring the health of large-scale ecosystems. Berlin: Springer-Verlag, 1995:5–31.

Rapport DJ, Hilden JE, Roots EF. Transformation of northern ecosystems under stress: Arctic ecological changes from the perspective of ecosystem health. In: Stress and disturbance in northern ecosystems. Springer-Verlag, 1996 (in press).

Rapport DJ, Regier HA, Hutchinson TC. Ecosystem behavior under stress. Am Naturalist 1985;125:617–640.

Rapport DJ, Whitford WG. Common properties of ecosystems under stress. 1997 (in preparation).

Regier HA, Baskerville GL. Sustainable redevelopment of regional ecosystems degraded by exploitive development. In: Clark WC, Munn RE, eds. Sustainable development of the biosphere. London: Cambridge University Press, 1986: 75–103.

Repetto R. Earth in balance sheet: incorporating natural resources in national income accounts. Environment 1992;34:12–20.

Schindler DW. Experimental perturbations of whole lakes as tests of hypotheses concerning ecosystem structure and function. Oikos 1990;57:25–41.

Schmidheiny S. Changing course: a global business perspective on development and the environment. Cambridge, MA: MIT Press, 1992.

Smith WH. Air pollution and forests: interactions between air contaminants and forest ecosystems. New York: Springer, 1981.

Yazvenko SB, Rapport DJ. A framework for assessing forest ecosystem health. Ecosystem Health 1996;2:41–55.

CHAPTER 4

Answering the Critics

David Rapport

W E HAVE already touched upon some of the issues raised about the appropriateness of the health metaphor and its proper limits. Many critics base their arguments against ecosystem health on a fundamental misconception about the purpose of metaphor. These critics claim that, because ecosystems are not organisms, health has no meaning at the ecosystem level.

Lackey (1996) provides a convenient summary of his objections (shared by many others) of difficulties with the concept of ecosystem health. The first of his objections rests squarely on the assumption (incorrect in our view) that a metaphor should be discarded if it does not apply to all situations. Lackey concedes that the notion of ecosystem health is superficially appealing, but is flawed for the following reasons.

Ecosystems are not organisms and do not behave like organisms. Although we would agree with this comment, we argue that this reason is insufficient by itself to discard the concept of ecosystem health. Clearly ecosystems become dysfunctional under stress (Karr 1993), which gives rise to the validity of the analogy of ecosystem as patient (Somerville 1995). The health metaphor suggests a focus on methods of early detection of system dysfunction, possible interventions, and preventive strategies for maintaining health.

No indices of ecosystem health exist nor is developing any possible. Arguing along a similar vein, Kelly and Harwell (1989, P. 13) state that the health metaphor is unsuitable because ". . . unlike indicators of adverse human health effects, there are no comparable integrative, simple measures or indices that show the effects of disturbances on ecosystems." Such assertions of a lack of objective measures for the impact of disturbance on ecosystems are simply false. David Schindler's painstaking work on the transformation of a well-buffered boreal lake to a highly acidified one documented in detail the impact of this source of stress on the structure and function of this ecosystem (Schindler 1990). In fact, numerous ecosystem-level experiments have yielded indicators of ecosystem transformation under stress (for a review, see Carpenter et al 1995). A number of indicators have already been incorporated into guidelines for ecosystem health (see, for example, Rapport, Gaudet, and Calow 1995; Whitford 1995; Karr et al 1986).

The use of this metaphor in political debates masks the value judgments that were made—judgments that are policy choices, not scientific choices. A judgment on health does involve an assessment of how the system is functioning relative to desirable states or goals. In essence, it draws upon both scientific judgments relating to function and societal values related to goals. To make this distinction clear in political debates requires being explicit about this duality. For example, if one determines that "the health of a particular lake is improving," one should explicitly include the values upon which the assessment rests. For example, one might say, "the health has improved since phosphorus loading has been curtailed, thus reducing undesirable summer alga blooms and enhancing habitat for sports fisheries." Such a statement makes clear both the objective measures of change in the condition of the lake and the societal values that may be affected (Hartig and Zarull 1992).

The roles of science and scientists in defining ecosystem or ecological health are contentious. To categorize something as healthy requires an implicit determination of the desired or preferred state. The example provided by Lackey (1996, P. 66) to illustrate this point is somewhat dubious. He speaks of purposeful changes in land use—for example, choices between a pristine forest, a highly productive dairy pasture, a fertile field of corn, or a bustling university campus. He then raises the question, "Which ecosystem is healthiest?" Posing the question in this manner may be a "red herring." It is analogous to asking, Which is healthiest: a dog, a cat, or a rabbit? Clearly, each ecosystem has a potential for health or degradation. The true question involves an examination of the degree to which ecosystems may become impaired, with a resultant loss of ecological services, as a result of stress from human activity and extreme natural events. Health assessments are valid only when comparing like with like, not when comparing apples and oranges. For the sake of argument, suppose that a stress, such as overgrazing, leads to the transformation of one type of system to another—for example, from a semi-arid grassland to a desert. Can one make any valid statements in this case with respect to ecosystem health? Is a semi-arid grassland any less healthy than a desert? Of course not. In this case, however, this issue is not the critical question. Rather, we must ask whether the grassland transformed by overgrazing is as healthy as the initial grassland. The answer is straightforward: Even the healthiest of deserts does not support a herd of cattle and, from a rangelands perspective, this transformation is clearly unhealthy.

Kelly and Harwell (1989, P. 13) suggest another objection to the ecosystem health concept. Their objection asserts:

> Attempts at an analogy of ecological health to human health have not been satisfying, in part because the exposure of ecosystem to stress is very complex . . . in part because ecosystems are more diverse and more complex than the human metabolic systems . . . and in part because ecosystems are much less internally coordinated and less able to respond to stress by controlled compensatory mechanisms that engender homeostasis of the systems. If ecosystems truly could be seen as superorganisms, with a few key components or processes reflecting the state of being of the ecosystem superorganism, then ecological response and recovery predictions in principle could be as reliable as human health check-ups and prognoses. But the reality is that predictive ecology lags far behind because ecological systems are less robustly defined, their dynamics thus being inherently less tractable, and their state not so easily fully characterized.

While this critique suggests (correctly, in our view) that systematic assessment and diagnosis of ecosystems are likely more complex than the comparable operations in human health, we believe that its assertion that the assessment of human health is a simple system approximating an exact science is far from the mark. Furthermore, complexity of the ecosystem is not a sufficient reason to abandon the effort to characterize dysfunction and relate its appearance to probable causes. While one may agree that ecology, owing to its complexity, may never be a fully predictive science, case studies such as those of Schindler (1990) and others (see, for example, Hildén and Rapport 1993; Rapport et al 1995) have provided a foundation for a better understanding of the complex mechanisms governing the response of ecosystems to both single and multiple stresses. These studies have revealed the importance of temporal lags, spatial displacements, synergistic and antagonistic effects of stresses, ambiguities in the use of single indicators, and a variety of biological, physical, and chemical mechanisms that mediate the impacts of stress on ecosystems.

Minns (1992, P. 110) concentrates his critique on the difficulty in setting standards for ecosystems because of a lack of consensus on the expected values of various indicators in a healthy system. Again, recent empirical studies (see, for example, Elmgren 1989; Hildén and Rapport 1993; Schindler 1990) serve to build consensus and establish expected values. Obviously if "norms" or "standards" could not be established, health assessments and environmental management would become very arbitrary. Sprugel (1991) has shown that establishing the normal range of ecosystem behavior in forested systems requires very-long-term data. In addition, one must admit that many alternative, wholly different communities are possible and "natural." Nevertheless, it is still quite possible to identify conditions that fall far outside that range—that is, clear signs of degradation of ecosystems that are not within the purview of the development and evolution of healthy systems (Whitford 1995; Gunn 1995). In examining the response of boreal lakes to acidification, Schindler suggests it is often difficult to "confidently distinguish between natural variation and low-level effects of perturbations on ecosystems" (Schindler 1987). His own 1990 studies indicate, however, that sufficient long-term data can differentiate between pathological states and norms. If "norms" cannot be established, not only is the practice of ecosystem health in doubt, but all approaches to ecosystem management are also questionable.

Another criticism stems from the fact that the "health paradigm" has come under attack for various reasons within the health sciences; critics suggest that this fact indicates that it provides a poor model to emulate. Traditional biomedical models undoubtedly have major shortcomings, particularly in their attribution of diseases to proximal causes within a simplistic, linear causal reductionist framework (Ehrenfeld 1995; Levins et al 1994). These limitations are now well recognized within the evolving practice of health care, however. The health concepts we draw upon explicitly recognize that the simple biomedical model is being replaced with a complex, multicausal model.

INTERFACING SOCIETAL VALUES AND SCIENCE

One major criticism of ecosystem health, particularly from the viewpoint of the natural sciences, is that it incorporates societal values in both setting goals and evaluating performance. Is this factor really a weakness of the field, or is it one of its main strengths? In the area of environmental management, we maintain that it is impossible to avoid the question of societal values; the challenge lies in bridging the present gap between the social and natural sciences. To ignore societal values is to ignore the potential to involve society in the discussion and in the solutions to ecosystem health problems. If solutions to ecosystem health challenges require public support, as they surely do, then it is critical to build a framework for analysis that explicitly includes a prominent role for societal values.

Values—that is, the degree of importance attached to ecological services derived from ecosystems or landscapes—will vary among individuals and interest groups. They are also very much related to the temporal and spatial scales on which environmental analysis is conducted. For example, individual values generally prevail on local scales, community values prevail on regional scales, and societal values dominate on still larger scales. Ecosystems and landscapes are regional in extent and generally involve time frames that span more than one generation. Thus, the community, which has a longer time horizon than the individual, represents the best "match" in terms of relating values to ecosystem outcomes.

Values are critical determinants of acceptable ecosystem management practice, and the decisions on which ecosystem transformations and resulting dysfunctions are permissible represent crucial issues. Working through the political process, societal values motivate incentives, regulations, and laws to manage stress on ecosystems. Ultimately, the subjects that people care about and the knowledge of how human activity influences these properties motivates decisions.

One must, however, be clear about one thing: neither ecosystem health nor any science has a mandate to determine societal values. Instead, science ought to articulate options by providing knowledge in a form that the public understands and that is pertinent to its concerns. In essence, scientists should serve as "independent counsel"—providing advice on assessments of existing conditions, likely future options, potential interventions, and other issues. To be effective, one must remain cognizant of societal values (which, in turn, change in response to new issues and information) and relate the findings of "how the system functions" to those aspects about which people care.

DISTINGUISHING AMONG HEALTH, INTEGRITY, AND SUSTAINABLE DEVELOPMENT

In policy statements, three "buzz word" phrases frequently appear in efforts to advance goals that are meant to achieve a viable future: ecosystem integrity, ecosystem health, and sustainable development. All are seemingly facets of the same "supergoal"—that of a viable environment. All have the problems of precise definition. And all apparently relate closely to one another. A failure to clarify the relationship between ecosystem health, integrity, and sustainable development may lead to unwarrented criticism. What, if any, distinctions can be made among these concepts? What special role does ecosystem health play in this triad? In addition, are these concepts good substitutes for one another?

Part of the problem reflects the newness of these terms—none has been in existence for more than a decade, and all of them are now being defined and refined in their own right. Thus, in some ways, it may be pointless to engage in a discussion of which terms are best. As the question has already been raised (see, for example, Gallopin 1995), however, a few comments with respect to both similarities and differences may prove useful.

No single or simple answer exists for the question of whether the terms are complementary or competitive. One view is that ecosystem health and ecosystem integrity are complements—a view enshrined in the conclusions of the outcome of the Rio Summit, which contained the phrase "to safeguard the health and integrity of the world's ecosystems" (United Nations 1992). A second perspective notes that ecological integrity ought to be the primary goal, for it is closely rooted to the natural evolution of life systems (Woodley et al 1993). A third view claims that ecosystem health is the prerequisite for sustainable development (Gallopin 1995).

The underlying supposition of all three concepts is that they seek to define conditions essential for the survival of future generations. In this similarity, the concepts at least are broadly oriented toward the same goal. Thus, embedded in the concept of sustainable development, is the principle that development must meet ". . . the needs of the present without compromising the ability of future generations to meet their own needs" (World Commission on Environment and Development 1987, P. 43). Embedded in the concept of ecological integrity (Woodley et al 1993) is the notion that natural evolution of ecosystems represents the best of all possible worlds and that it enhances the diversity and complexity of ecosystems, which provide immeasurable benefits to humankind. This natural "integrity" serves as a cherished goal worthy of safeguarding. The concept of ecosystem health stresses resilience, productivity, and vigor of life systems—the characteristics that confer viability and sustainability in terms of the biophysical, socioeconomic, and human health dimensions.

Are these concepts really at odds, and if so in what respect? We would first argue that distinctions such as those described above can differentiate between these concepts both in theory and in practice. We would hasten to add that, in environmental policy, finding common ground is far more important than dwelling on any differences. The means often determines the ends, however, and the proponents of each approach would undoubtedly draw attention to the primacy of the concept they advocate. For ecosystem health, for example, we would emphasize its focus on measures of the planet as patient, advancing methodologies to detect dysfunction at an early stage and to relate adverse changes to potentially modifiable causal factors. For sustainable development (World Commission on Environment and Development 1987), the primary goal is not the health of the system, but the sustainment of economic activity without compromising options for future generations. In essence, economic growth appears to constitute the primary goal, the hope being that wise use of technology can accommodate both sustained economic growth and environmental soundness. This possibility is hotly contested (Arrow et al 1995; Ayres 1995; Myers 1993). Many authors have claimed that the very notion of sustainable development is at best a pipe dream and at worst an oxymoron.

The conflicting opinions about the feasibility of sustainable development are based on different perceptions. Those opposed to the notion tend to view the Earth as finite but with a large but ultimately limited carrying capacity; this capacity may already be exceeded. According to this view, sustainability of the biosphere is not compatible with further economic growth unless the growth is defined so as not to drain and deplete the Earth's natural capital (such as soils, water, air, forests, and fisheries). In contrast, those promoting the feasibility of sustainable development see the Earth as an expandable universe, thanks to the powers of technology. They suggest that the planet can accommodate continuous growth through technological interventions. The latter advocates point to the apparent triumphs of technology—for example, crop yields have been bolstered and have averted, at least in most areas of the world, the gloomy predictions of Thomas Malthus. In this view, the "precautionary principle" errs on the side of being far too conservative, sacrificing immediate gains in economic growth for highly dubious future gains associated with preservation of present resources (Maddox 1995).

The notion of ecological integrity takes a very different tack in assessing the relationship of sustainable development and ecosystem health. Ecological integrity gives primacy to conservation—not of species, however, but of ecosystem organization. The argument here is seemingly straightforward: The goal is to preserve the natural integrity of the Earth's ecosystems—that is, the full complement of species and interrelationships that would prevail in the absence of human influences. More realistic circumstances recognize that, except for some remote ecological reserves, practically all of nature has been transformed by human activity. Thus, the goal might be to manage these systems so as to safeguard the functional complexity. Achieving this goal requires not the precise configuration of the system that would have been in place in the absence of human activity (in many cases, it would be difficult to determine the nature of such a system), but the protection of an "equivalent" degree of complexity and ecosystem function (Karr 1993; Woodley et al 1993).

ECOSYSTEM HEALTH: THE LAST FRONTIER OF MEDICINE

The notion of health has been the guiding force in the practice of medicine, irrespective of whether we speak of "Eastern" or "Western" medicine or "native (indigenous peoples) medicine." It is connected to, but not synonymous with, longevity. This concept is defined as the maintenance of "full function" appropriate to the stage of life; thus, it incorporates the avoidance of dysfunction. This condition or "resource" enables one to have the best possibilities to realize one's aspirations and goals.

Health at the level of the individual is the concept with which everyone is intuitively familiar. Health at the level of the community (public health) is a natural extension of this idea, particularly given that the analysis of health prospects and disease prevention is sometimes best addressed at the level of the whole community—for example, in the case of communicable, water-borne, or air-borne diseases.

Adapting the notion of health to other species—that is, to a larger entity, loosely defined in terms of ecological communities or ecosystems—represents a stretch, however. It requires that one accept the concept of ecosystem (or planet) as patient. This extension can be viewed from several perspectives. From the individual's point of view, the boundary conditions have direct implications for personal health, at least on a probabilistic basis. As discussed previously, considerable evidence suggests that global and regional changes are directly linked to personal health risks (Epstein 1995; Last 1993; McMichael 1993; Patz 1996). From society's point of view, it is becoming increasingly clear that all social goals depend on maintaining the viability of the larger life systems. From the methodology's point of view, much is potentially transferable across the different levels: identification and classification of indicators of dysfunction; signs and syndromes; expert systems for diagnosis; the merits of prevention over cure.

To suggest that ecosystem health is the last frontier of medicine is to hypothesize that the natural progression of health concerns at the level of the individual and at the level of the individual's species has a validity in assessing the health of the assemblage of many species making up the Earth's ecosystems. Present efforts in medicine are strongly focused on curing the individual after the disease or "complaints" are already present. Relatively little effort is placed in examining the conditions that gave rise to the disease. The links between the environment and human health strongly suggest that the "upstream" cause of many human ills lies with environmental transformation. Eventually, the practice of human medicine and human health care must address the root causes—that is, the transformations in the ecosystem that drive many individual and public health problems. For example, once cholera has been identified in a populated area (for example, Mexico City), it is not sufficient to treat only the "end of the pipe"; rather, one must determine the origin of the disease. This inquiry must address not only the microbe, but also the conditions in which the microbes flourish. From this perspective, the cost-effective treatment of cholera might well shift from treatment of patients who have contracted the disease to waste management in the watershed whose conditions foster the propagation and spread of the pathogen.

Ecosystem health goes even further than what environmental management can draw from medicine (in terms of methods and approaches) and what medicine can draw from ecosystem science (in terms of causal mechanisms). It challenges us to identify the societal goals that are compatible with sustainable life systems, to identify and validate indicators of ecosystem function that are essential to its evolution and perpetuation, and to merge goals (societal values) with biophysical realities. These challenges require goal-directed behavior of environmental management, toward maintenance of ecosystems within the parameters of their natural variability and within the bounds of alternative, self-sustaining, "desirable" states. The articulation of ecosystem health goals and indicators of performance to achieve these goals is a necessity if the future for humankind is to be viable.

REFERENCES

Arrow K, Bollin B, Costanza R, Dasgupta P, Folke C, Holling CS, Jansson BO, Levin S, Maler KG, Perrings C, Pimentel D. Economic growth, carrying capacity, and the environment. Science 1995;268:520–521.

Ayres RU. Economic growth: politically necessary but *not* environmentally friendly. Ecol Econs 1995;15:97–99.

Carpenter SR, Chisholm SW, Kreb CJ, Schindler DW, Wright RF. Ecosystem experiments. Science 1995;269:324–327.

Ehrenfeld D. The marriage of ecology and medicine: are they compatible? Ecosystem Health 1995;1:15–22.

Elmgren R. Man's impact on the ecosystem of the Baltic Sea: energy flows today and at the turn of the century. Ambio 1989;18:326–331.

Epstein PR. Diseases and ecosystem instability: new threats to public health. Am J Public Health 1995;85:168–172.

Gallopin GC. Perspective on the health of urban ecosystems. Ecosystem Health 1995;1:129–141.

Gunn JM, ed. Restoration and recovery of an industrial region: progress in restoring the smelter-damaged landscape near Sudbury, Canada. New York: Springer-Verlag, 1995.

Hartig JH, Zarull MA. Towards defining aquatic ecosystem health for the Great Lakes. J Aquatic Ecosystem Health 1992;1:97–108.

Hildén M, Rapport DJ. Four centuries of cumulative impacts on a Finnish river and its estuary: an ecosystem health-approach. J Aquatic Ecosystem Health 1993;2:261–275.

Karr JR. Protecting ecological integrity: an urgent societal goal. Yale J Intl Law 1993;18:297–306.

Karr JR, Fausch KD, Angermeier PL, Yant PR, Schlosser IJ. Assessing biological integrity in running waters: a method and its rationale. Special Publication 5. Champaign, IL: Illinois Natural History Survey, 1986.

Kelly JR, Harwell MA. Indicators of ecosystem response and recovery. In: Levin SA, Kelly JR, Harwell MA, Kimball KD, eds. Ecotoxicology: problems and approches. New York: Springer-Verlag, 1989.

Lackey RT. Pacific salmon, ecological health, and public policy. Ecosystem Health 1996;2:61–68.

Last JM. Global change; ozone depletion, greenhouse warming, and public health. Ann Rev Public Health 1993;14:115–136.

Levins R, Awebuch T, Brinkmann U, et al. The emergence of new diseases. Am Scientist 1994;82:52–60.

Maddox J. Sustainable development unsustainable. Nature 1995;374:305.

McMichael AJ. Planetary overload, global environmental change and the health of the human species. Cambridge, UK: Cambridge University Press, 1993.

Minns CK. Use of models for integrated assessment of ecosystem health. J Aquatic Ecosystem Health 1992;1:109–118.

Myers N. The question of linkages in environment and development. BioScience 1993;43:302–310.

Patz JA. Global climate change and emerging infectious diseases. JAMA 1996;275:217–223.

Rapport DJ, Gaudet CL, Calow P, eds. Evaluating and monitoring the health of large-scale ecosystems. Heidelberg, Germany: Springer-Verlag, 1995.

Rapport DJ, Regier HA. Disturbance and stress effects on ecological systems. In: Patten BC, Jorgenson SE, eds. Complex ecology: the part–whole in ecosystems. Englewood Cliffs, NJ: Prentice-Hall PTR, 1995.

Schindler DW. Experimental perturbations of whole lakes as tests of hypotheses concerning ecosystem structure and function. Oikos 1990;57:25–41.

———. Detecting ecosystem response to anthropogenic stress. Can J Fish Aquatic Sci 1987;44(suppl):6–25.

Somerville MA. Planet as patient. Ecosystem Health 1995;1:61–71.

Sprugel DG. Disturbance, equilibrium and environmental variability: what is "natural" vegetation in a changing environment? Biol Conserv 1991;58:1–18.

United Nations. Adoption of agreements on environment and development: the Rio declaration on environment and development. UNEP.A/CONF/151/2/Rev.1. 13 June 1992.

Whitford WG. Desertification: implications and limitations of the ecosystem health metaphor. In: Rapport DJ, Gaudet C, Calow P, eds. Evaluating and monitoring the health of large-scale ecosystems. Heidelberg: Springer-Verlag, 1995:273–294.

Woodley S, Kay J, Francis G. Ecological integrity and the mangement of ecosystems. Ann Arbor, Michigan: St. Lucie Press, 1993.

World Commission on Environment and Development. Our common future. Oxford, UK: Oxford University Press, 1987.

PART II

Approaches to Assessing the Health of Ecosystems

Part 2 provides a comprehensive overview of ecosystem assessment. It begins by explaining the reasons for assessment, and then moves on to identify the participants in assessment. Next, it describes the system properties of concern. Finally, it examines the array of potential indicators of the vulnerability of the ecosystems to disruption, the onset of system pathologies, and the recovery process. The difference between whole-system evaluations and the demonstration of particular environmental impacts is stressed in the selection of quantitative and qualitative approaches.

CHAPTER 5

Environmental Assessment: By Whom, for Whom, and to What Ends?

Richard Levins

E NVIRONMENTAL assessment has become a major activity around the world, with numerous international and national agencies engaged in the monitoring of pollutants, endangered species, health status of populations, productivity, biodiversity, and other properties of ecological formations on different spatial and temporal scales. It has also turned into a business, supporting myriad firms whose only asset is their reputation for being able to produce assessments that satisfy their clients. Finally, local grassroots organizations have proliferated to articulate concerns about the environments of their communities as affecting their health or livelihood.

In addition, environmental assessment is becoming a profession, a trend that confers both advantages and disadvantages. It is certainly desirable to develop a battery of techniques and skills that make measurements more reliable and a professional culture that is aware of the major kinds of errors. Nevertheless, professionalization turns assessment itself into a commodity. In this sense, it acquires a market value that has no necessary relation to its human value and becomes available to those who can afford to buy it. As a result, its "success" is measured less by peer review and more by salesmanship, and the legal or administrative goals of the purchasers of assessment tend to narrow not only the scope of the assessment but also the intellectual horizons of the assessors. The specialization that accompanies professionalization is the enemy of vision.

Environmental assessment is not a neutral technical matter. Measurement is always loaded when its outcome affects people's lives. In the Middle

Ages, when the measurement of land was a life-and-death matter for peasants, the ownership of the measuring rods created intense conflict. In the landlord–tenant relationship, when payment took the form of a share of the yield, the ownership of weights and measures was a source of debate. Today, control over the measurement of environmental impact is contested territory.

TYPES OF ENVIRONMENTAL ASSESSMENT

Different kinds of assessment exist that are undertaken for different purposes by different actors. Each sponsor of assessment has some point of departure—that is, some preference as to the outcome—that may influence the scope of the assessment, the investigative tools used, and the way in which results are presented.

Assignment of Liability

Assignment of liability may be made when damage has occurred or is alleged to have occurred to the health, income, property values, or general well-being in a community. In this case, the relative importance of different factors in the environment is of primary interest. Consequently, the analysis aims at identifying the "main" effects of factors. Synergistic effects and other forms of interactive complexity represent inconveniences that make the task of assigning liability more difficult and therefore favor those who would like to avoid responsibility. Conflict may arise over the legitimacy of concerns for, and interpretation of, "interactive" as opposed to "main" effects.

Justification or Rejection of Proposed Change

Conflict may also arise when parties seek to justify or reject some proposed change in land or resource use, economic activity, or demographic policy. Sometimes the assessment comes only after a policy decision has already been made. In such a case, the study undertaken by the sponsors of the projected development is intended to sell the benefits and belittle the costs to the community. At best it can recommend ameliorative measures after the transformation has been carried out.

The techniques used for this justification or rejection emphasize cost/benefit analysis. Considering who receives the benefits and who pays the costs is important, however, as they are not always the same and are allocated nonrandomly. According to the environmental justice movement in the United States, the more harmful enterprises are often located in the poorest neighborhoods that have the least political power, especially Afro-American and Latino communities or on the reservations of the Native American peoples. This pattern is repeated in other countries—for example, the tribal areas of India, the outer islands of Indonesia, and the rainforests of South

America and the Philippines are the foci of the least restrained extraction and contamination. On an international scale, advances in environmental protection in more affluent countries are often accompanied by the export of the more destructive activities to the poorer countries.

While the enterprise proposing some project is concerned only with the probabilities of actionable harm from that project, the recipient community must consider the impact on a larger domain. Because a small probability per opportunity multiplied by many opportunities can approach inevitability, it is necessary to ask, What is the net risk or harm to people in this community of the ensemble of industrial activities, even when each one confers only a small risk by itself? Certain questions—Is this development necessary? Will it really improve people's lives? Are there better alternatives?—would not be asked in the context of instrumental assessment. Thus, in the case of justifying or rejecting some type of change, conflict focuses on the legitimacy of concerns about equity and justice in the assessment.

Planning for Resource Use

Plans for the use of a resource may also affect one's perspective. Individuals or organizations that would benefit from the use of the resource have a stake in minimizing the expected impact. Therefore, they tend to prefer to limit the study to those processes that are closest to the direct effects of their activity. On the other hand, individuals or communities whose health, income, or way of life may be injured would prefer to trace the consequences of the proposed changes to their ultimate ramifications. The former strategy is reductionist in outlook, the latter more "holistic." Thus, disputes may arise over differences in the philosophy of science as well as the facts about particular impacts.

In addition, the variability of outcomes may serve as a source of conflict. The average or expected outcome of an action is not an adequate indicator when even rare events may have drastic consequences. Major disasters have been associated with the simultaneous failure of more than one more or less reliable protection, for example.

Several months before the meltdown at Chernobyl, Russia, the administrator of the nuclear power plant gave an interview in which he described the various safety systems and calculated that an accident would not be expected more than once in 10,000 years. The chilling aspect of this claim is not that it may have been wrong, but that it may have been right: Given the existence of 1000 reactors, a probability of 1 disaster per 10,000 years per plant in Europe means an average waiting time of 10 years. Thus, the more common a production system, the more stringent the requirements must be to maintain the same level of safety.

In looking at the aggregated effects of industrial activity, we must ask such questions as, What has been the rate of violation of protective norms per

100,000 gallons of petroleum shipped or tons of pesticide produced? The time needed to correct dangerous or harmful impacts is an essential ingredient of any assessment. If risks accumulate through technical failures, and human error and deliberate violations of legal norms at rate a are detected, challenged, and removed at rate b, then at any one time a/b harmful events will occur despite the existence of any regulations. Thus, a/b becomes a parameter of the system from the viewpoint of the recipients of a projected development, albeit not from the viewpoint of the developer. Such an index includes not only technical terms but also such issues as the responsiveness of regulatory agencies to neighborhood complaints, the delays within agency administrations, and the responsiveness of the producer of the effect— considerations that are not easily quantifiable.

Thus, conflict may arise over the significance attached to unlikely but disastrous events, and over whether probabilities are considered one project at a time or for the entire ensemble of potentially dangerous activities.

Evaluation of Overall Ecosystem Health

Evaluating the health of an ecosystem as a whole is undertaken to determine whether interventions are needed. This type of assessment is linked to sustainability—the capacity of a system to maintain itself in the face of various external impacts and its own internal dynamic and to be exploited without undermining that capacity. It is likely to include a much broader range of questions and larger temporal and spatial scales than the assessment of individual proposed projects; in addition, it will probably take into account human concerns that are not readily quantifiable. For instance, the protection of the soil and vegetation of a region cannot be separated from the preservation of rural life as a whole. Likewise, "preservation" cannot mean quick-freezing the communities, but rather preserving its norm of reaction, its pattern of response to change, and its autonomy as a self-directed participant. Thus change is not by itself an indicator of ill health, as discussed in Chapter 6. In this type of assessment, conflict arises between "hard" and "soft" indices, with participants debating the legitimacy of qualitative, aesthetic, and ethical concerns.

Reconciliation of Ecology and Economics

Since the report of the Brundtland Commission (WCED 1987) was issued, a goal of environmental assessment has been the reconciliation of ecology and economics. The economics in question is only of one kind—a neoclassical model of a private and increasingly corporate market system. It assumes that "growth" is necessary and that the more growth the better; thus, it suggests that the invention of new consumables makes for prosperity. Unless something is proved to be toxic, it may be produced without asking whether it represents a wise use of resources. Economic efficiency is the appropriate criterion for technical choices, and the noncash desirables of society—such as health, culture, tranquility, justice, and equality—must be assigned economic values so as to make decisions about them.

Although this viewpoint has certainly dominated in recent years, it is by no means the only way of formulating the relationship between human society and the rest of nature. Macro-level assessment of the state of the species, the strategies for posing and achieving human goals, and the menu of development choices are all issues that would be excluded from the world of project assessment, policy consultations, and legislation. This type of assessment will survive mostly in academic research and in challenges coming from civil society. It is therefore important that the macro- and micro-assessors have at least overlapping communities and that the preparation of assessors include exposure to the broader questions. Conflict here takes the form of an opposition between short-term (practical) and long-term (theoretical) assessment.

PARTICIPANTS IN ENVIRONMENTAL ASSESSMENT

Various actors are engaged in evaluating the same environment, with different interests in the outcome, criteria for determining a satisfactory state of the community or ecosystem, ways of conceptualizing the problems, and preferred methods of evaluation.

Government Agencies

Government agencies may be interested in foreign exchange, taxable income, or gross domestic product. Members of the governing group may have direct interests or interests that overlap with those of the developers. Alternatively, a governing party may want to win political support from those who expect to benefit from some environmental policy or development project. In most places, a public policy is committed in varying degrees to the protection of the environment. When this commitment conflicts with the development goals, however, the protective policies are rarely rigorously enforced. Consequently, one form of environmental conflict arises over the resources made available for enforcement.

Developers

The developers, of course, see the change as an investment. As a result, they want to proceed as quickly as possible and with a minimum of constraints, regulations, or environment rehabilitation costs.

Environmental Assessors

Those who carry out the studies may be employed directly by government or the developers or be contracted as a consulting firm. Their interests are more complex. The economic viability of consultants depends on their ability to present their client's case. They must compete with other consultants for the contract. On the one hand, they do not want to add costly measurements that would undermine their competitive position. On the other hand, they cannot afford to miss something that may become an issue during public hearings. If they can show that an additional measurement is likely to come up in the review process, they can benefit from the larger contract needed to carry out this step. If they investigate ecological relationships that will not affect the decision or might raise new doubts as to its advisability, however, they merely add to the project costs and risk their reputation. Thus, consultants must navigate through complex issues subject to many influences.

The scientists who carry out the study have interests that overlap with but are distinct from the other parties to a dispute. To some extent, they identify with their consulting or development company, government agency, or non-governmental organization. They also have professional pride that becomes an economic factor—after all, their reputation for competence will affect their future employment. Many of them entered their profession out of concern for environmental protection so that ethical considerations weigh heavily with them. As a result, they may be pulled in different directions by the demands of prudence, loyalty, integrity, and humanitarian concerns. At the same time their only real asset is "credibility"—a reputation that is easily lost if their recommendations seem too outrageously beyond the domain of the potentially acceptable.

The Community

The community that receives the proposed development will not be homogeneous. Some members of the community may benefit as local representatives of the industry, as brokers of jobs, as potential employees, or as politicians. Others may find their livelihoods threatened by commerce from outside, lose their land, inhale the fumes, or see their way of life disrupted. They may undertake their own assessments of the ecosystem, hire their own experts, or find allies among the many environmental organizations that can offer technical assistance.

Grassroots Organizations

Grassroots assessment uses people's intimate and detailed knowledge of their own circumstances and the possibilities of frequent and widespread observation. The technologies used by such grassroots organizations must be the least costly. They have to be robust in the face of errors in sampling, handling of specimens, or departures from the analytic protocols, and they place a premium on indicator variables that anyone can see. The objective is not really precise measurement, but rather the qualitative demonstration that things are getting better (or worse) or are likely to become better (or worse). The result is a decision as to the direction of public efforts rather than the precise calibration of numerical values. Chapter 8 explores the role of grassroots assessment along the U.S.–Mexico border.

THE INEVITABILITY OF CONFLICT

Because of the disparate perspectives and organizations involved in environmental assessment, there is no objectively "correct" way to assess, even when technically correct procedures for particular measurements exist. Some measurements may be very sensitive to small immediate changes; others may detect only cumulative or very strong impacts. What is good assessment depends on the goals of the assessor and of the sponsor of the assessor.

As a result, conflict often surrounds environmental assessment. The evaluation then becomes only one element in a network of partly congruent and partly conflicting interests. It does not by itself determine the outcome of a conflict but can play two different types of roles. First, where strongly held conflicting beliefs do not correspond with real conflicting interests, it may be possible to reconcile the conflict by discovering measures that would be compatible with all parties' interests. Second, where the interests are in fundamental conflict, the assessment can confer additional power on one or another party to the dispute.

The technical elaboration of standards for assessment does not remove conflict from the assessment process but rather represents part of that conflict. Seemingly neutral concepts such as efficiency are not without problems. In agriculture, we can talk about seed harvested/seed planted (the "manifold return" of the Bible) in terms of yield per hectare, yield per person-day of labor expended, yield per unit of water or other resource consumed, energy efficiency, value added per unit input, or return on investment. In addition to efficiency, our criteria may include the following: equitable benefit; improvement of productive capacity of the soil, the pest-controlling capacity of the fauna, or the atmospheric-detoxifying capacity of the flora; empowerment of women farmers; production of a nutritious and secure diet; buffers against economic or environmental disasters; steady employment; and preservation of a culture.

Employment is a particularly virulent source of conflict. While reduction of labor costs is part of the "efficiency" of management, farm laborers might argue that the more employment created per hectare, the better. If a social commitment to provide subsistence for everybody exists, then retrenchment of employment merely shifts the costs from the employer to the society as a whole. Creating more jobs is socially "efficient" though it might appear inefficient in terms of profit.

Considerations of the cultural sustainability of the community, its capacity to rebound from specified types of perturbation, and the control exerted by community members over their own conditions of interchange with the outside may all be included or excluded in the assessment. Sometimes variables are omitted because of technical difficulties related to measurement or the costs of data collection. In such cases, research strategies for the development of assessment methods are a necessary part of the development of the field as a whole. Qualitative methods are designed specifically to be inexpensive and accessible to the least affluent parties to an assessment. They permit the inclusion of variables that are not easily measured, they can include variables that have very different physical form, and they provide a framework for grasping conceptually the dynamics of what would otherwise be overwhelmingly complex systems.

CONCLUSION

It is impossible to eliminate the partisan nature of assessment. Even mediation of interests does little more than ratify existing relationships of power through the idiom of "realism"—that is, the possibilities of attaining specific goals given existing power relationships. The burgeoning field of "negotiation" studies and mediation is most effective when conflicts of interest are more imagined than real, and the discovery that interests are not incompatible permits meeting the needs of all parties. It is also necessary, however, to acknowledge the conflictive nature of the process and to use that knowledge in evaluating alternatives.

As the field of environmental health and assessment develops, it also becomes more professionalized. This trend creates pressures to restrict the scope of the assessments, limit the time scale, constrain the number of outcome variables that are considered, standardize methods, eliminate social aspects, and hide the conflict of interests so that the process will appear more "objective."

Professional assessment also requires a client, and not all interested parties have equal resources to hire technical services. Therefore, a danger exists that the profession will become linked by economic and social ties to the more affluent and powerful parties in a dispute. That trend might mean sacrificing the appreciation of the full biosocial complexity for a more narrowly defined economic practicality. An awareness of this conflict and the constraints on objectivity is a necessary part of the preparation of ecological assessors.

REFERENCE

World Commission on Environment and Development (WCED). Our common future. Oxford, UK: Oxford University Press, 1987.

CHAPTER 6

The Ecosocial Dynamics of Rural Systems

David Barkin
Richard Levins

W HEN we consider the health of human ecosystems, sustainability in the narrow sense is not adequate for our investigation. We must also examine how the system meets human needs, provides sustenance, and maintains cultural identity while preserving and creating knowledge. These criteria are closely related, however, so that sustainability, economic viability, and democratic empowerment are actually interdependent in the long run.

From the point of view of a qualitative systems analysis of rural ecosystems, the crucial elements are the networks of positive and negative feedbacks. These networks keep a system recognizable despite variation in quantitative indices. They also drive its evolution, as stability in one part of a system is achieved through or leads to changes in other parts. In this chapter, we will first look briefly at premodern ecosocial systems and then discuss how they have been transformed or replaced very rapidly in recent times.

An analysis of the crisis of rural society from a systems perspective also suggests prerequisites for preventing or retarding the "modernizing" trajectory that is wreaking havoc on rural life throughout the world. We propose the need for modifying the prevailing dual structure of wealth and poverty, creating a viable means for rural communities to contribute to the construction of complex and healthy ecosocial systems.

PREMODERN ECOSOCIAL SYSTEMS

On a scale of millennia, no human ecosystems are truly sustainable. In pre-modern times, however, the changes were often much slower than those we see in today's world, unfolding on a scale of centuries rather than decades. Births barely compensated for mortality, so that populations grew very slowly. Productive capacity was alternately reduced and restored with fluctuations in climate and shifts in technology and demography, so that long-term trends were usually imperceptible. The slowness of change did not mean that these societies were stagnant. Over the course of centuries, human activity altered the vegetation cover, the species composition shifted, traditions evolved, and knowledge grew and sometimes became obsolete.

The notion of peoples cultivating the soil "as their ancestors had for millennia" is sentimental nonsense. Most peoples at one point came from somewhere else. The migrations have been documented for many cases—the spread of the Caribs from South America into the Antilles, the Aztecs into the valley of Mexico, the Bantu peoples into South Africa, the Thai and H'muong from China into southeast Asia, the Maori to New Zealand, and the Grand Fleet from Tahiti to Hawaii a thousand years ago. Indeed, humans settled the tundras only some 4000 years ago and penetrated the rainforests more recently.

Peoples also tell stories accounting for the acquisition of their major crops and techniques, and many deities have traditionally been associated with the teaching of the crafts. Traditional knowledge changes through its transmission to apprentices and in the act of its application, even though its bearers may describe the process as "preservation."

Furthermore, peoples have altered their environments either deliberately or inadvertently so that no habitats are completely free of human influence. With the spread of humanity, the large mammals that neolithic people once hunted declined. On a scale of centuries, productivity of the Mesopotamian soils, the Central American forests, and the Mediterranean coast have all changed as deforestation, grazing, erosion, salinization, and eutrophication from human activity changed production conditions. Plants and animals were domesticated and transported, and each change in demography and the environment enabled new pathogens to reach endemic status and adapt to human populations.

Even before the European conquest, the indigenous peoples had a history that included social and technical change. For instance, the Anasazi culture of Chaco Canyon developed suddenly, flourished for some 250 years, and then vanished quickly around 1150 C.E. Between the fourteenth century and the coming of the conquistadors, the Aztecs evolved from a roughly egalitarian kinship-based community to a highly stratified empire. Archeologists' examination of sequences of skeletons from premodern communities show changes in stature and tooth wear that attest to changing diets, differing lifespans, and the waxing and waning of diseases. Sometimes change can occur very rapidly, as when the adoption of the horse by the peoples of the North American plains transformed whole societies within a few decades or when the final deforestation shattered the culture of Easter Island.

Nonetheless, before the eruption of European capitalism on a world scale, the processes of homeostasis in human ecosystems usually predominated over those of directional change for long periods of apparent stability (Wolf 1982). The homeostatic web included pathways of diverse land use, regulating both production and productive capacity. Agronomic pathways existed through which plant–plant, plant–animal, and animal–animal interactions stabilized production in the face of fluctuating environments; demographic pathways kept populations in a rough equilibrium with production and were themselves fine-tuned by adjustments of production to need and reproduction to production. Through epidemiological pathways, victims of infectious disease died or became immune, reducing the number of susceptible individuals in the population to halt the spread of the disease and leaving a residue of experience as to how to reduce exposure and contagion. Social-economic pathways redistributed production according to need to compensate for the very local uncertainties of the hunt or harvest. Processes of collective learning from experience were applied as new problems arose and were confronted; the knowledge gained in this way was then incorporated into the culture. Social-cultural interactions preserved the cohesion of the population, allowing for redistribution of food and community labor for public works. Thus, the human homeostatic web included sharing of knowledge and solidarity in encounters with the outside.

Cumulative changes finally reached thresholds where diverse—and even complementary—demands on human activity became incompatible. At this point, the vulnerability to stressors increased, allowing external events (such as invasions or droughts) to become the obvious triggers of transformations; these transformations were typically brief and were followed by seemingly stable successor production systems. Thus, the time scale for archeologically defined "cultures" is usually several centuries, rather than millennia or decades.

A diversified resource base and feedback processes maintained the demography, food supply, disease patterns, social cohesion, knowledge base, and technology—not in a static fashion, however, but within bounds that permitted continuity. For most peoples, most of the time it seemed that the way things were was also how they had to be. This social homeostasis included several feedbacks and buffers.

Food Diversity

The diversity of food sources served as a hedge against the uncertainty of the environment. Although all crops are roughly adapted to the regions where they are grown, they differ in their optimal temperatures and their water requirements. In dry years, maize may suffer but wheat does better in the United States. In wet years, the opposite is true. Maize is a thirsty crop, while wheat becomes more subject to fungal diseases with excess water. Insect pests usually affect only a few species of plants, so an outbreak generally harms only part of the harvest. When plants are grown together, the decline in one type of plant gives greater access to light or minerals for others, allowing the latter to flourish.

Polycultures have had other beneficial effects. Legumes may fix nitrogen that benefits grasses, for example, and deep-rooted species with symbiotic fungi (mycorrhizae) mobilize phosphorus, potassium, and other minerals. Crop mixtures support more predators of the pests, and they may confuse or frustrate herbivores looking for their preferred host plant. Low, dense vegetation such as sweet potatoes suppresses weeds, while taller plants slow wind movement and shade crops that suffer from too much direct sunlight. Corn provides mechanical support for bean vines and can divert fruitworms from more vulnerable peppers. Cassava and cocoyams can be left in the ground until needed, which allowed human farmers of the premodern era to cope with more urgent tasks.

Mixed land use also allowed for beneficial ecological interactions among plots. For example, forests provide directly harvestable products, regulate and stabilize the flow of water, serve as a refuge for the natural enemies of pests, ameliorate the microclimate of the nearby open land, and support wildlife that can be hunted. Savannas support livestock that provide meat, dairy, wool, leather, traction, and manure; they also maintain insect pollinators of crops and serve as a buffer between cultivated areas.

Rotations of cultivation with grazing (sometimes carried out by different peoples sharing the land) permitted the coexistence of grasses with legumes that otherwise would be "shaded out." For premodern society, the edges of forests and open land provided transition zones especially rich in animal and plant species. The land was rich enough in decomposer organisms to recycle the community's waste. Uncultivated land of the time contained a large number of useful plants in low numbers that served medicinal, dyeing, construction, and other nonfood uses.

Crop Adaptation

Crops adapted to the local conditions of weather, soils, and farming practices. These local varieties would then be compared and exchanged, with more desirable varieties replacing the less desirable ones. Because these varieties would eventually adapt to their new local conditions, cycles of opposing divergence and convergence guided the evolution of domesticated plants.

Labor Diversity

Diverse land use and productive processes required many different kinds of labor. These labor requirements suited different ages, knowledge, skills, strengths, and inclinations, so that people could actively contribute to society for most of their lives.

Food Storage

Food was stored in various ways. For example, food might be preserved in physical form in jars and storerooms, in biological form as pigs and cattle, or in the form of social obligations through which supplies were redistributed in compensation for local fluctuations in rainfall, animal abundance, or pests. The technical difficulties and high cost of transport and storage made food self-sufficiency necessary for most peoples through most of history.

Land Preservation

The land was preserved through rest periods. When agricultural production declined due to soil exhaustion or the accumulation of pests, a plot could be abandoned to fallow or a village moved, allowing the land to recuperate.

Variation in Food Production

The effort devoted to food production varied with need. In periods of drought, more land was cultivated; when productivity increased, smaller areas were sown. In periods of food shortage, peoples might devote more time to hunting. A surplus could be consumed only directly or shared in ways that intensified social ties, but generally could not be accumulated for increased production later. This particular negative feedback distinguishes subsistence economies from capitalist economies; in the latter, production is determined by prices and only somewhat influenced by need. In capitalist societies, increased productivity may spur expansion to take advantage of profit opportunities while diminished productivity may lead to the decision that a particular enterprise is not profitable; an accumulated surplus may be invested in further production. In subsistence economies, need drives production, unlike today, when need is increasingly created to absorb production.

Population Growth

Population growth was usually extremely slow. As noted earlier, births just compensated for or barely exceeded mortality in the premodern era. Regulation occurred through several mechanisms: social arrangements that managed marriage, sexuality, and the duration of lactation; abortion and infanticide; adoption; and less controlled feedbacks, as when the contagion rates of disease increased when more nonimmune children lived in close proximity. Each ecosocial production system had its own flexible but real carrying capacity. As this limit was approached, processes of reduction increased. When population density did increase beyond manageable levels, communities might divide and reproduce their original conditions elsewhere or adopt new productive activities from a repertoire of knowledge.

Productive activity requires extensive knowledge. A common error is to expect "simple" societies to support only simple knowledge and limited thought. On the contrary, in the absence of sophisticated tools people depend even more on a very sensitive grasp of their surroundings and nuances in the behavior of their environments. In the premodern age, this knowledge was maintained in people's memories and transmitted both by direct practice and by repeated telling. It included knowledge of usable resources, dangers, techniques and skills for resource management and the protection of resources, diagnosis of disease and healing procedures, the people's own history, artistic expression, rituals, rules of cooperative labor for public works (such as water management), and rules for other social interactions.

The knowledge of premodern peoples was created in the only way that knowledge is created anywhere: from experience reflected upon and shared in the light of previous knowledge. What is unique about modern science is not any epistemological distinctiveness, but rather its organization as a historically created moment in the division of labor in which people, resources, and institutions are set aside in particular ways for the explicit purpose of organizing experience to find out and to give a self-conscious attention to ways of doing something and avoiding error.

Because change during premodern times usually occurred slowly, the accumulated knowledge was more or less sufficient and was reinforced by its successful application. Thus, older generations appeared more valuable to the young and social relations were reproduced through apprenticeship and observation. Today's call for the preservation of local cultures is not a demand to freeze them as immobile exhibits in an anthropological museum, but rather to maintain the integrity of their evolution—a condition in which "the past has a vote, not a veto" (Kaplan 1934).

Contacts with the Outside World

Local rural ecosystems were almost never in complete isolation. People have always traveled and exchanged goods through trade, gifts, ritual transfers, plunder, or other means, and these exchanges were accompanied by the flow of ideas and genes. Each culture usually related to the others around it on its own terms, and even conquest resulted in the exaction of tribute without necessarily destroying the integrity of the conquered people. Thus, external factors impinged on the internal workings of the community through internally determined pathways. The essential dynamics of these systems included the existing patterns of contacts and the peoples' own interpretations of these contacts, allowing the external to become internalized into larger regional systems made up of a mosaic of habitats and cultures.

THE EMERGENCE OF A CRISIS OF RURAL LIFE

With the rise and spread of capitalism, processes of disruption overwhelmed the networks of stabilizing feedbacks (Wolf 1982). Over the last few centuries and with the recent intensification of its penetration into all corners of the world, the previous system of loosely linked but relatively autonomous eco-social systems was thrown into crisis. The result has been a worldwide crisis of rural life. Although trade, diversification of sources, and specialization permit a flow of goods that can compensate for production failures or ecological limits on what can be produced, the resulting "interdependence" often becomes a very limited, one-way street. Food power becomes a weapon of coercion against the weak, and economic fluctuations can have a more drastic impact on people's access to food than any weather extremes or outbreaks of agricultural pests. A strong state or international system is needed to work against the market forces to compensate for the harmful effects of inequality in world trade while still realizing benefits from the exchange (Barkin et al 1990).

Although the processes that have emerged differed greatly from country to country and even within each country, their results have been remarkably similar. The colonization of America, Africa, and southeast Asia gave rise to a never-ending series of displacement, appropriation, and expropriation; breathtaking booms in new commodities were followed by crashes that left ghost towns, stranded populations, and depleted resources in their wake. As successive waves of colonizers laid claim to the most highly productive lands, the use of land evolved from its historic vocation—producing basic products for human and social survival—to the present emphasis on crops that promise a profit to their owners. For more than 500 years, the first peoples of the Americas and their successors were repeatedly forced to seek refuge in ever more marginal conditions, in ever more fragile ecosystems. Such a process broke the homeostatic feedbacks in several ways.

First, displaced to new habitats or with their traditional lands invaded by oil, mining, and lumbering industries, people found themselves without knowledge suitable to the new situations or a structure of social relations that would keep it viable, such as the rules that organize cooperation, transmission of authority, and the appropriation and distribution of value and surplus. This shortcoming made them vulnerable to the power and prestige of technological packages newly introduced by the dominant force. These new technologies gave short-term results but were soon revealed as destructive for the production base and social relations. They replaced person-to-person ties within the community with links between individual community members and outside advisors, suppliers, purchasers, and eventually employers and politicians. Unequal access to outside resources promoted inequalities of wealth and power within the community and undermined the relative democracy and customary power structures of traditional kin-based societies that had been held together by shared experience and mutual dependence. It also weakened the position of women and diminished the credibility of aged members of the society. Thus, the long-term process of pushing indigenous peoples to increasingly marginal lands emerged as one of the most important factors in the loss of cultural identity in developing countries.

Second, the dynamics of population were altered. First came depopulation, caused by previously known and newly introduced diseases, hunger, massacres, and despair. In the Americas, the indigenous population may have declined by as much as 90% in the centuries following the conquest by Europeans. With this decline came a loss of traditional knowledge. The number of Pawnee in North America was reduced from thousands to hundreds, for example, and their physicians died faster than they could replace themselves by training their successors. Later, their healers had to go to anthropology libraries to retrieve information about how to use the medicine kits. Agricultural knowledge was also lost or became irrelevant.

Despite such traumatic depopulation, people lived in crowded conditions. When populations are squeezed into limited, less productive, or more fragile areas, it becomes impossible to maintain an adequate rotation of crop and fallow; likewise, the uncultivated resources of the forest that once provided diversity and emergency food cannot be carefully utilized. The dual needs for production and restoration became increasingly incompatible as shorter fallow periods depleted productivity and increased weed and pest problems. Shortfalls could be made up by selling trees or other resources or by working for wages for the encroaching businesses. Thus, the poor became compelled to degrade their own means of production so as to survive.

If traditional methods of production and organization are not followed, they also lose their standing as a guide to survival. In turn, the bearers of that knowledge lose their social stature. Eventually the feedbacks that once linked reproduction to production capacity were broken, permitting rapid population growth that became another symptom and an exacerbating cause of the crisis. The new demography changed not only population density, but also the growth rate, age distribution, distribution of resources among age groups, and gender-based division of labor.

Meanwhile, in the areas from which the indigenous peoples had been displaced, commercial enterprises attempted to apply European/North American production methods to extract maximum returns from monocultures as quickly as possible. Along the Mediterranean coast, for example, depletions that took centuries in ancient times now ripened in decades.

By the mid-twentieth century, rural entrepreneurs using state and corporate resources began to shape a nascent scientific/technical tradition that would soon be known as the "green revolution." Displacing agronomists who had worked in the peasant tradition, technical staffs applied agrochemicals, machinery, and nonrenewable energy sources to increase productivity for a limited set of crops. Responding to the neo-malthusian specter, policy makers urged the multilateral financial and development institutions, such as the United Nations Food and Agriculture Organization (FAO), the International Monetary Fund, and the World Bank, to expand the reach of the green revolution. Insisting on the need to extract greater volumes from commercial farms, development practitioners focused their efforts on promoting agricultural development to those social groups best prepared to respond: those who were located in the environments most suitable for intensified irrigated farming and who were most easily integrated into modern institutional settings, including the elitist political structures and the credit system. These organizations, in turn, responded by giving their full support for the rising strata of technically "progressive" and commercially aggressive rural entrepreneurs. The support was partly motivated by personal and family ties to that sector, by the conviction that modernization was urgent and that stagnation was the only alternative to the green revolution package, and by the need to establish a political support system in countrysides where the entire class structure was threatened. Thus, the rapid changes in productive technology and economic arrangements were seen as a way to stabilize the political system.

In addition, the transformation sometimes represented a conscious and deliberate onslaught against a sophisticated local technology. In 1930s Mexico, for example, the Cardenas government strengthened a local extension and research system; the system had been attempting to help peasants who received new lands under the agrarian reform improve their ability to raise crops within traditional diversified milpa productive systems. After determining the political implications of such autochthonous agriculture, Norman Borlaug and a group from the Rockefeller Foundation worked to dismantle this system (Jennings 1988).

The developmentalist idea that only one pathway of modernization exists has had devastating impacts on rural economies and ecologies, closing off serious discussion of alternatives and accepting the harmful impacts of this pattern of development as the inevitable cost of progress. Levins (1986) provides a critique of this developmentalism trend.

The technologies introduced in the modern age also devalue the products and the resources used for their production. When the price of goods is affected by the dynamics of trade, business cycles, fashion, and competition, not only are the goods themselves made less valuable, but the value of all inputs and factors that go into their production also declines. The knowledge systems that serve their production, the bearers of that knowledge, and the entire society that produced them also become devalued unless they can be transferred to other production systems.

Ready access to credit and control of the most fertile lands allowed modern farmers with the appropriate resources, connections, and fortunate locations to join the green revolution. These farmers could employ the most modern technologies to increase the production of the most valuable crops and livestock to new heights. In turn, the political and social structure gave them access to more land through colonization schemes and large-scale hydroelectric and irrigation projects and to the distribution channels that allowed them to realize the profits that eluded other groups.

The profligate use of water, energy, and agrochemicals was a logical response to the developmental incentives that stimulate output by offering subsidized prices for key agricultural inputs. Little thought was given to the long-term impact of the new "input package" on soil, biota, or water quality. The health risks to the workers and consumers were addressed only belatedly and partially. No importance was attached to the objections that such advances would further impoverish the majority of farmers for whom credit was rarely available. Instead, scarce resources for research and technical assistance were channeled from traditional farming groups to the "modernizers." For their part, investors seeking rapid amortization of their investments made economically rational, but ecologically and socially devastating decisions that called for the most rapid extraction and minimum conservation and rehabilitation efforts.

This pathway of development has transformed and continues to transform the ecosocial structure and dynamics of rural society in two major ways.

First, the internal redundancy, with its buffers against adversity, and the networks of sustaining negative feedbacks have been undermined. They have been replaced by stronger destabilizing, polarizing, and eroding pathways and positive feedbacks, respectively. Poverty sends young people to work for wages away from the farm. The resulting labor shortage on village lands leads to neglect of public works and the purchase of inputs such as herbicides to compensate for unavailable hand labor. The fallow is shortened to use more of the land for production. This strategy increases weed problems and further exacerbates the labor shortage. In addition, the purchases of inputs increase the demand for cash and promote the shifting to crops for market; the monoculture then reduces the productivity of its gardens. Finally, the young people return from the cities and oil fields with their wages, buy land, hire the labor of fellow villagers, and ignore traditions.

Second, the entire rural economy has been inserted more completely into the larger systems of the rural–urban dynamic and the global economy. The dominant pathways have become the links to the outside that make the rural communities flowthroughs of the larger economies. The complex criteria that guided the practical decisions of relatively homogeneous communities have been reduced to simpler economic calculations within an increasingly differentiated population or else removed entirely from the control of the local communities. The long-term trend toward trade practices that place peasants at a disadvantage relative to agricultural businesses when prices are determined, and agriculture as a whole at a disadvantage relative to industry and finance, also undermines the effectiveness of the decision-making powers that remain in the village.

THE DYNAMICS OF RURAL POVERTY

Underlying this crisis in rural life is rural poverty—the historical consequence of existing systems of economic organization that continue to discriminate against direct producers in favor of producers in other sectors. The direct producers are not endowed with comparable amounts of accumulated capital that would raise the productivity of their land and labor. Thus, they are at a disadvantage compared with other sectors of the economy. They are also in a weak competitive position relative to farmers in other parts of the world, lacking access to the technical, financial, and institutional systems of support that have protected those farmers during different historical periods from threatening competition. Some of the pathways that exacerbate the rural crisis are discussed below.

Discriminatory Macroeconomic and Sectoral Policies

Production and export taxes, complex systems for controlling foreign exchange and trade (for example, through overvalued exchange rates and protected tariffs for industrial products), and price controls on various commodities are among the most common tools used to extract some surplus from rural producers. In the post–World War II period, new techniques for fiscal and monetary regulation were added to the toolbox, promoting transfers from rural communities to finance industrialization and systems of wage control that involved commodity regulatory mechanisms. The high costs and arbitrary impacts of these programs left the crops produced by the rural poor (and sometimes even the wealthier producers) seriously "disprotected."

Other facets of the public policy agenda have had contradictory effects—for example, increasing output while exacerbating the social inequalities that characterize most rural societies in the developing world. The green revolution brought significant productivity increases to those groups that were able to gain access to the technical know-how and finance the input package (see Jennings 1988 for a discussion of green revolution research strategies). Similarly, public investment in irrigation and colonization schemes to expand the productive frontier generally promoted large-scale commercial agriculture that was amenable to mechanization (Barkin 1972; Hecht 1985). Such programs not only had devastating effects on the environment, but also were socially destructive. Local populations were relocated or even exterminated when the complex ecosystems rapidly became unproductive, even for the new colonizers.

Inadequate and Polarized Land Tenure Systems

Unequal access to land and insecure tenure arrangements represent major obstacles to maintaining or improving environmental quality and productivity. Land ownership in much of the developing world remains highly concentrated despite numerous attempts at land reform. Throughout Latin America, for example, the greater number of small farms (increasing 2.2% per year in the post–World War II period) and shrinking plot sizes created a peasantry that is being pushed and pulled "away from being primarily farm producers and toward increasing integration into the labor market" (de Janvry et al 1989, 406–407). The privatization of communal lands of indigenous peoples not only disrupts communal means of environmental protection but also frequently represents an intermediate stage in the concentration of land into the hands of external investors. The recently recognized communal land title in some Andean regions, for example, is nullified by its limitation to the surface of the land; subsoil (oil and mineral) rights are granted to outside investors whose extractive activities destroy the hunting, fishing, and farming potential of the area.

Antipeasant Bias in Development Institutions

Resources are systematically denied for "peasant" approaches to problem solving and social organization. This bias is readily evident in the differential manner in which peasant and commercial products prices are manipulated by regulatory agencies, and the types of decisions about the import of basic commodities that impact small-scale, rain-fed agricultural zones more negatively than the larger, irrigated farming sector. As a result, many seemingly scale-neutral innovations of the green revolution and biotechnology have been transformed into mechanisms to further social polarization.

Unequal Distribution of Income and Political Power

The system of regional or provincial bosses (*caciques*) frequently represents a major obstacle to progress for the rural poor. Although this "bossism" may take numerous forms, its effects are remarkably similar and reminiscent of the stories told about manorial lords in the Middle Ages. Whether it arises from the use of wealth to consolidate political power, a political/military power that forcefully accesses wealth, or a consolidation of an initial intermediary role brokering links to the outside wealth and power, a power hierarchy— sometimes tied to political parties, and extending from the local communities—often plays a role in determining the availability of desperately needed aid packages, work projects, and welfare programs.

Inappropriate Employment Policies

Throughout the developing world, one of the most serious problems facing planners is the creation of remunerative employment. Traditionally an important source of livelihood for large parts of the population, agricultural employment has been declining precipitously in recent decades. This change reflects the incorporation of new labor-saving technologies into commercial agricultural sectors, which diminish the share of labor in this area while stranding workers in the peasant sector for want of better alternatives.

The opening of economies to international competition complicates matters in three ways. First, traditional productive activities become unprofitable as imported consumer goods displace locally produced goods and the vendors themselves find it more profitable to import than to produce. This trend often is accompanied by a change in food consumption patterns: the imports are often not suitable for local production, but the prestige that power and wealth confer encourage a disdain for local traditional foods that ultimately undermines the market for local produce. As a consequence, production shifts to specialty products for an international market in which local producers must overcome near-monopolies of outlets and credit. Second, foreign investment brings in new technologies and raises the scale of production, reducing the rate of job creation below social needs. Third, a secular decline affects trade of raw materials and food crops relative to industrialized products. As a result, heroic efforts to increase agricultural production are required just to maintain current levels of poverty.

These trends are common to all primary producers. National fisheries and deep-sea fishing are plagued by problems of overharvesting, for example, while coastal ecosystems are menaced by contamination; commercial demands lead governments to transfer rights from traditional fishing communities. Foresters face competition from imported wood products even as they are forced to intensify their cutting beyond the capacity of the woods to support the new levels of extraction.

Pressures Against Local Cultural Institutions

As peasant farmers have been transformed into "proletarianized" workers, they have assumed all of the responsibilities of such groups but gained none of the privileges that might come from having a steady income in return for productive work (Barkin 1985). This individualization of the labor force is notable in many rural communities where inherited systems of mutual self-help and voluntary labor to construct projects of collective interest are rapidly disappearing without adequate replacements. The authority of the community is being eroded and replaced by new forms of authoritarian imposition.

Migration and the Feminization of Poverty

Women's role in rural society has changed dramatically in recent decades. With the proletarianization of the labor force and the greater difficulty experienced in satisfying social needs with on-farm and rural community production, the typical family has had to develop complex survival strategies that involve migration and greater participation in wage labor. Even though more women are now wage laborers and follow such migratory patterns, a worldwide tendency exists toward the formation of rural households without men. Unlike in the past, when women's dominant role was household management and child-rearing, increasing numbers of women throughout the world now bear the additional burdens of actually providing for the basic subsistence needs and other income needs for their families. Moreover, these new duties have not ended discriminatory practices that limit women's access to education or economic opportunities.

As the environment has degraded, life in the rural sector has become more difficult, making women's tasks more difficult. Deforestation, for example, has meant that the search for firewood requires longer treks, younger trees are sacrificed on steeper slopes, and sometimes the number of cooked meals that can be prepared is reduced. Similarly, the tasks of ensuring water supplies are becoming more arduous. Such an overload affects household nutrition, as family plots where fruits and vegetables were cultivated and small farm animals reared on household and garden wastes are frequently renounced to pursue other activities. The replacement of diversified agriculture by single crops also creates a very uneven seasonal pattern of labor that is incompatible with child care and therefore tends to exclude women or sabotage their other activities.

Urbanization and Rural Poverty

Urbanization in the developing world is creating networks of densely settled areas, whose growth is fed largely by rural migrants. Increasingly, rural families count on the cities—and even international migration—for their very subsistence. Off-farm income is now an integral part of rural incomes. Collective decisions may send a son to work as a farm laborer in California and a daughter to the new sweatshops of urban industry or the burgeoning tourist hotels in the capital, even as another son continues to fish and another daughter tends the family garden plot. Thus the structure of the rural household has become more diffuse geographically and subject to a greater diversity of influences. The skills acquired through this strategy have not released the family from poverty, however, but represent resources of great potential if altered circumstances allow their productive use.

As the urban areas expand, they make enormous demands on resources and require places to deposit their waste; these needs are not accompanied by any corresponding improvement in these areas' ability to address the problems of the majority of poor people (Hardoy et al 1992). The megacities of the developing world are a historically new environment, vulnerable to fluctuations in the prices of food and fuel and to the invasion of new diseases. Nevertheless, they also represent potential sites for food production if areas can be reserved for such purposes. This approach would create productive employment, reduce transportation costs for food, provide places for recycling some of the urban waste, and reduce the density of urban settlement. The Cuban agricultural scientist Magda Montes, for example, has been circulating for discussion a proposal that the future of Cuban agriculture should plan for rural, suburban, and urban production (Montes 1994, unpublished).

Given the deteriorating employment situation and the discrimination against small-scale rural producers, it is unsurprising that environmental degradation is proceeding apace. People are being forced into the labor force even as real wages and rural incomes are declining. Increasing numbers of individuals must take refuge in peasant communities, and are obliged to resort to destructive techniques for their survival. From this perspective, then, corrective action requires a new program of productive job creation in rural areas to increase income, improve living standards, and protect the environment.

THE DUAL ECONOMY: A MODERN ALTERNATIVE FOR SUSTAINABLE DEVELOPMENT

Global integration is creating opportunities for some individuals, but nightmares for many more people. In this juxtaposition of winners and losers, a new strategy for rural development is required—one that revalues the contribution of traditional communities and their production strategies. In the present world economy, the vast majority of rural producers in the developing world cannot compete in international markets.

The crisis of rural ecosocial systems—their ill health—is systemic and cannot be remedied by improving the productivity of particular crops or teaching better marketing skills. Cultures cannot survive merely by having folk festivals and ethnological museums, nor will poverty be cured by the distribution of surpluses from abroad. Unless insulated in some way, these cultures' traditional products have ready markets only within the narrow confines of poor communities suffering a similar fate.

These marginal rural producers, however, offer an important promise: They can support themselves and make important contributions to the rest of society. Peasants and indigenous communities must receive assistance intended to enable them to continue living and producing in their own regions. Even by the strictest criteria of neoclassical economics, this approach should not be dismissed as inefficient protectionism, as most of the resources involved in this process would represent a minor or no opportunity cost for society as a whole.[1]

In the context of the present dynamic of "modernization," the most viable alternative model for such cultures is a dual economy. By recognizing the permanence of a sharply stratified society, the country will be in a better position to design policies that recognize and take advantage of these differences to improve the welfare of people in both sectors. A strategy that offers succor to rural communities—a means to make productive diversification possible—will make the management of growth easier in those areas developing links with the international economy. Even more importantly, such a strategy will offer an opportunity for the society to actively confront the challenges of environmental management and conservation in a meaningful way, with a group of people uniquely qualified for such activities.[2]

This alternative revalues the contributions of traditional production strategies of land use diversity, local recycling of crop residues, integration of farming and rural artisanship, recognition of diverse forms of cooperation in maintaining productivity and sharing its benefits, and the application, development, and consequent regard for indigenous and peasant knowledge and their carriers. It would also link the local economy to the world economy, albeit on terms limited by the communities themselves with a view toward sustainability.

The idea of a dual economy is not new. Unlike the present version that permeates our societies, putting the rich and poor into confrontation, the proposed approach calls for creating structures so that the segment of society that chooses to live in rural areas finds support from the rest of the nation in implementing an alternative regional development program. The new variant starts from the inherited base of rural production, improving productivity through agroecology techniques. It also involves incorporating new activities that build on the cultural and resource base of the community and the region for further development. Because it requires very specific responses to a general problem, this approach depends heavily on local involvement in design and implementation. While the broad outlines are widely discussed, the specifics require investment programs for direct producers and their partners.

Such a rural economy would be buffered against the uncertainties of weather and pests and have only limited dependence on the world market. Diverse land uses would jointly modulate the flow of water and microclimate, resist invasions by pests, preserve fertility of the land, provide useful products for consumption, home industry, and exchange, offer useful occupations throughout the year, and protect biodiversity through stewardship of both cultivated and reserved ecosystems. Where the ecosystems have already suffered major degradation, an ecologically sound restoration program could be undertaken as a positive contribution to the national economy. The relative economic autonomy would retard the emigration of the rural population toward the cities and provide a material base to maintain the integrity of the local cultures, their knowledge, and their governance mechanisms.

The political base for a dual economy might be sought in the traditional governing and social structures that have survived more or less intact as organs of community cohesion and self-government, the newly proliferating nongovernmental organizations (NGOs) that support local economic and social initiatives, the regional and national level indigenous rights movements that have been working to secure land rights, and urban and international allies. Unlike the dominant economic system, where success is measured in terms of efficiency, growth, and profitability, the dual rural economy pursues multiple goals of adequate production, security, ecological sustainability, and (increasingly) equity, cultural survival, and empowerment. Rather than a simple transfer of resources to compensate groups for their poverty, this strategy requires an integrated set of productive projects that offer rural communities the opportunity to generate goods and services that will contribute to raising their living standards while simultaneously improving the environment in which they live.

Under the dual economy, this sector of society would be linked to the rest of the society through its diversity of unique products and services, each offered in small quantities to the rest of the nation. In addition, it would receive compensation for its preservation of the watersheds, soils, and biological riches. It would also benefit through tourism, through the efforts of a small and perhaps fluctuating population flow back and forth to the cities (but only insofar as this group finds employment), and through its political and cultural participation in national life.

The modeling of a dual system and the assessment of its state of health would take several factors into account:

- The adequacy and reliability of the system of resource acquisition, production, and exchange for maintaining the community.

■ Redundancy and the co-occurrence of multiple and partly interchange-able land uses, crops, sources of livelihoods, and markets that function as buffers against the uncertainties of natural and social processes.

■ The multiple feedbacks that allow adjustment to the inevitable changes of circumstance.

■ The cross-links among natural, social, economic, and political processes that both stabilize and destabilize the semi-autonomous sectors in their demography, production, and social coherence.

■ The sinks that absorb the major impacts of the external and protect the vital aspects of the community.

■ The traditional and new aspects of community democracy that allow the collective intelligence to focus on problems of common concern.

■ Indicators of equality within the community, especially those that create common interests and allow the democracy to truly solve problems rather than existing as merely the format for the exercise of differential power on behalf of conflicting interests.

■ The long-term processes that make the stability only relative but that allow peoples to choose the directions and rates of change.

NOTES

1.   Many analysts dismiss peasant producers as working on too small a scale and with too few resources to be efficient. While it is possible and even necessary to promote increased productivity, consistent with a strategy of sustainable production (as defined by agroecologists), the proposal to encourage them to remain as productive members of their communities should be implemented under existing conditions. In much of Latin America, for example, if peasants cease to produce basic crops, the lands and inputs often cannot be transferred to other farmers for commercial output. The low opportunity costs of primary production in peasant and indigenous regions derive from the lack of alternative productive employment for the people and lands in that area. Although the people would generally have to seek income in the "informal sector," their contribution to national output would be meager. The difference between the social criteria for evaluating the cost of this style of produc-tion and the market valuation is based on the determination of the sacrifices that society would make in undertaking one or the other option. The theoretical basis for this approach harks back to the initial essay of W. Arthur Lewis (1954) and subse-quent developments that find their latest expression in the call for a "neostructural-ist" approach to development for Latin America (Sunkel 1993).

2.   Much of the literature on popular participation emphasizes the multifaceted contri-bution that the productive incorporation of marginal groups can make to society. (Friedmann 1992; Friedmann and Rangan 1993; Stiefel and Wolfe 1994). While very

little has been written about specific strategies for sustainability in poor rural communities, much of the experience recounted by practitioners with grassroots groups (see, for example, Glade and Reilly 1993) is consistent with the principles enunciated by theorists and analysts like Altieri (1987).

REFERENCES

Altieri MA. Agroecology: the scientific basis of alternative agriculture. Boulder, CO: Westview, 1987.

Barkin D. Global proletarianization. In: Sanderson S, ed. The Americas in the new international division of labor. New York: Holmes & Meier, 1985.

———. Los beneficiarios del desarrollo regional. Mexico City: Sep–Setentas 1972.

Barkin D, Batt R, DeWalt B. Food crops versus feed crops: the global substitution of grains in production. Boulder, CO: Lynne Rienner, 1990.

de Janvry A, Sadoulet E, Young LW. Land and labour in Latin American agriculture from the 1950s to the 1980s. J Peasant Studies 1989;16:396–424.

Friedmann J. Empowerment: the politics of alternative development. New York: Basil Blackwell, 1992.

Friedmann J, Rangan H. In defense of livelihood: comparative studies on environmental action. West Hartford, CT: Kumarian Press, 1993.

Glade W, Reilly C, eds. Inquiry at the grassroots: an Inter-American Foundation reader. Arlington, VA: Inter-American Foundation, 1993.

Hardoy J, Mitlin D, Satterthwaite D. Environmental problems in Third World cities. London: Earthscan Publications, 1992.

Hecht SB. Environment, development and politics: capital accumulation and the livestock sector in eastern Amazonia. World Develop 1985;13:663–684.

Jennings B. Foundations of international agricultural research. Boulder, CO: Westview, 1988.

Kaplan M. Judaism as a civilization: toward a reconstruction of American-Jewish life. New York: MacMillan, 1934.

Levins R. Science and progress: seven developmentalist myths. Monthly Review 1986;38.

Lewis WA. Economic development with unlimited supplies of labour. 1954. Republished in: Agarwala AN, Singh SP, eds. Economics of underdevelopment. New York: Oxford, 1963.

Rosset P. Rural report: the greening of Cuba. NACLA Report on the Americas 1994;28:37–41, 44.

Rosset P, Benjamin M. The greening of the sea of revolution: Cuba's experiment with organic agriculture. Sydney: Ocean Press, 1994.

Stiefel M, Wolfe M. A voice for the excluded: population participation in development: utopia or necessity? London and Atlantic Highlands, NJ: Zed Books and UNRISD, 1994.

Sunkel O. Development from within: toward a neostructuralist approach for Latin America. Boulder, CO: Lynne Rienner, 1993.

Wolf E. Europe and the people without history. Berkeley, CA: University of California Press, 1982.

CHAPTER 7

Assessing Ecosystem Health
Across Spatial Scales

Yrjö Haila

E VALUATION of the health versus the unhealth of ecosystems makes sense and is important in systems modified by human activity. Major problems of scaling then arise, as ecological processes and human activities usually have widely differing characteristic spatial and temporal extensions. The notion of ecosocial complex is one potential tool that can be used to obtain matching scales. An "ecosocial complex" is a system of material flow from a natural source to a sink taking place under the influence of human social activity.

To elaborate on this idea, this chapter first reviews general problems of scaling in ecology and in ecosocial processes. In the former domain, particularly in population and community research, the main issues seem to be adequately charted; in the ecosocial domain, however, rigorous distinctions between different spatial and temporal scales are difficult to draw and would easily become unnecessarily limiting. Therefore, the scales of "ecosocial complexes" must be defined on the basis of processes that maintain and reproduce ecological systems. In the spatial domain, the main entities are relatively homogenous habitat patches. These patches are embedded in landscapes, on the next level up, while the internal structure of patches is reproduced through small-scale succession cycles, on the next level down. Habitat patches are incorporated into "ecosocial complexes"—for instance, in forestry—through human influence on two levels:

■ On the overall landscape structure making up the upper-level constraint on processes within single habitat patches (area proportions of different habitat types and forest age classes, size distribution of patches as well as distances among them, and so on).

■ On structural characteristics and succession processes within patches.

The continuation or disruption of natural processes on these two scales can thus provide criteria for assessing ecosystem health in the spatial domain. The "historicity" of ecological systems brings a strong contextual element into the assessment of ecosystem health. When the human perspective is fixed, however, the time frame must be fixed as well. As a result, the apparent relativism loses strength, and it becomes necessary to make ethical commitments explicit.

DEFINING THE CONTEXT: ECOSYSTEM HEALTH AND ECOSOCIAL COMPLEX

"Ecosystem health" can be regarded as a heuristic metaphor for characterizing the condition of ecosystems that have been modified by human activity. It does not make particular sense to apply this term to completely natural ecological systems. To make this application, we would need a context-free definition of "health" versus "unhealth"—an impossibility because health and unhealth go in pairs. For instance, a healthy population of smallpox implies unhealth in the population of humans. Human-modified systems include a context for health assessment. In contrast, an ecosystem totally uninfluenced by humans would not be "healthy" or "unhealthy," but rather a natural system in a particular stage of its temporal trajectory. This perspective explains how we view the Martian environment, for example.

The adoption of this metaphor implies that assessments of ecosystem health must focus on how human activity and natural processes integrate; in other words, these factors must be analyzed together. In this chapter, the term "ecosocial complex" will be used as a conceptual tool for this purpose. It refers to any system of material flow from a natural source to a sink taking place under the influence of human social activity. An ecosocial complex must be reasonably bounded to be identifiable, but the degree of boundedness is relative and ultimately a practical question. What matters is that consideration of a particular ecosocial complex as separate from the processes in its surroundings increases our understanding of the dynamics of the complex.[1] In forestry, for instance, a single family farm using timber as a part of its subsistence economy constitutes an ecosocial complex, whereas a plot of corporate forest managed as a part of the company's overall strategy does not; in the latter case, an ecosocial complex might be defined on the scale of the corporation.

That ecosocial complexes are "socially mediated" means that the social activity of human subjects defines some of the variables that determine the dynamics of change in those complexes. For instance, removal of timber is a basic characteristic in the dynamics of managed forests; this development is, of course, completely due to human activity. How timber removal actually happens depends on the management schemes, technology, and cutting practices. In addition, variations exist in the social organization of timber removal—that is, who does the job, who provides machinery and tools and gives instructions, what the source of the knowledge base is, where the timber goes, what its uses are, who gets the profits, and so on. Such social configurations have an integral role in the dynamics of change of the entire complex; they are not merely background factors influencing the complex from the "outside."[2,3]

This chapter explores the problems of scaling in assessing ecosystem health. That scaling is a problem follows from the fact that ecosystems—let alone ecosocial complexes—are not spatially well-defined units. In the next section, we review problems of scaling in ecology, particularly from the perspective of bounding ecosocial complexes. We then ask what particular ecological processes might offer diagnostics for ecosystem health. The subsequent section addresses the paradoxes of historicity in nature: as nature is constantly changing, can any real criteria for health be given? The concluding section discusses ethical implications of such assessments.

SPATIAL SCALING OF ECOSOCIAL COMPLEXES

Scaling has emerged as an important issue in ecology in recent years (Delcourt et al 1983; Wiens et al 1986; Wiens 1989; Hoekstra et al 1991). Space and time in ecology are clearly not a priori, uniform Newtonian concepts; rather, they are constituted by the multitude of ecological processes themselves (Haila and Levins 1992). The interest in scaling coincided with an interest in what has come to be called "hierarchy theory" (Pattee 1973; Allen and Starr 1982; Salthe 1985; O'Neill et al 1986; Grene 1987). The connection is obvious: A hierarchical view of biology implies that different levels of the biological hierarchy scale differently in space and in time. The issue of biological hierarchy has two horns—ontological and methodological—that are partially related to one another. The ontological side refers to the organization of the world: complexity that implies distinction between different scales is an ontological feature of the world (Salthe 1985; O'Neill et al 1986; Wimsatt 1994). Regardless of what the world is "really" like, scales arise as a consequence of methodological decisions—that is, as a mechanical function of the "observation window" used in research (Rosen 1977; Wiens 1989). In addition, ideas of scaling and hierarchical levels may play a strategic role in research methodology by providing heuristics for grasping complexity (Levins 1973; Wimsatt 1974; Grene 1987; Dyke 1988).

O'Neill et al (1986) drew a distinction between two aspects of ecological hierarchy—namely, structural and functional. This distinction is ambiguous as "structure" and "function" really develop together.[4] Nevertheless, the distinction may prove useful, as it indicates dynamic complications in ecological systems. The main point is that no one-to-one mapping occurs between units in structural and functional realms.[5]

Structure is equivalent to function on another level of organization; for instance, trees provide structures for forest birds for nesting and seeking shelter but also maintain ecosystem functions that are vital for the birds by producing leaves that are consumed by herbivorous insects eaten by birds. In an interlevel analysis, the distinction between structure and function fades away. This merging of these two facets relates to the important distinction made by Salthe (1985), who identified the "focal level" on the one hand, and "initiating conditions" and "boundary conditions" on the other.

In population and community ecology, scales are phenomenologically fairly easy to define: they can, by and large, be identified with the spatial extensions of the entities studied (Wiens 1981, 1989; Haila 1990; Hoekstra et al 1991). In ecosystem ecology, adequate scaling presents a more difficult problem because the entities studied do not have unambiguous spatial extensions. In the traditional "Odumian" ecosystem approach, this problem was simply skipped over by pretending that a "community" is a bounded entity inside which the energy flow can be described using aggregated measures—that is, by assuming that the internal structure does not matter (see, for instance, the classic energy flow diagram in Odum 1959, Figure 11). This simple picture has shaped the image held by generations of ecologists and, through innumerable textbook reproductions, has grown into an icon among environmentalists.

Peter Taylor (1988) traced the background of Odum's ecosystem concept to the "technocratic optimism" of the post–World War II era. Viewing ecosystems in terms of cybernetic, self-regulating feedback loops was compatible with the aim of efficient ecosystem management. Depiction of ecosystems as cybernetic systems sacrifices their internal structure, however, and arbitrary internal aggregations must be used instead. "Redescription, or bookkeeping, of the measurements on an ecosystem has frequently been used as if it provided a representation of the ecosystem's dynamics" (Taylor 1988, 230). Every ecologist knows that ecosystems are not closed, as both energy and matter flow through them, but the use of aggregated measures "creates" boundedness.[6]

The boundaries of ecosocial complexes are even more relative than those of ecosystems because of the multidimensionality of human activity. The spatial and temporal scales of ecological processes and human activity never match completely. Instead, social connections established through communication and trade create ties between different societies independently of "natural" ecological boundaries or even purposefully across such boundaries. Trade is initiated to effect the exchange of locally available products for locally unavailable ones. A "society" composed entirely of self-subsistent small producers is a dollhouse ideal that, if imitated, quickly turns into a nightmare.[7]

Because of the continuous historical change that characterizes human societies, ecosocial complexes must be appropriately scaled in time as well. Basic distinctions can be drawn using the characteristic rate of historical change on each level as a criterion (see Braudel 1984). For instance, the present sociohistorical context of forestry has taken several centuries to unfold, reflecting the gradual formation of the capitalist world system (Wallerstein 1974, 1994). Fluctuations in the demand for particular wood products and, thus, the growth and decline of particular branches of forest industries have typically stretched over several decades. Major organizational changes in forest management, such as the adoption of monoculture forestry, have taken several years or a few decades. Changes in cutting practices, such as the spread of chain saw in the 1950s and mechanical harvesters in the 1980s, have occurred on a time scale defined by technical innovation and spread—that is, a few years.

The decision on where to draw boundaries in this series of temporal scales—that is, which processes to include "within" the complex and which to stabilize as boundary conditions—is arbitrary and depends on the purpose of the analysis. For instance, one might include cutting methods in the complex and "freeze" the present structure of production as a boundary condition; this decision seems fair given a time scale of a few years. Alternatively, one might include both cutting methods and the structure of production in the complex and consider the development of the world market to be a boundary condition; this rationale seems fair given a time scale of a decade or two.

Change will continue to occur—the goal of preserving ecosystem health cannot mean its discontinuation. Maintaining the status quo is simply impossible in a thermodynamically open system such as human society (Dyke 1992). Historical change arises continuously on several time scales, and what has happened previously serves as a constraint for what can happen at present (Braudel 1984). The problem of bounding social systems that are continuously changing was crystallized by Dyke as follows:

> In many cases, to think of systems in terms of stacked or nested levels we have to revert to a rigid conceptualization of whole and part we really are not happy with. . . . For example, once you have a national currency and a national banking system, the stability of individual enterprises cannot be

specified completely without talking about the stability of the national economy. But, likewise, the stability of the national economy depends, in part, on the stability of the individual firms. . . . Firms are parts of nations, but nations are also parts of firms in the sense that national banking, taxation, and so forth must be internalized by the firm in determining the conditions of its operation—so no unequivocal stacking is possible. (Dyke 1988, P. 66)

The solution is to search for phase separations. This task may be somewhat easier in the case of ecosocial complexes than in purely sociohistorical processes because ecosocial complexes "lean" on nature. Consequently, the nature-based processes provide standards for scaling only if we can identify them. Earlier in this chapter, we argued that it is fruitful to focus on critical reproductive processes. Although understanding dynamics ultimately requires analysis carried out on the level of ecosocial complexes, a focus on ecosystem processes may suffice for diagnostics. The first task then is to identify schematically those critical ecological entities and processes that maintain those entities.[8]

Patchiness is a universal feature in nature, but patches of any particular type should be defined from the point of different organisms. For instance, the canopy of an alder consists of a large number of separate patches for gall-mites but is continuous for a foraging warbler. Often patches are formed by subtle environmental variations that are barely observable to ecologists. For instance, variation in litter composition on the forest floor in the Finnish taiga creates patches for carabid beetles and spiders (Niemelä et al 1992)[9] and variation in foliage nutrient content in Australian eucalypt forests creates patches for arboreal marsupials and birds (Braithwaite et al 1988, 1989). Other types of patches are readily detectable—for instance, vegetated plots in desert environments, ponds in the tundra, or meadows created by thermokarstic processes inside continuous taiga on the permafrost lowlands of central Siberia.[10] On a small scale, patches are usually ephemeral—mushrooms, animal carcasses, and downed logs decay, and deciduous leaves drop. On another level of scaling, alpine mountaintops raising above forested lowlands or riparian habitats bordering water bodies and rivers can be regarded as habitat patches that have a more stable configuration modified only by slow geological processes.

The composition of species assemblages within patches is influenced by a variety of factors. Some are deterministic, such as selective predation pressure exerted by keystone predators (Paine 1966) or competition for light and nutrients among plants (Tillman 1988). Other factors are contingent, such as accidents of dispersal and colonization and local variation in physical conditions. Although the species composition in patches can sometimes be attributed to fairly deterministic "assembly rules," this goal seems a distant one even for plants because of the complexity of their species-specific responses (Austin and Gaywood 1994). Variation in the composition of plant assemblages across patches is, of course, amplified in assemblages of animals.

Dynamics of the patch assemblages are influenced by three major factors. First, organisms inhabiting a patch often modify the conditions there. Social insects such as ants and termites provide well-known examples; other cases involve organisms that eat up ephemeral resource patches, such as dung beetle larvae that consume their preferred habitat, cow droppings (Hanski 1987). Second, population dynamics in patch systems are often governed by a balance of local immigration and extinction, or "metapopulation dynamics" (Levins 1979, Gilpin and Hanski 1990). Although this phenomenon may not be as universal as sometimes assumed (Harrison 1994), empirical demonstrations that support it are also accumulating (see, for example, Hanski et al 1995). Third, either internally or externally triggered fluctuations in the conditions within patches induce successional change. Succession can proceed in highly deterministic phases, conditionally upon the set of species already present in the patch, or unpredictably with changes caused by the invasion of new species. In human-modified landscapes, invasion of habitat patches by new elements from the outside "matrix" is a universal disturbance type (Saunders et al 1989; Haila et al 1993).

Successional change integrates single patches into patchworks on a larger "landscape" scale. An interplay of scales occurs here: activity within a patch is conditional upon the activity in its surroundings. Particular complexity is created by the two-dimensionality of the system: both area proportions and specific configurations of different habitat types matter (Bascompte and Solé 1995). This complexity is studied in "landscape ecology," a subfield of ecology that has formulated interesting theoretical principles but provided relatively few empirical demonstrations of these principles as yet (Forman and Godron 1986; Turner 1989; Wiens et al 1993).

In a heterogeneous landscape, what happens in a particular patch depends on what happens in the surroundings. The assessment of ecosystem health becomes complicated by a kind of "relativity principle" that might also be called the "paradox of site": ecological processes are tied to sites, but the activity in a single site is seldom critical. This paradox has practical significance, for instance, in land use planning when decisions affect the fates of particular plots of land. Only seldom can it be claimed that irreplaceable values are attached to particular plots, even though each plot has value as a part of the overall landscape diversity.

THE RANGE OF SCALES: THE TAIGA AS AN EXAMPLE

In this section, the northern boreal forest, known as the taiga, is used as an example to relate ecosystem health to this mixture of scales (for overviews of the history and dynamics of the taiga, see Birks 1986; Delcourt and Delcourt 1991; Shugart et al 1992; Haila 1994). The taiga is largely a product of the Ice Age, and dramatic changes have occurred in the extension and composition of typical taiga biotopes in the course of climatic fluctuations over the last few million years. Such long-term, large-scale change is ongoing and has given rise to extant ecological patterns. These dynamics constitute the background that serves as a constraint for the present.

Long-term evolutionary processes cannot serve as a basis for ecosystem health, however; we cannot influence climatic fluctuations on an Ice Age scale. Instead, the realm of natural processes that can be related to ecosystem health is on the "meso-scale"—that is, processes on the level of populations and ecosystems (Holling 1992a).[11] Suitable conditions for ecological entities on the meso-scale are created by successional processes—irreversible but essentially repeatable change in structure and composition of ecosystems. In a way, an ecosystem could be considered a "vehicle" for ecological succession (Holling 1992a).

Wildfires do not burn uniformly, but rather leave behind pockets of old vegetation. Nested "within" these stand-level processes are small-scale succession cycles that Whittaker and Levin (1977) dubbed "mosaic processes." These localized changes in conditions follow, for instance, the death and decay of individual organisms. Mosaic processes are so ubiquitous in all ecosystems that they appear to be continuous; when considered in an appropriate spatial and temporal scale, however, they are revealed as recurring cycles—similar to forest growth following wildfires, for example.

Population responses to environmental change are usually not linear. Nonlinear processes have "junctures," or stages in which the state of the system undergoes great change accompanied by only a slight change in system parameters (Levandowsky and White 1977). Canopy closure is an important qualitative juncture in forest succession, for example, and has a clear effect on both undergrowth and ground-floor invertebrates such as ants (Punttila et al 1991, 1996) and carabid beetles (Niemelä et al 1996). The dynamics of pest outbreaks often show a nonlinear population response to forest succession; the spruce budworm is a classic case (Peterman et al 1979; Holling 1992b).

On a yet smaller scale, dynamics are driven by the substitution of dead individual organisms by new ones. Tree replacement may occur individual by individual, particularly in fire refugia, in which case small-scale heterogeneity is created by soil disturbance and woody debris creating favorable conditions for seedling establishment (Kuuluvainen 1994). Decaying wood provides a habitat not only for organisms that actually do the decaying, but also for other soil organisms; for instance, springtails use tree stumps as refuges against drought (Setälä and Marshall 1994). Likewise, food-web interactions in the soil create a structure that controls nutrient availability for microbes and higher plants (Bengtsson et al 1994).

CRITERIA OF ECOSYSTEM HEALTH

The qualitative junctures typical of succession processes offer diagnostic criteria for forest ecosystem health. Different types of junctures are important for different sets of species. "Continuity" is a key concept in the relationship between specialized populations and their required habitat (Esseen et al 1992; Haila 1994). Where the successional processes are frequent enough to guarantee continuity of critical habitats that are ephemeral at every single site, a broad suite of species thrives. This situation can be modeled using the metapopulation idea of Levins (1969). The role of different types of "mosaic phenomena" is particularly important. Specialized organisms often depend upon particular habitat elements that are continuously reproduced by small-scale successions.

Contingent events probably have a role in shaping successional communities as well. An important type of stochastic variation relates to habitat selection: variation appears in the small-scale distribution of forest organisms such as birds and ground arthropods in the taiga that is not directly correlated with external habitat features (Haila and Järvinen 1990; Haila et al 1993, 1994, 1996; Niemelä et al 1996). This finding implies that the populations are "floating" across the landscape. Landscape structures in the surroundings determine the abundance of each species in the regional species pool (Haila 1994).

This idea of junctures can be applied to ecosocial complexes in forests by assessing the frequency of critical successional junctures in managed forests compared with those in natural forests. If the patterns of occurrence are similar, then the original set of species is probably provided with the habitats it requires, and the ecosocial complex in question can be deemed healthy. The picture that arises can be presented as the following hierarchical schema: In spatial ecological organization, the main entities are relatively homogenous habitat patches. Patches are embedded in landscapes, on the next level up, and the internal structure of patches is reproduced through small-scale succession cycles, on the next level down. Thus, we can apply the driadic scheme of Salthe (1985) to this situation: Patches constitute a "focal level" that is constrained by "boundary conditions" on the landscape scale, and the dynamics stem from "initiating conditions" on the mosaic scale. As Salthe (1985, 111) noted, "an entity exists as a result of the interaction of particular values of initiating and boundary conditions." By implication, an abrupt change in either boundary conditions or initiating conditions can push the entity on the focal level into extinction. Thus, diagnostic criteria for patch continuity can be sought on these two scales.[12]

In silvicultural forests with efficient fire suppression, the human influence dominates the spatial dynamics completely (apart from occasional wind-throws). This statement is true for both of the diagnostic scales identified above. It is well established that management changes the structure of forests on the landscape (Mladenoff et al 1993; Syrjänen et al 1994). This change affects wildlife such as deer (Mladenoff and Stearns 1993) and gray wolf (Mladenoff et al 1995)[13] as well as forest birds (Järvinen et al 1977; Virkkala 1987, 1991). On the other hand, that fairly subtle changes in the within-stand structure of forests are important for trends in, for instance, forest bird populations has been demonstrated in Finland using long-term census data (con Jaartman 1973; Järvinen and Väisänen 1978; Haila et al 1980; Helle and Järvinen 1986). Specialized forest insects are sensitive to subtle changes in stand structure (Heliovaara and Väisänen 1984; Väisänen et al 1993; Siitonen and Martikainen 1994). An interplay also exists between stand structure and specific habitat elements. For instance, Kaila et al (1996) demonstrated that species assemblage found on decaying birch snags is largely determined by the surrounding patch's structure.

In agreement with Salthe (1985), one can also speculate that the focal level reflects changes on these two levels in different ways. Changes in boundary conditions often induce a more rapid response than changes in initiating conditions. For instance, habitat loss leads to the extinction of a particular population more quickly than changes in its reproductive parameters.

Thus, we can derive feasible criteria for ecosystem health in the spatial domain by using patches as the basic starting point and by paying attention to landscape-level constraints on the one hand and population-level initiating conditions on the other hand. These criteria determine the dynamic continuity of the patches. Boundary conditions are landscape structures that affect patch dynamics from the outside—often in a quite literal sense—and initiating conditions are the small-scale successional processes that produce, and reproduce, niches for the entire suite of organisms residing within habitat patches. The definition of a "patch" remains, however, arbitrary and depends on the ecology of the system.

In an analysis of these processes on the level of ecosocial complexes, the first requirement is to distinguish between the scales of boundary conditions and initiating conditions. In a phenomenological sense, this task is a simple one. In forestry, for instance, internal stand structure is created by cutting practices, and stand configuration is created by management schemes—surely separate processes. More important, however, is that the distinction between cutting practices and management procedures may carve into different ecosocial complexes, as these processes are partially uncoupled from one another. This separation has political relevance. The modification of cutting practices might require, for instance, that increasing autonomy be allocated to the lumberjacks who do the actual work, whereas the modification of management procedures might require the use of more bureaucratic methods. On the other hand, both corporate forestry and modern (neo)colonial forestry include a tendency to merge the scales together so that only one, uniform type of silviculture is left and controlled by the corporation/(neo)colonial administration.[14]

The present taiga was once covered by continental ice. When the temporal scale is sufficiently lengthened, all aspects of the biosphere appear temporary,[15] a principle known as "contextuality" in ecosystem health assessment (see also Norton 1992). The standards that can be applied are contextual. If they are, then the "contextuality" of standards is contextual as well. In other words, the world as it exists now is associated with an element of absolute value because it is the only living world that we actually know to exist.[16] The world as it exists now is a result of contingent history, but it is the world upon which human existence depends.

On the other hand, the time scale of biological adaptation and evolution relates primarily to long-term constraints of extant ecological systems and, hence, ecosocial complexes. We must also formulate ideas on how to preserve favorable initiating conditions that provide such systems with dynamics. A relevant temporal scale can be found by focusing on constitutive processes. In her essay "Paradoxes of historicity," Marjorie Grene wrote:

> The biological does not constitute the cultural, but it establishes the area within which the cultural becomes, and remains, possible. Human beings have not, of course, always known very much or very accurately about the complex underlying order of the living on which their own lives depend. But their being within nature, not simply against or above it, has certainly been a constituent part of most people's view of their own nature. And given what we know about the biological mechanisms on which any life, including human life, depends, we need, in our philosophical reflection on the question what it is to be a person, to take reasoned account of the existence and character of these foundations. Not historicity cut off from life, nor life devouring and denying historicity, is what we need to think about, but historicity as one life-style, one peculiar to our kind, as the kind of animals we happen to be. (Grene 1978, 36)

As Grene points out, the possibility of the "historicity" of individual persons is created and guaranteed by nature; humans go through a life cycle of continuous change but function within largely given natural preconditions. The same is true of ecosocial complexes: they change continuously but neverthe-less exist in a possibility space created by a natural background. This back-ground, however, is always and everywhere modified by previous human activity. Natural and cultural processes merge together. As a result, temporal scaling becomes much more complicated in the case of ecosocial complexes than in the case of human individuals: the life cycle of individuals covers an unambiguous time span but, as discussed earlier, different time scales of social processes are intermingled. We need some idea of a "meso-scale" in which crucial reproductive processes of ecosocial complexes take place. What might this "meso-scale" be?

This task is one of identifying constitutive processes of ecosocial com-plexes.[17] Ecosocial complexes arise through human productive practices. Pro-ductive practices, on the other hand, cannot reach equilibrium because, similar to the economy as a whole, they are based on a continuous through-put of energy and materials. Energy must come from an external source because it flows through the human productive system, according to thermo-dynamic principles, and new materials are needed constantly because old materials incorporated in human-made structures wear down and are dissolved in the surroundings (Dyke 1992).[18]

This outline is quite straightforward. First, it is very questionable to allo-cate a constitutive role to nature in large-scale historical processes. An idea of adaptation on the level of whole societies is a scaling error (see, for instance, Giddens 1981). Nature does not indicate which of the possible pathways of social development will be taken in a given situation, although natural con-straints certainly exclude some options (Haila and Levins 1992). The case is analogous with thermodynamic principles: the principles define what is impossible, but they do not indicate which possibility will turn into reality (Dyke 1992).

On a focused scale, the constitutive role of nature is almost a common-place. No fisheries exist without fish, no forestry without forest, no agriculture without productive soil. Analyzing this aspect further presents us with two tasks. First, we should identify the natural preconditions of particular pro-ductive practices, tied to time and space. Second, we should figure out how such specialized practices merge into their larger socioeconomic environ-ment. Once again, these tasks present us with two different scales. The lower scale consists of units of local productive activity, and the larger scale com-prises processes that are preconditions for this activity. An example from Yakutia demonstrates this interaction of scales in a particularly clear way. Herding in Yakutia is dependent on an environment that represents a sharp contrast to the surrounding habitat, and the sustainability of herding requires that both the local units (alas patches) and the sum total of grasslands on the regional scale are in viable shape.

The sharp contrast between the subsistence base (grassland) and the surrounding habitat (coniferous forest) in Yakutia could be surmised to be an exception—that is, the ecological background of human productive activity would usually be found in a lesser contrast with the dominant environmental type, as is the case with forestry in the taiga. Nevertheless, the traditional Yakut way of life has certainly merged with the taiga ecosystems as well as hunting and fishing cultures; traditional herding was no "monoculture." Variation across cultures in this regard seems a priori important: the tighter the coupling of a productive practice with a given type of habitat, the more vulnerable the society supported by that practice if the environment changes.

In the modern society, ties between habitats and human productive practices tend to be looser. Natural conditions do not define the profitability of particular ways of using nature, but rather economic and social changes and world market conjunctures. This effect has become visible through a particularly abrupt and violent process in many parts of the former Soviet Union,[19] but the same situation is observed all over the world, particularly in developing countries (see, for example, Martinez-Alier 1995). The social and economic determination of ecosocial processes brings ethical questions into a sharp focus.

AN EMPHATIC "NO!" TO SHALLOW ANTHROPOCENTRISM

Whenever choices must be made, ethical considerations must be taken into account. When conditions change and established ways of acting and thinking become outdated, ethical considerations are unavoidable.[20] Both of these statements certainly hold true for the ecological situation today. Although the arguments presented above may seem uncomfortably relativistic, in fact it is not the case at all. The role of nature in human existence is at issue. The apparent relativism makes ethical considerations explicit. Once this illumination occurs, it brings the entire range of criteria used in human valuation of the state of the world into the picture; by no means does this transformation imply selfish anthropocentrism (Sagoff 1985, 1992; Haila and Levins 1992; O'Neill 1993).

Another liberating consequence of the emphasis on the contextuality of ecosystem health is that a dualism of "normal" versus "pathological" (Canguilhem 1991) can be avoided. The issue is as follows: The adoption of external criteria would mean that some physiognomic features of extant ecosystems (or postulated past ecosystems) were named "normal" and elevated into an ideal model. This process would create an artificial distinction between "natural" and "unnatural" that supports only shallow moralism. As shown by Canguilhem, such a distinction in the medical field is socially defined; in fact, it is impossible to find unambiguous clinical criteria for "normal" human health (Rapport 1995b). Essentialistic criteria such as "normalness" do not matter, but rather specific criteria that relate to the dynamic processes and are sensitive to context and standpoint.[21]

Taking the term "ecosystem health" as a metaphor does not diminish its value in environmental considerations—a point made by Callicott (1995) and Rapport (1995a, 1995b), for example. As in medical ethics, metaphors can be used in environmental ethics for the purpose of deriving general principles. The "precautionary principle" (O'Riordan and Cameron 1995) is a good example of the outcome of this process. Of course, the contents of specific metaphors and the recommendations derived from them are questions of political controversy—but this case holds true, anyway. Preparing for uncertainty is one specific implication of the precautionary principle (Levins 1995). Another is the need to involve and empower local residents in the productive practices that actually determine the dynamics of any particular ecosocial complex (Saunders et al 1993).

In this chapter, we have connected ecosystem health to ecosocial complexes to underline the basically social character of environmental policy. This trend converges toward "political ecology," which draws upon the old tradition of political economy; that is, this field is sensitive to the social dimensions of environmental degradation and aims to enhance the position of those lacking power in the present conditions (see, for example, Martinex-Alier 1995). A more precise elaboration of these connections must be left to a later occasion.

NOTES

1. Theoretically, this idea draws upon nonlinear dynamics (see Dyke 1988, 1994). I prefer the term "complex" over "system" because of the close affinity of traditional systems theory with Newtonian systems in which, in the words of Robert Rosen (1986, 39), different categories of causation "are isolated into independent mathematical elements of the total dynamics." In contrast, a "complex system" is said to possess "a multitude of simple system descriptions, which cannot be combined into a single 'master description' of this type" (Rosen 1986, 41).

2. Although economy is often the most important factor defining the dynamics, the whole picture is not so simple. Ecosocial complexes are also greatly influenced—for instance, by local knowledge and tradition, status of the workers, division of labor in the whole society, and decision-making structures (social features that are contingent upon previous history).

3. Another useful conceptual clarification may be to distinguish the attributes "ecosocial" and "socioecological" from one another. "Socioecological" is more comprehensive and helpful for analyzing the relationships to nature of whole societies or modes of production, such as pastoralism (Taylor 1992; Taylor and Garcia-Barrios 1995).

The emphasis in "ecosocial complex" is more focused toward the interpenetration of particular elements of nature and productive practices. The boundary between "ecosocial" and "socioecological" defined in this way remains fuzzy, however.

4. I owe this emphasis to a comment by Chuck Dyke (see also Dyke 1988).

5. This relationship, however, may partially be an artifact of ecological research traditions. Population/community ecology and ecosystems ecology have developed largely separate from one another, as has often been noted (see, for example, O'Neill et al 1986; Brooks and Wiley 1987).

6. "Odumian" pictorial representation of ecosystems has its own history, from rough composite schemes to more and more elaborate energy circuits (Taylor and Blum 1991). Howard T. Odum used energy circuit models to describe society–nature systems, but in such applications, arbitrary boundary decisions and the loss of internal dynamics are fatal.

7. This point has an empirical edge in the historical fate of various "Robinsonades." A recent example is the Lykov family (a couple and four children), who were motivated by their rigid religious conviction to escape from the worldly society and live without any contact with other human beings in the Siberian taiga on the Sayan mountains. The tragic impossibility of the project became apparent when the family was found by a geological expedition in 1978, 33 years later (Peskov 1993).

8. We never know what nature is "really" like. As Evelyn Fox Keller (1992, 74) pointed out, however, the effectiveness of some of our tools based on science relates to the relationship between theory and reality. The world presents us with a "residual reality" that, although being "vastly larger than any possible representation we might construct," gives standards for eliminating beliefs that are blatantly wrong.

9. The effect is most likely due to an abundance of prey in patches of deciduous litter as compared with patches of needle litter; this relationship has been confirmed with a manipulation study (Matti Koivula with co-workers, in preparation).

10. This phenomenon is widespread in the Lena Valley in Yakutia, for example. These meadows, called alas in Yakutian and Russian, vary in area from tens of hectares to tens of square kilometers; they provide a historical explanation that the cattle- and horse-herding Yakut people have been able to settle deep into the taiga zone successfully. This case illuminates how ecosocial complexes can support socioecological systems (herding) in a fundamentally alien environment (the taiga).

11. Even individuals—the actual actors and interactors in ecological processes—could be included in the "meso-scale," but this consideration is not critical for this general argument.

12. The effect of air-borne pollution, such as acidification, has been excluded from consideration here because an extensive literature deals with these questions. Empirical evidence on the negative effect of acidification is naturally available from boreal forests as well.

13. David Mladenoff (personal communication) points out that the increase in wolf populations in recent years in northern Minnesota is due to human-controlled change in the forest landscape, as well as to diminishing persecution of these

animals. Forests on the landscape scale constitute an ecosocial complex, and management practices adopted in forests of different ownership status (such as national, state, industrial, or private forests) vary. Wolves in Minnesota live in human-dominated landscapes and are surely no indicators of "pristine wilderness" (see Mladenoff et al 1995; Haila 1995).

14. This instance exemplifies what Chuck Dyke has called an ethos of "sterilization" (in his manuscript "Political ecology of sterilization," 1995).

15. There is a tragic dimension in the death of human individuals, and an analogous dimension is created by the precarious existence of ecosocial complexes in conditions undergoing natural change. The disappearance of human settlement from, for example, Greenland or the high Alps during the Little Ice Age bears evidence for this statement.

16. This point has an analogy in the debate about "postmodernism" and the "demise of grand histories" proposed by, for instance, Jean-Francois Lyotard. Most travelers on the postmodernist bandwagon have not noted that, if "grand histories" are dead, then the world as we know it today requires even more concern, ethical and otherwise (see Lyotard 1992).

17. I use the term "constitutive" in a weaker sense than Grene, meaning not as "giving rise to" but as "creating the possibility for."

18. This effect is strong enough to render the neoclassical ideal of "equilibrium" completely unrealistic, as pointed out by Mirowski (1995).

19. Berman (1996) describes this process in the case of Chukotka. In this example, reindeer husbandry ran into serious difficulties after the collapse of "modern" transport and communication systems that were based on heavily subsidized helicopter connections and, taking a longer historical timeframe, on heavily subsidized mining industries (gold and tin, in particular).

20. See Dewey (1939); Juha Hiedanpää brought this point to my attention.

21. "Wilderness" is sometimes used as a substitute for "normal" in reference to the state of nature, but this terminology is misleading (Haila 1995); the term can rather be understood as another metaphor (Callicott 1992).

REFERENCES

Addicott JF. The population dynamics of aphids on fireweed: a comparison of local populations and metapopulations. Can J Zoology 1978;56:2554–2564.

Allen TFH, Starr TB. Hierarchy: Perspectives for ecological complexity. Chicago: University of Chicago Press, 1982.

Austin MP, Gaywood MJ. Current problems of environmental gradients and species response curves in relation to continuum theory. J Vegetation Sci 1994; 5:473–482.

Bascompte J, Sol RV. Rethinking complexity: modeling spatiotemporal dynamics in ecology. Trends Ecol Evol 1995;10:361–366.

Bengtsson J, Setälä H, Zheng DW. Food webs and nutrient cycling in soils: interactions and positive feedbacks. In: Polis G, Winemiller K, eds. Food webs: pattern and process. London: Chapman and Hall, 1995:30–38.

Berman DI. Chukotka in postcommunism. In: Sepp M, ed. Strangers in the Arctic. Ultima Thule and modernity. Pori: Pori Art Museum and FRAME, 1996: 104–111.

Birks HJB. Late-Quaternary biotic changes in terrestrial lacustrine environments, with particular reference to north-west Europe. In: Berglund BE, ed. Handbook of Holocene palaeoecology and palaeohydrology. New York: John Wiley & Sons, 1986:3–65.

Braithwaite LW, Binns DL, Nowlan RW. The distribution of arboreal marsupials in relation to eucalypt forest types in the Eden (NSW) woodchip concession area. Australian Wildlife Research 1988;15:363–373.

Braithwaite LW, Austin MP, Clayton M, Turner J, Nicholls AO. On predicting the presence of birds in Eucalyptus forest types. Biol Conserv 1989;50:33–50.

Braudel F. The perspective of the world. Civilization and capitalism, 15th–18th century. Vol 3. London: William Collins Sons, 1984.

Brooks DR, Wiley EO. Evolution as entropy. Toward a unified theory of biology. 2nd ed. Chicago: University of Chicago Press, 1988.

Callicott JB. The value of ecosystem health. Environ Values 1995;4:345–361.

Canguilhem G. The normal and the pathological. New York: Zone Books, 1991.

Costanza R, Norton BG, Haskell BD, eds. Ecosystem health. New goals for environmental management. Washington, DC: Island Press, 1992.

Delcourt HR, Delcourt PA. Quaternary ecology. A paleoecological perspective. London: Chapman & Hall, 1991.

Delcourt HR, Delcourt PA, Webb T. Dynamic plant ecology: the spectrum of vegetational change in space and time. Quaternary Sci Rev 1983;1:153–175.

Dewey J. Theory of valuation. In: Neurath O, Carnap R, Morris CW, eds. International encyclopedia of unified science. Vol. II(4). Chicago: University of Chicago Press, 1939:1–67.

Dyke C. The world around us and how we make it: human ecology as human artifact. Adv Human Ecol 1994;3:1–22.

———. From entropy to economy: a thorny path. Adv Human Ecol 1992;1:149–176.

———. The evolutionary dynamics of complex systems. A study in biosocial complexity. Oxford, UK: Oxford University Press, 1988.

Esseen P-A, Ehnström B, Ericson L, Sjöberg K. Boreal forests—the focal habitats of Fennoscandia. In: Hansson L, ed. Ecological principles of nature conservation. London: Elsevier, 1992:252–325.

Forman RTT, Godron M. Landscape ecology. London: John Wiley and Son, 1986.

Giddens A. A contemporary critique of historical materialism. London: Macmillan, 1981.

Gilpin M, Hanski I, eds. Metapopulation dynamics: empirical and theoretical investigations. London: Academic Press, 1991.

Grene M. Hierarchies in biology. Am Scientist 1987;75:504–510.

———. Paradoxes of historicity. Rev Metaphysics 1978;32:15–36.

Haila Y. Natural dynamics as a model for management: is the analogue practicable? In: Sippola A-L, Alaraudanjoki P, Forbes B, Hallikainen V, eds. Northern wilderness areas: ecology, sustainability, values. Rovaniemi: Arctic Centre Publications 7, 1995:9–26.

————. Preserving ecological diversity in boreal forests: ecological background, research, and management. Annales Zoologici Fennici 1994;31:203–217.

Haila Y, Toward an ecological definition of an island: a northwest European perspective. J Biogeography 1990;17:561–568.

Haila Y, Hanski IK, Niemelä J, Punttila P, Raivio S, Tukia H. Forestry and the boreal fauna: matching management with natural forest dynamics. Annales Zoologici Fennici 1994;31:187–202.

Haila Y, Hanski IK, Raivio S. Turnover of breeding birds in small forest fragments: the "sampling" colonization hypothesis corroborated. Ecology 1993;74:714–725.

Haila Y, Järvinen O. Northern conifer forests and their bird species assemblages. In: Keast A, ed. Biogeography and ecology of forest bird communities. Hague: SPB Academic Publishing, 1990:61–85.

Haila Y, Järvinen O, Väisänen RA. Effects of changing forest structure on long-term trends in bird populations in SW Finland. Ornis Scand 1980;11:12–22.

Haila Y, Levins R. Humanity and nature. Ecology, science and society. London: Pluto Press, 1992.

Haila Y, Nicholls AO, Hanski IK, Raivio S. Stochasticity in bird habitat selection: year-to-year changes in territory locations in a boreal forest bird assemblage. Oikos 1996;75.

Haila Y, Saunders D, Hobbs R. What do we presently understand about ecosystem fragmentation? In: Saunders DA, Hobbs RJ, Ehrlich P, eds. Nature conservation 3: reconstruction of fragmented ecosystems. Chipping Noreton: Surrey Beatty and Sons, 1993:45–55.

Hanski I. Nutritional ecology of dung- and carrion-feeding insects. In: Slansky F Jr, Rodriguez JG, eds. Nutritional ecology of insects, mites and spiders. New York: John Wiley & Sons, 1987:837–884.

Hanski I, Pakkala T, Kuussaari M, Lei G. Metapopulation persistence of an endangered butterfly in a fragmented landscape. Oikos 1995;72:21–28.

Harrison S. Metapopulations and conservation. In: Edwards PJ, May RM, Webb NR, eds. Large-scale ecology and conservation biology. Oxford, UK: Blackwell Scientific, 1994:111–128.

Haukioja E, Neuvonen S, Hanhimäki S, Niemelä P. Birch leaves as a resource for herbivores: seasonal occurrence of increased resistance in foliage after mechanical damage of adjacent leaves. In: Berryman AA, ed. Dynamics of forest insect populations. Patterns, causes, implications. New York: Plenum Press, 1988:163–178.

Heliövaara K, Väisänen R. Effects of modern forestry on northwestern European forest invertebrates: a synthesis. Acat Forestalia Fennici 1984;189:1–32.

Helle P, Järvinen O. Population trends of North Finnish land birds in relation to their habitat selection and changes in forest structure. Oikos 1986;46:107–115.

Hobbs RJ, Saunders DA, eds. Reintegrating fragmented landscapes. Towards sustainable production and nature conservation. New York: Springer Verlag, 1993.

Hoekstra TW, Allen TFH, Flather CH. Implicit scaling in ecological research. On when to make studies on mice and men. BioScience 1991;41:148–154.

Holling CS. Cross-scale morphology, geometry and dynamics of ecosystems. Ecol Monographs 1992a;62:447–502.

———. The role of forest insects in structuring boreal landscape. In: Shugart HH, Leemans R, Bonan GB, eds. A systems analysis of the global boreal forest. Cambridge, UK: Cambridge University Press, 1992b:170–191.

———. Resilience and stability of ecological systems. Ann Rev Ecol Syst 1973;4:1–23.

Järvinen O, Kuusela K, Väisänen RA. Effects of modern forestry on the numbers of breeding birds in Finland in 1945–1975. Silva Fennica 1977;11:284–294.

Järvinen O, Väisänen RA. Recent changes in forest bird populations in northern Finland. Ann Zoologici Fennici 1978;15:279–289.

Kaila L, Martikainen P, Punttila P. Dead trees left in clearcuts benefit saproxylic Coleoptera adapted to natural disturbances in boreal forest. Biodivers Conserv 1996;5.

Keller EF. Secrets of life, secrets of death. Essays on language, gender and science. New York: Routledge, 1992.

Kuuluvainen T. Gap disturbance, microtopography, and the regeneration dynamics of boreal coniferous forests in Finland: a review. Annales Zoologici Fennici 1994;31:35–52.

Laine K, Niemelä P. The influence of ants on the survival of mountain birches during an *Oporinia autumnata* (Lep., Geometridae) outbreak. Oecologia (Berlin) 1980;47:39–41.

Levandowsky W, White BS. Randomness, time scales, and the evolution of biological communities. Evol Biol 1977;15:69–161.

Levins R. Preparing for uncertainty. Ecosystem Health 1995;1:47–57.

———. Evolution in communities near equilibrium. In: Cody ML, Diamond JM, eds. Ecology and evolution of communities. Cambridge, MA: Belknap Press, 1975:16–50.

———. The limits of complexity. In: Pattee H, ed. Hierarchy theory. The challenge of complex systems. New York: George Braziller, 1973:111–127.

———. Extinction. In: Gerstenhaber M, ed. Some mathematical questions in biology. Lectures on mathematics in the life sciences. Vol. 2. Providence: American Mathematical Society, 1970:77–107.

———. Some demographic and genetic consequences of environmental heterogeneity for biological control. Bull Entomol Soc Am 1969;15:237–240.

———. Evolution in changing environments. Princeton: Princeton University Press, 1968.

Loucks OL. Evolution of diversity, efficiency, and community stability. Am Zoology 1970;10:17–25.

Lyotard J-F. The postmodern explained to children. Correspondence 1982–1985. London: Turnraround, 1992.

Martinez-Alier J. Political ecology, distributional conflicts, and economic incommensurability. New Left Rev 1995;211:70–88.

Mirowski P. The realms of the natural. In: Mirowski P, ed. Natural images in economic thought. Markets read in tooth and claw. Cambridge, UK: Cambridge University Press, 1994:451–483.

Mladenoff DJ, Sickley TA, Haight RG, Wydeven AP. A regional landscape analysis and prediction of favourable gray wolf habitat in the northern Great Lakes region. Conserv Biol 1995;9:279–294.

Mladenoff DJ, Stearns F. Eastern hemlock regeneration and deer browsing in the northern Great Lakes region: a re-examination and model simulation. Conserv Biol 1993;7:889–900.

Mladenoff DJ, White MA, Pastor J, Crow TR. Comparing spatial pattern in unaltered old-growth and disturbed forest landscapes. Ecol Applic 1993;3:294–306.

Niemelä J, Haila Y, Halme E, Pajunen T, Punttila P. Small-scale heterogeneity in the spatial distribution of carabid beetles in the southern Finnish taiga. J Biogeography 1992;19:173–181.

Niemelä J, Haila Y, Punttila P. The importance of small-scale heterogeneity in boreal forests: diversity variation in forest-floor invertebrates across the succession gradient. Ecography 1996;19.

Norton BG. A new paradigm for environmental management. In: Costanza R, Norton BG, Haskell BD, eds. Ecosystem health. New goals for environmental management. Washington, DC: Island Press, 1992:23–41.

Odum EP. Fundamentals of ecology. 2nd ed. Philadelphia: W. B. Saunders, 1959.

O'Neill J. Ecology, policy and politics. Human well-being and the natural world. London: Routledge, 1993.

O'Neill RV, DeAngelis DL, Waide JB, Allen TFH. A hierarchical concept of ecosystems. Princeton: Princeton University Press, 1986.

O'Riordan T, Jordan A. The precautionary principle in contemporary environmental politics. Environ Values 1995;4:191–212.

Paine RT. Food web complexity and species diversity. Am Naturalist 1966;100:65–76.

Pattee HH, ed. Hierarchy theory: the challenge of complex systems. New York: Braziller, 1973.

Payette S. Fire as a controlling process in North American boreal forest. In: Shugart HH, Leemans R, Bonan GB, eds. A systems analysis of the global boreal forest. Cambridge, UK: Cambridge University Press, 1992:144–169.

Peskov V. Taezhnij tupik. Moscow: Actes Sud, 1992.

Peterman RM, Clark WC, Holling CS. The dynamics of resilience: shifting stability domains in fish and insect systems. In: Anderson RM, Turner BD, Taylor LR, eds. Population dynamics. Oxford, UK: Blackwell, 1979:321–341.

Punttila P, Haila Y, Pajunen T, Tukia H. Colonization of clearcut forests by ants in the southern Finnish taiga: a quantitative survey. Oikos 1991;62.

Punttila P, Haila Y, Tukia H. Ant communities in taiga clearcuts: habitat effects and species interactions. Ecography 1996;19.

Radkau J. Wood and forestry in German history: in quest of an environmental approach. Environ History 1996;2:63–76.

Rapport D. Ecosystem health: exploring the territory. Ecosystem Health 1995a;1:5–13.
———. Ecosystem health: more than a metaphor? Environmental Values 1995b; 4:287–309.

Rosen R. The physics of complexity: Ashby memorial lecture. In: Trapple R, ed. Power, autonomy, utopia: new approaches toward complex systems. New York: Plenum Press, 1986.
———. Observation and biological systems. Bull Math Biol 1977;39:663–678.

Sagoff M. Has nature a good of its own? In: Costanza R, Norton BG, Haskell BD, eds. Ecosystem health. New goals for environmental management. Washington, DC: Island Press, 1992:57–71.
———. Fact and value in environmental science. Environ Ethics 1985;7:99–116.

Salthe SN. Evolving hierarchical systems. New York: Columbia University Press, 1985.

Saunders DA, Hobbs RJ, Margules CR. Biological consequences of ecosystem fragmentation: a review. Conserv Biol 1991;5:18–32.

Schulze E-D, Mooney HA, eds. Biodiversity and ecosystem function. New York: Springer Verlag, 1993.

Setälä H, Marshall VG. Stumps as a habitat for Collembola during succession from clear-cuts to old-growth Douglas-fir forests. Pedobiologia 1994;38:307–326.

Shugart HH, Leemans R, Bonan GB, eds. A systems analysis of the global boreal forest. Cambridge, UK: Cambridge University Press, 1992.

Siitonen J, Martikainen P. Occurrence of rare and threatened insects living on decaying *Populus tremula:* a comparison between Finnish and Russian Karelia. Scand J Forest Res 1994;9:185–191.

Slatkin M. Competition and regional coexistence. Ecology 1974;55:128–134.

Steele JH. Some comments on plankton patches. In: Steele JH, ed. Spatial pattern in plankton communities. New York: Plenum Press, 1978:1–20.

Syrjänen K, Kalliola R, Puolasmaa A, Mattsson J. Landscape structure and forest dynamics in subcontinental Russian European taiga. Annales Zoologici Fennici 1994;31:19–34.

Taylor AD. Large-scale spatial structure and population dynamics in arthropod predator–prey systems. Annales Zoologici Fennici 1989;25:63–74.

Taylor PJ. Reconstructing socio-ecologies: system dynamics modeling of nomadic pastoralists in sub-Saharan Africa. In: Clarke A, Fujimura J, eds. The right tools for the job: at work in the twentieth century life sciences. Princeton: Princeton University Press, 1992:115–148.
———. Technocratic optimism, H. T. Odum, and the partial transformation of ecological metaphor after World War II. J Hist Biol 1988;21:213–244.

Taylor PJ, Blum AS. Pictorial representation in biology. Biol Philosophy 1991;6,:25–134.

Taylor PJ, Garcia-Barrios R. The social analysis of ecological change: from systems to intersecting processes. Soc Sci Info 1995;34:5–30.

Tilman D. Plant strategies and the dynamics and structure of plant communities. Princeton: Princeton University Press, 1988.

Tuomi J, Niemelä P, Mannila R. Leaves as islands: interactions of *Scolioneura betuleti* (Hymenoptera) miners in birch leaves. Oikos 1981;37:146–152.

Turner MG. Landscape ecology: the effect of pattern on process. Ann Rev Ecol Syst 1989;20:171–197.

Väisänen R, Biström O, Heliövaara K. Sub-cortical Coleoptera in dead pines and spruces: is primeval species composition maintained in managed forests? Biodivers Conserv 1993;2:95–113.

Virkkala R. Spatial and temporal variation in bird communities and populations in north-boreal coniferous forests: a multiscale approach. Oikos 1991;62:59–66.

―――. Effects of forest management on birds breeding in northern Finland. Ann Zoologici Fennici 1987;24:281–294.

Vuorisalo T, Walls M, Niemelä P, Kuitunen H. Factors affecting mosaic distribution of galls of an eriophyid mite, *Eriophyes laevis*, in alder, *Alnus glutinosa*. Oikos 1989;55:370–374.

Wallerstein I. Development: lodestar or illusion? In: Sklair L, ed. Capitalism and development. London: Routledge, 1994:3–20.

―――. The modern world-system. Capitalist agriculture and the origin of the European world-economy in the sixteenth century. London: Academic Press, 1974.

Whittaker RH, Levin SA. The role of mosaic phenomena in natural communities. Theor Popul Biol 1977;12:117–139.

Wiens JA. Landscape mosaics and ecological theory. In: Hansson L, Fahrig L, Merriam G, eds. Mosaic landscapes and ecological processes. London: Chapman & Hall, 1995:1–26.

―――. Spatial scaling in ecology. Functional Ecol 1989;3:385–397.

―――. Scale problems in avian censusing. Stud Avian Biol 1981;6:513–521.

Wiens JA, Addicott J, Case T, Diamond J. The importance of spatial and temporal scale in ecological investigations. In: Diamond J, Case T, eds. Community ecology. New York: Harper & Row, 1986:145–153.

Wiens JA, Crawford CS, Gosz JR. Boundary dynamics: a conceptual framework for studying landscape ecosystems. Oikos 1985;45:421–427.

Wiens JA, Stenseth NC, Van Horne B, Ims FA. Ecological mechanisms and landscape ecology. Oikos 1993;66:369–380.

Wimsatt WC. The ontology of complex systems: levels of organization, perspectives, and causal thickets. Can J Philosophy 1994;20:207–274.

―――. Complexity and organization. In: Schaffner KF, Cohen S, eds. Proceedings of the meetings of the Philosophy of Science Association, 1972. Dordrecht, Netherlands: Reidel, 1974:67–86.

Zackrisson O. Influence of forest fire on the North Swedish boreal forest. Oikos 1977;29:22–32.

CHAPTER 8

The Efforts of Community Volunteers in Assessing Watershed Ecosystem Health

Cynthia Lopez
Geoff Dates

EMPOWERING COMMUNITY GROUPS

The purpose of this chapter is to describe the process whereby community members become interested and involved in assessing the health of local rivers, watersheds, or ecosystems. Volunteers from affected communities can serve a crucial role in monitoring and protecting ecosystem health. Throughout this chapter, we emphasize their perspective of watershed ecosystem health and its assessment.

The conditions that provoke the interest and formation of a local water monitoring group do not necessarily create an automatic interest in the health of a watershed or ecosystem. Rather, community-based monitoring groups typically begin by identifying common interests, elucidating organizational goals and objectives, and forming specific questions about a water body, such as a river. As monitoring is conducted to answer these questions, groups gain more knowledge and new questions arise. Such groups seek training and information from a variety of sources. The overall process leads not only to their own empowerment, but also to a broader understanding of, and concern for, watershed or ecosystem health.

In the sections that follow, we explore what ecosystem health means to these people, as compared with definitions that appear in the literature, and how these individuals' perspectives change as their own efforts, knowledge, and organizations evolve. Through several case studies, we describe the types of problems and conditions that may provoke and encourage these individuals to coalesce and act to monitor and protect local waterways. We review the indicator measures available to them and how volunteers and others use the resulting data.

During their education and empowerment process, local water monitoring groups may seek assistance from government agencies, other monitoring groups, or support organizations. This help may enhance the longevity of a monitoring group, as well as benefiting those who provide the assistance.

Information for this chapter was obtained from interviews with community organizers of nascent and established water monitoring groups from the United States and from the U.S.–Mexico border region. We also used documents provided by these groups, government publications, and literature published by support organizations. Lastly, we relied upon our experience with the River Watch Network (RWN).

RWN is a national nonprofit organization that provides technical and organizational support to newly forming community-based volunteer monitoring groups. Although it supports advocacy groups that need assistance in gathering information about local rivers, RWN is not an advocacy organization. Rather, it serves as an intermediary, bringing disparate groups together during the initial formation of local monitoring groups. RWN helps to resolve conflicts between parties before confrontation occurs. Hence, the examples we present in this chapter reflect situations where conflict and confrontation, both of which are inherent when community-based advocacy groups demand change, have been minimized.

WATERSHED ECOSYSTEM HEALTH

Before we explore volunteer water monitoring and its relationship to watershed ecosystem health assessment, we must define ecosystem health. For this definition, we turn to government and academic literature. The U.S. Environmental Protection Agency defines watershed as ". . . a geographic area in which water, sediments, and dissolved materials drain into a common outlet—a point on a larger stream, a lake, an underlying aquifer, an estuary, or an ocean" (U.S. EPA 1984). An ecosystem has been defined as ". . . a functioning interacting system composed of one or more living organisms and their effective environment, both physical and biological . . . and may include its spatial relations" (Fosberg 1963). For our purposes, a watershed ecosystem includes the land, the water flowing over and under its surface, and the myriad life forms, including humans, that inhabit each area.

This system is highly complex. Land, water, and life form a system of interrelationships, including synergistic and antagonistic relationships. These relationships and their effects change over time and space. In terms of time, a river's ability to use nutrients, generate dissolved oxygen, and assimilate organic matter will vary annually, seasonally, and even daily. In terms of space, these same processes will vary from upstream to downstream. For example, pristine mountain streams may be naturally nutrient-poor compared with lowland rivers. Thus, adding nutrients to the mountain stream may have significantly greater effects than adding the same amount to a large coastal river. These relationships and effects are poorly understood by professional researchers, and even more poorly understood by community volunteers with limited training who attempt to assess local conditions or impacts.

Aquatic ecosystem health is a U.S. national goal that the federal Clean Water Act defines as ". . . the chemical, physical and biological integrity of the Nation's waters" (U.S. Water Pollution Control Act 1972). Maintenance of this integrity is supposed to occur through compliance with water quality standards adopted by the states. These standards contain numerical or narrative criteria that define "acceptable" levels of ecosystem characteristics (such as dissolved oxygen) or pollutants (such as fecal coliform bacteria). However, most standards are limited to a few water column characteristics and pollutants, and in our opinion do not assess integrity, much less ecosystem health.

Other definitions of ecosystem health include notions of self-sustenance, ecological functions and processes, or biological diversity. These vague concepts do not necessarily provide clear benchmarks or guidance as to the specific field measures that can be used to provide an indication of ecosystem health.

Even if ecosystem health were well defined, it would be impossible to measure, estimate, or assess every organism, interaction, or characteristic of a watershed ecosystem. Hence, "indicator" measures are used. An indicator has been defined as ". . . a characteristic of the environment that, when measured, quantifies the magnitude of stress, habitat characteristics, degree of exposure to the stress or degree of ecological response to the exposure" (Hunsacker and Carpenter 1990).

In effect, indicators serve as substitutes for a broad range of characteristics, each of which could be measured, but at prohibitive cost and time commitment. Because these indicators represent the specific items that volunteers or professionals measure to assess watershed ecosystem health, decisions about which indicators to monitor are critical to these groups' efforts. In the next section, we focus on indicators recommended by government agencies (ITFM 1994) and professionals (Cairns 1993; Rosgen 1994; Plafkin et al 1989). These indicators represent a "menu" from which local volunteers may choose when developing their respective monitoring plans.

INDICATORS OF WATERSHED ECOSYSTEM HEALTH

Although a watershed includes terrestrial, aquatic, and human components, our discussion focuses on the aquatic system with some consideration of human health impacts. Terrestrial components are discussed only as they influence the aquatic system. As a comprehensive review of watershed ecosystem health indicators and approaches lies beyond the scope of this chapter, we will present merely broad categories of indicators and selection guidelines.

Physical Indicators

To analyze a watershed ecosystem, one must understand the physical characteristics of the river network. These characteristics change over the course of the river (from upstream to downstream) and over time in ways that influence all other aspects of the river system (Vannote 1980; Leopold 1994). Physical characteristics influence how the river adjusts its boundaries, which in turn affects habitat conditions for river organisms. Physical characteristics may be monitored for two reasons: to assess changes in the stream channel that affect human use of adjacent lands, and to analyze habitat conditions that affect aquatic and riparian life.

Physical factors that influence stream channel changes include the river's width, depth, velocity, flow, channel slope, channel material roughness, sediment load, and sediment size (Rosgen 1994). Factors that influence habitat conditions include the bottom composition, in-stream cover, embeddedness, velocity, scouring and deposition, channel alteration, pool/riffle/run composition, bank stability, bank vegetative stability, and stream-side cover (Plafkin et al 1989). Temperature and turbidity are also important factors.

Monitoring physical indicators involves estimating or measuring them in the water column, river channel, or riparian zone. Volunteers typically use visual estimates, field measurements, or water sample collection methods for this purpose. Monitoring physical indicators presents a challenge, however, because their high natural seasonal and annual variability makes it difficult to distinguish natural from cultural influences. Physical indicators can be used to assess adverse effects on the river and adjacent lands. Low flows can cause increased contaminant concentrations, higher temperatures, and reduced habitat. Channel movement and sedimentation can change habitats, threaten property, or reduce water clarity (Plafkin et al 1989).

Interpretations of the results of indicator measurements usually depend on establishing some sort of stream channel or habitat reference conditions with which actual conditions are then compared. Rosgen (1994), for example, developed a stream channel classification system that organizes streams into relatively homogeneous types according to their physical characteristics. These classifications are used to assess channel stability and predict future changes.

For habitat conditions, Vannote (1980) posits a "river continuum concept" where the river is a continuous gradient of physical conditions to which the biological community responds in predictable ways along the continuum. Plafkin and co-authors (1989) suggest locating an "ideal" site within an eco-region and comparing results at monitored sites with this ideal. We are not aware of any state water quality standards that address the concept of stream channel changes and habitat conditions in any sort of quantitative manner.

Chemical Indicators

Chemical indicators in the water column provide information on biological processes. Many indicate conditions essential (or toxic) for life. The presence of some chemicals may affect human use of the water. Most monitoring agencies and volunteer groups employ chemical indicators to assess water quality. Some of the more common indicators measured include oxygen supply (such as dissolved oxygen and oxygen demand), ionic strength (such as pH, salinity, acid neutralizing capacity, conductivity, and total dissolved solids), nutrients (such as nitrogen and phosphorus), potentially hazardous chemicals, and chemicals affecting odor and taste.

Volunteer or professional monitoring of chemical indicators may involve measuring concentrations in the water column directly, such as by immersing a hydrolab directly in a river. Alternatively, it may involve the collection of water column samples and their analysis for various chemicals in the field or laboratory. Samples of aquatic life and sediments can also be collected and analyzed for specific chemicals. Monitoring chemical indicators is a challenging task for volunteers and professionals alike, for several reasons. Not all toxic chemicals are known. Given resource constraints, it is not feasible to measure even the known chemicals of concern. The chemical concentrations measured in the water column are not necessarily available to, or taken up by, living organisms. Natural daily, seasonal, and annual variabilities also make it difficult to obtain enough samples to enable the investigator to determine whether results are due to natural or cultural influences.

In many cases, chemical indicators may be used to assess adverse effects. For example, a low oxygen supply can adversely affect aquatic animal respiration, organic waste decomposition, and chemical oxidation. Ionic strength can cause corrosiveness, reduced potability, eye irritations, toxicity, and reduced habitat suitability. The presence of hazardous chemicals can lead to toxic reactions among exposed human and aquatic populations. Such chemicals may also bioaccumulate in food. Chemicals that affect odor and taste can cause food and water to become unattractive or unpalatable to users. Of course, the magnitude of these adverse effects depends upon the concentration of the agents in the water and the degree of exposure and sensitivity of the ecosystem to them (ITFM 1994).

Interpretation of resulting measurements usually involves a comparison to water quality criteria and/or standards. However some indicators, such as nutrients, typically do not have numerical criteria. Furthermore, background levels vary from upstream to downstream and across different ecoregions. Toxicological information may be unavailable for some chemicals. In addition, by the time that monitors detect changes in the indicators, the ecosystem may already have sustained long-term damage. For these reasons, chemical monitoring alone is not a reliable predictor of ecosystem health. Nevertheless, measurement of chemicals in the water column is useful in diagnosing specific causes of change in aquatic life or as part of human health assessments (Cairns 1993).

Biological Indicators

Biological indicators assess the numbers, health, ecological functions, or life cycles of living organisms. Living organisms are considered "integrators" of the many effects caused by changes in physical and chemical characteristics. For this reason, they are considered one of the best indicators of ecosystem health. In fact, the U.S. Environmental Protection Agency (EPA) and other organizations are moving toward using biological indicators of ecosystem health (ITFM 1994). Common biological indicators include macroinvertebrates such as aquatic insects, worms, and mollusks; fish; microorganisms such as *Escherichia coli* bacteria; phytoplankton; periphyton; aquatic plants; and zooplankton.

In most cases, monitoring biological indicators involves collecting living organisms from the water column or river bottom and counting and/or identifying them. These samples may be subjected to environmental stress under controlled laboratory conditions and their response to the stressor assessed.

Monitoring biological indicators may involve several types of assessments. For example, bioassays may be used to determine the toxicity of a substance by comparing substance and placebo effects on similar test organisms. Field measurements of individuals or populations of organisms, particularly sensitive species, may be conducted to determine the effects of contaminant exposure. Exposure effects are revealed through analysis of the individual organisms or of their overall abundance and production. The community structure of one or more whole taxonomic groups may also be examined to determine species diversity, relative abundance, relative dominance, biomass (total mass of living matter in a given unit of area), types of organisms present, tolerance to specific stressors, size classes, and feeding strategies. Alternatively, investigators may measure the ecological processes of a community, such as decomposition and productivity, rather than the organisms themselves. Biological indicators may be combined into an integrated biological "index" that may respond to a number of different stressors (Cairns 1989).

Volunteer and professional monitors may use these indicators to assess adverse effects on the environment. For example, overabundance of macroinvertebrates or aquatic plants can clog water supply intakes and wastewater treatment plant outfalls. Macroinvertebrate abundance, community structure, and function indicate food availability for game fish, alteration from pollution, and general ecosystem status. Toxicity testing of macroinvertebrates may provide evidence of toxic conditions. The presence of noxious fish species can crowd out indigenous or other desirable species. Fish abnormalities may be repugnant to anglers, indicate the presence of parasitic infections, or demonstrate the presence of toxins (ITFM 1994). If humans are exposed to microorganisms, they may develop diseases, such as gastrointestinal disease. Phytoplankton may cause undesirable tastes, odors, toxins, and colors. Their abundance, community structure, and function can indicate food availability for herbivores, alteration from pollution, and primary productivity status. The presence of periphyton and zooplankton indicate similar conditions. Lastly, overabundance of aquatic plants may create displeasing or difficult conditions for recreation (ITFM 1994).

Again interpretation of results usually requires establishing reference conditions, with which actual conditions are then compared. The upstream–downstream gradient described by Vannote's river continuum concept, for example, may be used as a reference to distinguish natural changes from those due to human influences (Vannote, 1980). Also, an "ideal" reference community of benthic macroinvertebrates within an ecoregion may be compared to monitored sites. For interpreting measures of microorganisms, water quality standards provide numerical criteria for indicator bacteria, such as the fecal bacteria *E. coli*. These criteria are based on an analysis of an "acceptable" level of risk of gastroenteric disease caused by human contact with water containing these bacteria. Of course, the terms "ideal" and "acceptable" are subject to individual interpretation. Based on their own ideal, volunteer monitoring groups may find that state criteria or standards are not adequately protective of health.

Human Health Indicators

Human health, well-being, and quality of life may be affected by disturbances in watershed ecosystems. Watershed degradation can adversely affect human health through various illnesses caused by water-borne disease agents. Toxic chemicals, many potentially mutagenic and carcinogenic, can harm the nervous, immune, cardiovascular, and reproductive systems of humans of all ages (National Research Council 1993).

Because humans cannot be placed in experimental laboratory settings, researchers examine the occurrence of disease within defined populations, and in the presence and absence of exposure to contaminants, while controlling exposures to other disease causing agents. The discipline of assessing human health impacts using these methods is called "epidemiology" (defined by Cairns [1993] as the study of causes, distribution, and control of disease in populations). Monitoring of human health indicators involves the study of a representative sampling of a population, followed by assessments of its exposure to the disease agent. The presence of disease can be determined in many ways, including via collection of biological samples such as blood and/or tissue, physician examinations, or symptom reports. Once disease agents are found in water via water sampling and analysis, exposure to these agents can be assessed from behavior reports, observations, occupational or medical records, or biological samples. Biological samples can be analyzed not only for evidence of disease, but also for indicators of prior exposure to suspected disease agents. After exposure and disease status have been determined, unexposed and exposed groups are compared to determine whether they have a different statistical "risk" of getting disease. Resulting measures are measures of association (two variables are associated if one is more or less common in the presence of the other) and are not indicators of causation. Nevertheless, the stronger the association, if a dose–response relationship exists (every resulting increase or decrease in dosage or exposure is matched by a concomitant increase or decrease in the response or disease) and results are replicable, the greater the evidence for a causal relationship.

When humans cannot be studied, animal investigations may be substituted. Wild animals may be studied in the field. The presence of disease may be determined by examining the animal or its biological samples. Exposure assessments are also typically based on biological samples or observations. Animals may be used in toxicology studies conducted in the laboratory as well. In this case, adverse health outcomes are assessed by exposing the animals to varying doses of the contaminant. The animals are then sacrificed to determine whether any adverse effect has occurred. If animals manifest disease upon exposure to a contaminant across species, it is assumed that humans may also respond adversely to the contaminant. The data gathered from such studies are then extrapolated to humans to establish levels considered "safe" for the average human, given likely exposures.

Studying humans, wildlife, and laboratory animals is difficult. It is time-consuming, costly, and an imposition, and may give uncertain results. In addition, subtle, rare, or long-term effects are difficult to detect in such investigations. Subtle effects, for example, may prove difficult to discern and measure. A large sample is needed to detect rare disease occurrences. To detect long-term effects, the population must be followed for a long period of time.

Although it can measure only association (not causation), epidemiologic research is considered the strongest evidence of potential human effects, as the appropriateness of extrapolating animal results to humans is less certain. However, long-term effects of contamination may not be detected in time to prevent widespread health problems. Nevertheless, human health data can be valuable in that their collection complements other aspects of watershed monitoring and volunteers become more aware of the link between environmental and public health. Clearly, the human health monitoring process has the potential to generate public interest in ecosystem health assessment.

Human Use and Perception Indicators

Human use of watershed resources responds to environmental degradation. For example, deterioration of water quality may result in water use restrictions, consumption warnings, and limits on water-based recreational activity. Other common human use indicators measure changes in commercial fishing, drinking-water consumption, recreational use, industrial use, aesthetics, transportation, general economic well-being, and future use. Monitoring of these indicators typically involves inventorying the quantity of use, determining resource quality, and attaching economic values to the resources used or determining their preservation costs.

Interpretation of the results usually involves a comparison of the quantity and quality of the resource used with known stocks. Increasingly, the use is assessed against some sort of benchmark of sustainability. For example, current fish harvest rates may be compared with replenishment rates to determine the potential for depletion. In some cases, the value of the resource is stated in dollar terms, allowing the contribution and costs to the regional economy to be quantified.

Human perceptions of environmental quality and quality of life may play significant roles in watershed resource management decisions. If resource users are considered to be "customers," then their perceptions of the quality of the "product" represent important feedback for resource managers. Common indicators of perceptions of environmental quality include overall satisfaction with current conditions, property values, and resource use.

Human perceptions may be monitored via opinion surveys, questionnaires, or other tools that assess human use of watershed resources. Interpreting the results usually requires a comparison of subjective and objective assessments.

Human use or perception indicators are limited by their reliance upon the assignment of a monetary value to resources. Some resources are not amenable to such a valuation. In addition, current generations may inappropriately estimate the net present value of such resources, to the detriment of future generations.

Selecting Indicators

Obviously, no professional or volunteer program can monitor all of the possible indicators of watershed ecosystem health. Instead, monitoring groups must select the indicators that yield the most valuable information for the least expenditure of time and money. This goal poses a challenge even for professional, well-staffed, and well-funded programs.

Recently, the U.S. EPA and the U.S. Geological Service formed the Intergovernmental Task Force on Monitoring (ITFM) to recommend improvements in U.S. water quality monitoring. The ITFM developed professional guidelines to help monitoring organizations select indicators; this guidance is applicable to both professional and volunteer groups. The ITFM recommendations may be consolidated into two suggestions. First, groups should develop clear monitoring goals and objectives connected to data users' resource management goals and objectives. Second, they should select indicators that address the monitoring goals based on scientific and program considerations.

Clear monitoring goals and objectives should be tied to explicit and generally accepted ecosystem conditions or objectives to be achieved and maintained. Groups should gear their monitoring activities so that they determine the extent to which current conditions meet these objectives. The objectives should be stated in terms of the specific ecosystem characteristics or "indicators" that will be measured. For example, state water quality standards represent a set of objectives that define acceptable conditions for regulatory purposes, with states then attempting to manage waters to achieve these conditions. Their water quality standards typically define acceptable levels of bacteria, turbidity, dissolved oxygen, pH, temperature, and EPA priority toxic pollutants. In addition, some states use biological criteria that describe community structure and function. As noted earlier, however, most state water quality standards focus on a limited set of water column characteristics that may not reflect the most important features of a watershed ecosystem. No state water quality standards appear to define watershed ecosystem health criteria in terms of a comprehensive set of related physical, chemical, and biological characteristics and processes.

When groups consider scientific and program needs, they narrow the list of possible indicators to monitor. Examples of scientific considerations for each indicator include, but are not limited to, the following:

- Importance in maintaining a balanced biological community

- Relevance to stakeholders

- Measurability

- Quantifiability

- Sensitivity to the impacts being evaluated

- Natural variability

- Response to changes over time

- Ability to integrate effects over time and space

- Ability to provide timely information

- Broad applicability

Examples of program considerations include the following, for each indicator:

- Cost-effectiveness

- Difficulty of monitoring

- Measurability by a generally accepted method

- Ability to be explained to the target audience

Most support organizations, such as RWN, work with nascent local water monitoring groups to follow ITFM recommendations in selecting relevant indicators for their local monitoring programs. In the next section, we describe the status of community-based volunteer monitoring programs in the United States. We then provide five examples of established and nascent monitoring programs.

VOLUNTEER WATER MONITORING

We broadly define volunteer water monitoring as any monitoring effort that involves nonpaid community volunteers who collect information about local water bodies. Our discussion here focuses on the recent growth in community-based volunteer water monitoring groups. We describe the composition of these groups, the reasons they typically organize, their decisions to monitor, and the uses of their data.

In early 1993, the EPA distributed hundreds of standardized questionnaires to volunteer water monitoring groups throughout the United States. Much of the information about local water and river monitoring groups cited here is derived from the results of this survey (Rhode Island Sea Grant and U.S. EPA 1994; U.S. EPA 1994). According to the EPA (1994), "hundreds of thousands of volunteers" participate in local monitoring programs nationwide, and the majority of volunteer monitoring groups have been established during the past 20 years. Of the 517 programs identified by the EPA, more than 50% were established after 1990 (U.S. EPA 1994). In 1992, three times as many programs were founded as were formed in 1988 (Rhode Island Sea Grant and U.S. EPA 1994). The reasons for this growth are unclear. It may partially reflect the effects of the 1972 Water Pollution Control Act (also known as the Clean Water Act), which provided funding for volunteer efforts; it may also demonstrate an increasing willingness of government agencies to use volunteer data (Lee 1994). In 1989, the EPA allowed states to use volunteer data in their water quality assessment reports, as mandated by section 305(b) of the Water Pollution Control Act. Another possible reason for the growth of such groups may be the "value-neutral" nature of monitoring. No age, race, class, or gender requirements exist for organizing a volunteer monitoring group (U.S. EPA 1994).

The U.S. EPA concludes that most monitoring programs are grassroots movements. To support this conclusion, it cites results indicating that volunteer organizations monitor local locations, are small, low-budget efforts, and often receive support from local funding sources. The median annual budget for the groups surveyed is $4000, and the median number of volunteers is 25 (Rhode Island Sea Grant and U.S. EPA 1994). Volunteers report that the top three uses for monitoring data include education (building public awareness about natural resources and the threats they face), problem identification (acting as a "watchdog" for problems in the water and watershed), and local decision making. Other uses of the data include research, planning, restoration, enforcement, and legislation (Rhode Island Sea Grant and U.S. EPA 1994).

Volunteer programs are growing not only in numbers, but also in the complexity and sophistication of their monitoring activities, and in the number and type of water bodies monitored. Most groups begin by monitoring only one body of water. River monitoring is most popular, followed by lake, estuary, wetland, beach, and well monitoring. An increasing number of groups, however, are beginning to monitor multiple water bodies. As emphasized in the National Directory of Volunteer Environmental Monitoring Programs:

> In the past, volunteer monitoring programs tended to specialize in monitoring one specific type of water body. This pattern is changing as more and more volunteer programs are taking a whole-watershed approach to monitoring—for example, evaluating a lake in conjunction with its tributaries, outlet streams, and associated wetlands. Of the 517 programs in the database, 195, or 38%, monitor more than one water body type. (Rhode Island Sea Grant and U.S. EPA 1994)

Many groups may begin monitoring only a few parameters or indicators of interest, but over time may include additional indicators in their assessments. The most common parameters monitored across all water bodies are temperature, dissolved oxygen, and pH. For rivers, volunteers also commonly monitor or assess macroinvertebrate populations, water flow, nitrogen, debris, and habitats. They may monitor many other diverse parameters as well, ranging from bacteria to aquatic vegetation, wildlife populations, sediments, pesticides, and hydrocarbons. It appears that volunteer water monitoring programs are following a trend:

> . . . more and more programs are also recognizing the value of collecting [information other than water quality testing:] for example, information about local land uses, or shoreline vegetation, or recreational uses of a water body, or population of bird or amphibians. . . . [these] activities go beyond the boundaries of traditional water quality testing. (Rhode Island Sea Grant and U.S. EPA 1994)

Given the complexity of watershed ecosystems and the capabilities of volunteer monitoring groups, selection of indicators, methods, and sites is an important decision. In our experience, volunteers design programs ranging in sophistication from rigorous to haphazard. Some groups may simply decide to monitor and will not develop goals, objectives, or even a monitoring plan. In some cases, ready-made monitoring kits may be purchased without any clear purpose. Some groups may be unclear as to why they monitor certain indicators. Others may become involved in predesigned programs with preselected indicators, such as those run by the Izaak Walton League of America, Maryland Save Our Streams, and Massachusetts Water Watch Partnership, or state-run monitoring programs such as Texas Watch and Kentucky Water Watch. These organizations develop a study design process for their subgroups and provide them with "off the shelf" monitoring programs.

Other monitoring groups customize programs to address local issues, conditions, resources, and capabilities. Some may have the foresight to consider potential data users when first developing their programs. Support or advisory groups may be invited to guide volunteers in developing an organization, goals, objectives, and questions. In this case, indicators, study designs, methods, and quality assurance plans are based on the specific questions involved, data requirements, capabilities, and resources (River Watch Network 1995). Other tasks that may be customized include sampling, scheduling, pilot studies, data analysis, report preparation, and annual evaluation of the program.

Once they become interested in monitoring a local water body and decide upon indicators, community volunteers typically require training to proceed with their plan. They may obtain instructional manuals, hire consultants, or attend workshops to gain this knowledge. The rigor of the training they seek and the methods they use typically depend upon the intended use of the data gathered. Many instructional materials and support organizations exist solely to assist newly formed monitoring groups. Their objective is to help local volunteer groups become capable and efficient with their monitoring projects through educational, technical, or organizational assistance. They also hope to support local projects in becoming sustainable by establishing a body of well-trained volunteers who are capable of training other participants.

Contact between volunteer monitoring groups, agencies, and support organizations also positively affects and alters the agencies and advisors. Interaction with local river monitoring groups, for example, allows support organizations to refine their objectives, educational strategies, monitoring tools, and the need for new services. On the other hand, contact with regulatory agencies may result in conflict. Monitoring groups may expect action if they discover a violation based on a regulatory agency's standards. They may also want to monitor indicators that regulatory agencies do not recognize or for which regulatory standards have not been established.

The use of volunteer data also varies in terms of its rigor and controversy. Some data may be used for education and awareness-building purposes; other information may affect the legal and regulatory process. Other uses of the data include research, planning, restoration, enforcement, and legislation.

Data quality requirements for increasing education and awareness of community members are simple. The data must reveal processes and conditions that increase community members' awareness and understanding. At the other end of the continuum, data quality requirements for legal and regulatory proceedings will likely be complex and rigorous. The data must be of sufficient quality to persuade professionals in the field that a course of action that may affect people's livelihoods is supported by the data. In these cases, at a minimum, EPA-approved Quality Assurance Project Plans should be included in the monitoring efforts.

CASE STUDIES

From our interviews and work with local monitoring groups, several factors appear to influence community interest in local waterways. We group these factors into four general categories:

- Individual curiosity and desire to learn

- Encouragement and support from potential data users, such as state environmental agencies

- Environmental crises, such as toxic spills

- Threats to quality of life, or a way of life, such as the loss of a local recreation area

In this section, we describe five monitoring organizations and the issues that motivated each group, the indicators selected, and the data uses.

Southwest: Isleta Pueblo and the Rio Grande

The Native Americans of Isleta Pueblo live south of Albuquerque, New Mexico, and have a strong cultural link with the Rio Grande. The river is crucial to tribal ceremonies that require immersion and ingestion of the sacred waters. Over the past several years, tribal religious leaders and others intimately involved with the river, such as farmers, noticed a substantial degradation of the Rio Grande. Tribal leaders became concerned over the safety of human contact with river water during religious ceremonies (Sanchez, personal communication). Their concerns have been reinforced by others; in 1993, the Rio Grande was cited as the most endangered river in North America (American Rivers 1993). A New Mexico environmental organization quantified the threat:

> A third to half of the Rio Grande's original fish fauna have disappeared from several of the longer reaches. The silvery minnow is on the brink of extinction . . . [in] the Rio Grande Bosque, a number of species have already vanished including the grey wolf, grizzly bear, longnose gar, shovelnose sturgeon, and phantom and bluntnose shiners. (Amigos Bravos 1994)

As per the 1987 Congressional Amendments to the U.S. Water Pollution Control Act, the Pueblo submitted its own water quality standards, with human health criteria, to the EPA. The EPA approved these standards in December 1992. The Pueblo standards influenced the renewal of the city of Albuquerque's National Polluted Discharge Elimination System (NPDES) permit by requiring that the city's sewage effluent be of higher quality. Albuquerque objected to this restriction and sued the EPA for approving the Pueblo's standards. In a compromise decision, Isleta's standards were not fully applied, but nonetheless resulted in stricter NPDES effluent limitations being imposed on Albuquerque. As a result of the compromise, an intensive, two-year study is being performed to determine the "natural" ambient water levels of four metals. The results of the study, which were due in 1996, could still lead to stricter NPDES permit levels for Albuquerque. The study is being conducted cooperatively between Isleta, the New Mexico Environmental Department, the EPA, and the city of Albuquerque, acting through the U.S. Geologic Survey.

More recently, Isleta has joined with Sandia Pueblo, which is located upstream of Albuquerque, in an informal alliance. The two areas' Environmental Departments plan to develop a basic water quality and biological monitoring program that will incorporate Pueblo volunteers. This effort will eventually provide information regarding spatial impacts, both upstream and downstream from Albuquerque. Isleta and Sandia hope that other neighboring Pueblos will join this alliance, leading to a more comprehensive monitoring of the Rio Grande and its tributaries.

The U.S.–Mexico Border: The Rio Bravo River Watchers

Where the Rio Grande/Rio Bravo (RG/RB) serves to delineate the U.S.–Mexico border, community members from Ciudad Juarez and El Paso County coalesced in 1993 to form the binational Rio Bravo River Watchers (RBRWs). In creating this organization, they received assistance and technical support from RWN and the Texas Natural Resource Conservation Commission's Texas Watch Program. The group was formed in response to local threats to the environment, human health, and quality of life from residential, agricultural, and industrial pollution.

These pollutants threaten an entire way of life. Approximately 10% of residents from unregulated settlements on both sides of the border, called colonias, rely upon fish from the river for subsistence and to earn small amounts of cash. During the hot summers, colonia residents who do not have other recreation options swim in the river. In colonias located in the RG/RB floodplain, prevalence levels of hepatitis A and diarrheal diseases are as much as five times the U.S. national average (Sawyer et al 1989; Nickey 1994). The most likely cause of these high prevalence levels is resident exposure to water from wells or the river, which has been contaminated with disease agents harbored in sewage dumped into the river.

The RBRWs want to know the extent of river contamination from sewage and industrial waste. Their monitoring project has two purposes: to determine whether the river meets state water quality standards for fecal coliform bacteria, pH, temperature, dissolved oxygen, and total dissolved solids/conductivity, and to determine whether contact with the river poses any risks to human health.

Specifically, community volunteers monitor basic water quality parameters in the field along a 40-mile stretch of the river, upstream and downstream from the towns of El Paso and Juarez. They also transport water samples to their local laboratory, where the samples are analyzed for fecal coliform bacteria. The volunteers are monitoring conditions over several seasons to obtain a "profile" of water quality and fecal contamination across time and space. Their results will be compared with standards set by the Texas Natural Resource Conservation Commission (TNRCC) and the International Boundary and Water Commission (IBWC). A few of the RBRWs occasionally monitor colonia wells for the same indicators as part of an epidemiologic research project assessing the effect of water contamination on the residents' health. As part of the epidemiologic study, river and well water, plus river sediment samples, are analyzed by an outside laboratory for the presence of metals and semi-volatile and volatile organics every six months.

Because the RBRWs are trained and certified by Texas Watch and the RWN, they produce what are considered quality data; this information is used by TNRCC and IBWC. For example, the two agencies are using the data to produce a follow-up to their 1994 regional report on water quality in the Rio Grande/Rio Bravo (TNRCC 1994). The data are also used to determine whether the Rio Grande supports its designated uses, as mandated by the Water Pollution Control Act and as reported in the biennial Texas Water Quality Assessment 305(b) Report to Congress. These reports dictate pollution control priorities as well.

The RBRWs will soon add macroinvertebrates to their monitoring schedule. With training and support from the IBWC, they may expand upon their contribution to human health assessment by monitoring the RG/RB for the cholera vibrio. The addition of other indicators will permit a more comprehensive assessment of the RBRWs' local waterway. These groups are in close contact and share information with other volunteer water monitoring organizations in the RG/RB watershed. As yet, RBRWs have not implemented a watershed approach to their monitoring activities, although communication with other groups indicates movement in that direction. The sheer size of the watershed inhibits such an approach—the RG/RB drainage basin covers 335,550 square miles and spans three U.S. and five Mexican states (TNRCC 1994).

New England: The Connecticut River Watch Program

Citizens and environmental professionals in the lower Connecticut River watershed work together to monitor, protect, and improve the watershed's rivers through the Connecticut River Watch Program (CRWP). Currently, CRWP volunteers monitor the Connecticut River and two tributaries: the Mattabesset River and the Coginchaug River. The program is sponsored by the Middlesex County Soil and Water Conservation District.

The CRWP was formed in 1991 as a joint program of the Connecticut River Watershed Council, a four-state advocacy organization for the Connecticut River and the RWN. The program began in the Middletown area, with input and support from interested citizens, local environmental organizations, town government, and the Connecticut Department of Environmental Protection.

In 1992, the program expanded to address the river monitoring and community education needs of the Conservation District's federally funded five-year initiative to manage the nonpoint source (NPS) pollution problems of the Mattabesset River. This river has been targeted by the Connecticut Department of Environmental Protection as a priority for management of

NPS pollution. Sedimentation due primarily to urban development has caused the river to suffer from significantly degraded biological activity. Compounding these problems is a lack of public awareness and access.

Eventually, the CRWP will involve community members throughout the valley, with groups being active in all four states covered by the watershed. The goal is to involve these individuals in monitoring and protecting the physical, chemical, and biological integrity of the Connecticut River system by identifying and addressing water quality problems that affect the health and human use of the river ecosystem (Brawerman and Dates 1995).

The CRWP works to build public awareness of river resources and threats to human health and collects scientifically credible data. A team of technical advisors assists in data interpretation and provides direction for the CRWP; it consists of representatives from the state Departments of Environmental Protection and Public Health, the U.S. Geological Service, local universities, a local professional environmental laboratory, town land-use commissions, the RWN, and CRWP laboratory staff. Each year, approximately 60 individuals from affected communities participate in the program.

The CRWP monitors 40 sites on the Connecticut River's main stem and several of its tributaries, including the Mattabesset, for a number of indicators. Rivers are monitored to determine compliance with state water quality standards for fecal and enterococci bacteria, turbidity, total suspended solids, dissolved oxygen, pH, and benthic macroinvertebrates. Volunteers perform habitat assessments and shoreline surveys. The rivers are also monitored to assess the effects of NPS pollution and any health risks associated with water contact. For the Mattabesset, the CRWP established baseline conditions and now monitors changes resulting from the implementation of best management practices.

The CRWP operates under an EPA-approved Quality Assurance Project Plan, and the data it collects are of high quality. This information is being used by local communities to improve the health of their river resources. For example, the project has uncovered a number of bacterial contamination problems in area streams; towns are currently investigating the sources of this contamination. Bacteria data were also used to justify expansion of sewers in an area where septic systems had failed. The Connecticut Department of Environmental Protection uses the CRWP's data for planning purposes and includes it in its 305(b) Water Quality Report to Congress.

The CRWP is considered a model program for monitoring water ecosystem health. Although it is intended to be a long-term program, the future direction that the program will take remains uncertain. With five years of consistent water quality data collected and evaluated, the CRWP and local communities will revisit the program's goals, objectives, and study design, all in an effort to ensure that the CRWP responds to the needs of watershed communities and the state.

New England: The Mystic River Watch

Local volunteers formed the Mystic River Watch (MRW) partly in response to an environmental crisis and partly in support of the Mystic River Watershed Greenway Plan (Wahle, personal communication). In 1989, the Mystic River Watershed Advisory Council, led by Steve Golden of the National Park Service, produced the Greenway Plan. The plan summarized problems affecting water quality, land use, fisheries, public access, historic preservation, health, and safety in the watershed. It included recommendations for cooperative watershed protection efforts among the four affected river communities (National Park Service 1989).

In the summer of 1989, a very visible problem attracted the attention and concern of Mystic residents and the local business community: the estuary turned an unusual shade of green due to "an extensive algal bloom" that covered the estuary "in a dense mat" (Mystic River 1990). Concerned community residents suspected that road workers spraying hydromulch (sawdust with grass seed, fertilizer, and green coloring) on land near a highway bridge spanning the estuary had affected the waterway. Workers sprayed the hydromulch mixture on the bare ground immediately before a rainstorm. Residents hypothesized that an increase in algae growth resulted from hydromulch runoff. However, this hypothesis was never confirmed.

Given this crisis, the development of the Greenway Plan, and additional concerns about a local, chronically malfunctioning sewage treatment plant, the community offered strong support for a volunteer monitoring program. In 1990, approximately 40 community members were active volunteer monitors of the MRW during the summer months. They began checking levels of nutrients, such as phosphorus and nitrates, dissolved oxygen, pH, and fecal coliforms. During the academic years, the volunteers worked with local schools that performed the monitoring.

Although the estuary algal bloom crisis subsided within the first year of MRW activity, volunteers continued to monitor every two weeks during the summer until 1992. At many of their sites, they consistently found high levels of fecal coliform bacteria that exceeded the 200 colonies per 100 mL dictated by the water standard (Mystic River 1990, 1991). Fecal coliforms were particularly high after rains. During the third year of the organization's existence, volunteer interest declined, partly because of individual time constraints. Hence, MRW volunteers monitored less frequently, on a monthly basis. Today, a "skeleton" crew of volunteers conducts special project-oriented sampling. Possible future projects include examinations of the effects of contamination upon local shellfish or fish populations. The group is also interested in investigating the contribution of boating traffic to the presence and distribution of hydrocarbons in the estuary.

Originally, the MRW received training and support from the RWN. The organization also began with several highly motivated individuals who played crucial roles in its establishment. Unlike the RBRWs and the CRWP, however, the MRW did not receive substantial support from state agencies. Although the agencies may review MRW data, they do not use the data as do other state agencies that incorporate volunteer data in their 305(b) reports.

The Merrimack River Volunteer Environmental Monitoring Network

We include this case study as an example of a program design that has attempted to consider the watershed ecosystem, but has yet to be implemented. Its experience illustrates the challenge of designing a monitoring program for a large watershed with many well-established grassroots river and lake monitoring groups already in place.

The Merrimack River Watershed Initiative (MRWI) began in 1988 with government support to provide a forum for multiple participants to "develop a holistic approach to the protection and management of the watershed" (New England Interstate 1997). Participants define a watershed as consisting of "all the land which drains to a particular body of water." They consider their approach different in that it is resource-based (with the watershed considered to be the primary resource) and holistic in nature (New England Interstate 1997).

One project that the MRWI wanted to develop was a coordinated volunteer monitoring program. The organization asked the RWN, the Massachusetts Water Watch Partnership (MWWP), and the Merrimack River Watershed Council (MRWC) to design such a program. The resulting coordinated volunteer monitoring program is known as the Volunteer Environmental Monitoring Network (VEMN). It has two components: a watershed-wide support system and study design, and individual monitoring programs in sub-watersheds or river reaches.

The watershed-wide support system will consist of a team of people, businesses, organizations, and agencies that will provide various support services to existing and future monitoring groups in the watershed. It will include a watershed monitoring coordinator, consulting organizations (such as the RWN, MWWP, and Cooperative Extension), and technical advisors from state and federal agencies, universities, and businesses.

The study design includes a core set of assessments that the VEMN will encourage all groups throughout the watershed to undertake in their own localities:

- A visual assessment of basic watershed characteristics

- A health risk assessment that includes water sampling and analysis, and health data gathering

- A water quality standards assessment that includes water sampling and analysis for indicators listed in the New Hampshire and Massachusetts water quality standards

The design also includes optional assessments to answer tributary, river-reach, or issue-specific questions. For example, a tributary-specific question might concern the impact of a certain wastewater treatment plant on the river. Another issue might arise over the effectiveness of river restoration projects. When such a question arises, the team will identify indicators, sampling and analysis methods, site selection criteria, and quality assurance/quality control (QA/QC) protocols to answer it. For each type of study, the team will develop a range of sampling and analytical protocol monitoring options (from simple to complex) that address various data quality goals. Hence, a monitoring group can choose the level of sophistication it will undertake, according to its available resources and its data quality goals. These goals will be based on the expected users of the data and their anticipated applications of the information. The team will also provide hands-on training and written documentation (such as manuals) for use throughout the watershed, produce watershed-wide reports, and explore ways to provide long-term human and financial support for these programs.

The actual monitoring will be managed by new or existing sub-watershed organizations. These groups will monitor areas according to the watershed-wide study design, but will select the appropriate protocols to match their resources and data quality goals. In addition, they will maintain their autonomy and manage their own programs. Potentially these organizations could collect environmental information to address issues unique to their sub-watersheds.

Support and guidance will be provided by the watershed-wide support system team. The program is administered and managed by the MRWC, with the advice of a steering committee. Volunteer river monitoring will be coordinated with ongoing state, federal, regional, and local river monitoring, protection, and restoration efforts.

Case Conclusions

The examples of monitoring groups described here demonstrate that each group forms because of different threats, crises, and concerns, and levels of support from local agencies. In one case (that of Isleta Pueblo), conflict with a government body, the city of Albuquerque did arise. This confrontation likely occurred because Isleta Pueblo has standard-setting authority, unlike the other groups described. This authority provides the Pueblo with the power to question local standards, establish alternative standards, and enforce them. It may also adversely affect those interested in maintaining the status quo. Confrontation might potentially arise in other situations where a local group selects to monitor indicators that local or state agencies may not recognize as "legitimate." If volunteers consider the indicator to be important, and the agencies remain intransigent, conflict may arise.

In three of the cases, state and local agencies were supportive of volunteer activity, albeit to differing degrees. For example, the CRWP and MRWI could not exist without government support. The RBRWs would be unable to monitor as extensively without equipment loaned to them by the state. In these cases, volunteers monitor indicators recommended by their government supporters.

Every monitoring group assesses at least one physical characteristic, most commonly temperature. Some carry out habitat assessments that survey physical habitat conditions. Few volunteer monitoring organizations are using their data in epidemiologic studies. Among the examples described above, only the RBRWs currently allow their data to be used in an epidemiologic study (Lopez, personal communication). The VEMN, however, has added this type of study to its core list of recommended surveys. Although none of the groups described monitors wildlife populations, this type of assessment is common. For example, the Mad River Watch Program is in the process of implementing a wildlife monitoring program. An organization called "Adopt a Beach" located in Seattle, Washington, is monitoring bird populations and mortality in Puget Sound (U.S. EPA 1984).

Each group profiled here developed different goals, objectives, questions, monitoring strategies, and indicators. Examples of the most common questions posed by community volunteers include the following:

- Is our local water body polluted?

- How do we define "polluted"?

- What are the sources of pollution contaminating our river?

- What is the extent of contamination?

- Can the river support aquatic life?

- Is the water safe for human contact or consumption?

As monitoring is conducted to answer these questions, and the groups gain more knowledge, other questions may arise. For example, if the river is found to be contaminated with fecal coliform bacteria, is the source of contamination human or animal? What is its source? What impact does this contamination have on aquatic life? Even if the fecal coliform levels fall below the standard, is it safe to swim in the river? Is the standard truly health-based? It is precisely this process that leads to a broader understanding of, and concern for, the watershed and its health. The process may also help groups to understand their own limitations, as every question and every impact cannot be assessed given the complexity of a watershed ecosystem, the limits of volunteer manpower and time, and the potential for conflict.

CONCLUSIONS

Many benefits accrue to community members who organize to voluntarily monitor their local river, watershed, or ecosystem. Going through the process of organizing, seeking information, and monitoring enhances the volunteers' awareness, empowerment, skills, and ability to contribute to the prevention or identification of water contamination.

Such a development also benefits overworked local, state, and federal environmental and public health agencies. These agencies have the opportunity to put volunteer labor to work in monitoring more water bodies, more often, and for more parameters. They also develop a positive working relationship with a well-informed constituency. Of course, this process requires agencies to invest in training, managing, supporting, and communicating with these community volunteers.

Benefits also accrue to support organizations that may be funded or receive service fees to assist local, community-based volunteer monitors. With each group they aid, these organizations learn about different regional ecosystems and the applicability of various monitoring tools in these cases. They can subsequently use this knowledge to gauge the need for new services that might benefit their future clients.

Volunteer monitoring has its limits, however. As the case studies make clear, none of the groups systematically analyze ecosystem health. In fact, no federal, state, or local environmental agency currently conducts comprehensive watershed ecosystem health assessments. With the exception of the Connecticut program that monitors benthics and habitat, most programs focus on monitoring water column indicators. In short, they have fallen into the trap of defining ecosystem health as compliance with water quality standards that do not accurately describe reference conditions for ecosystem health assessment.

Taking a watershed ecosystem approach can overwhelm professionals and volunteers alike, because historic information on parameters of interest is limited, numerous stakeholders (businesses, municipalities, and individuals) with different needs and interests must be considered, and government agencies must coordinate activities (Goodno R, personal communication). Specific factors that inhibit a comprehensive ecosystem health assessment by volunteers include limits on volunteer time; difficulty in defining goals, objectives, and questions; trouble finding and targeting the appropriate data users; the need to combat doubts surrounding volunteer credibility; the choice between a confusing number of organisms and complex interactions in an ecosystem; and dealing with the limitations of a variety of indicators. Perhaps the most important limitation for volunteers is cost. The costs of monitoring, laboratory analyses, equipment, and rigorous quality assurance protocols can be particularly cumbersome.

Our work with community-based water monitoring groups has led us to observe other conditions outside the control of volunteers that prevent comprehensive ecosystem health assessments. First, most government agencies, professionals, and academic researchers do not understand and therefore cannot clearly define ecosystem health reference conditions, nor can they provide satisfactory guidance as to how to assess ecosystem health. For volunteers, the challenge of setting up a watershed ecosystem monitoring program is related to the lack of clarity provided by these professional "experts." Indeed, until ecosystem health assessment and watershed management theories are clearly stated in terms of ecologic reference conditions, volunteer groups have little incentive to monitor ecosystem health. Many such water monitoring organizations choose to assess compliance with water quality standards because clear benchmarks have been established. When they find violations, however, most state agencies do not have a visible, public process for enforcing regulations or dealing with violators. This situation may frustrate volunteers who have little power to force clean-up or remedial action, potentially leading to conflict or volunteer apathy. Some groups avoid this situation by working directly with local decision makers, such as road agents, public works directors, and individual landowners.

Second, a comprehensive ecosystem health assessment requires the labor of community-based volunteers. Yet professionals, academics, and many government agencies underestimate the capabilities of volunteer water monitoring groups, and are skeptical that they can produce data that meet fairly rigorous data quality requirements. The professionals assume that volunteer monitors cannot conduct their analyses using standard methods. As a result, agency and academic experts often recommend kits or methods to volunteer groups that they would never use themselves. Our experience has demonstrated that, with training and guidance, volunteer monitoring groups are capable of producing data of quality comparable to that of professionals. Community volunteers can be trained to monitor more complex chemicals or indicators that are typically the domain of experts.

In addition, individuals and agencies who are willing to use volunteer data may not clearly specify their data requirements or provide guidelines for data submission. Data quality requirements differ according to the proposed application of the information. For example, an agency may be adverse to pursuing enforcement action unless the data are of the highest quality. To satisfy data requirements, volunteers need guidance from the users of this information, so that the volunteers may then select the most appropriate methods and indicators given the proposed end-use.

Although a watershed approach is considered superior to the monitoring of single water bodies and the benefits are many, the barriers to such an approach are extensive. We believe volunteers can overcome these barriers when they receive the appropriate respect, support, and guidance from all stakeholders and experts.

ACKNOWLEDGMENTS

We would like to acknowledge the assistance of the following individuals who provided us with comments during the preparation of this chapter: Jack Byrne, Executive Director of the River Watch Network; Blane Sanchez, Water Quality Officer from the Pueblo of Isleta; Jane Brawerman, Director of the Connecticut River Watch Program; Lisa Wahle, Volunteer for the Mystic River Watch Program; and Ralph Goodno, Executive Director, Merrimack River Watershed Council.

REFERENCES

American Rivers. Endangered rivers of America: a report on the nation's 10 most endangered rivers and 15 most threatened rivers of 1993. Washington, DC: American Rivers, 1993:7–13.

Amigos Bravos. ¡Bravo! Annual Review. Taos, New Mexico: Friends of the Wild Rivers, 1994.

Brawerman J, Dates G. Connecticut River Watch Program 1994 final report. Haddam, CT: Middlesex County Soil and Water Conservation District, Dec. 1995.

Cairns J. A proposed framework for developing indicators of ecosystem health. Hydrobiologia 1993;263:1–44.

Dates G. River monitoring study design workbook. Montpelier, VT: River Watch Network, 1995.

Fosberg FR. The island ecosystem. Man's place in the island ecosystem, a symposium. Honolulu, HI, 1963.

Hunsacker CT, Carpenter DE, eds. Environmental monitoring and assessment program: ecological indicators. Research Triangle Park, NC: Office of Research and Development, U.S. Environmental Protection Agency, 1990.

Intergovernmental Task Force on Monitoring (ITFM). Strategy for improving water quality monitoring in the United States. Final report. Washington, DC, 1995.

Intergovernmental Task Force on Monitoring (ITFM). Strategy for improving water quality monitoring in the United States. Technical appendices. Washington, DC, 1994.

Lee V. The volunteer monitoring movement: a brief history. Proceedings of the Fourth National Citizens' Volunteer Monitoring Conference, Putting Volunteer Information to Use. U.S. Environmental Protection Agency, Office of Wetlands, Oceans, and Watersheds, Assessment and Watershed Protection Division. April 10–14, 1994.

Leopold LB. A view of the river. Cambridge, MA: Harvard University Press, 1994.

Lopez C. Personal communication with Lisa and Peter Wahle of the Mystic River Watch. Dec. 1995.

———. Personal communication with Blane Sanchez. Oct. 1995.

———. Personal communication with Ralph Goodno of the Merrimack River Watershed Council. Sept. 1995.

Merrimack River Initiative. Watershed connections. Merrimack River Initiative Management Plan. Wilmington, MA: New England Interstate Water Pollution Control Commission, March 1997.

Mystic River–Whitford Brook Watershed Association. Mystic River Watch program newsletter. Summer 1991.

———. Mystic River Watch program newsletter. Summer 1990.

National Park Service, North Atlantic Region, River and Trail Conservation Assistance Program. Office of Planning and Design. Golden S (manager), Tracy C (landscape architect). The Mystic Greenway Plan. Sept. 1989.

National Research Council, Committee on Pesticides in the Diets of Infants and Children, Board of Agriculture, Board of Environmental Studies and Toxicology, Commission on Life Sciences. Pesticides in the diets of infants and children. Washington, DC: National Academy Press, 1993.

Nickey LN. U.S.–Mexico border health and environmental issues. Testimony presented to the U.S. Congress and Texas State Legislature. Jan. 15, 1994.

Plafkin JL, et al. Rapid bioassessment protocols for use in streams and rivers: benthic macroinvertebrates and fish. Report EPA 444/4-89-001. Washington, DC: U.S. Environmental Protection Agency, 1989.

Rhode Island Sea Grant, University of Rhode Island, U.S. Environmental Protection Agency, Office of Water. National directory of volunteer environmental monitoring programs. 4th ed. EPA 841-B-94-001. Jan. 1994.

Rosgen DL. A classification of natural rivers. Catena 1994;22:196–199.

Sawyer JS, et al. Hepatitis A in a border community. J Environmental Health 1989;4:2–5.

U.S. Environmental Protection Agency. Volunteer Monitor. Spring 1994;6(1).

U.S. Water Pollution Control Act. Amendments of 1987. Public Law 92-500.

———. Amendments of 1972. Public Law 92-500, 86 STAT.816.

Vannote RL. The river continuum concept. Can J Fish Aquatic Sci 1980;37:130–137.

CHAPTER 9

Assessing Cumulative Health Effects in Ecosystems

Patricia A. Lane

CUMULATIVE EFFECTS: BACKGROUND AND DEFINITION

One of the most prevalent and pervasive forms of environmental illness comes through cumulative effects. In one form or another, these effects influence most parts of the planet. Although the seriousness of this ailment and the need to manage it have been recognized in the assessment literature and legislation for more than two decades, very little progress has been made in curing the disease. The best overall approach is, of course, prevention. When prevention comes too late, however, management and treatment are necessary. Indeed, in many parts of the world, it is already too late for prevention.

Cumulative effects are not amenable to the traditional environmental impact assessment that focuses on single projects (especially physical works) or a strategic environmental impact assessment that addresses the determination of environmental consequences of plans, programs, and policies. Strategic environmental assessment remains in its infancy, and at present is both poorly developed and poorly applied. It has also met with resistance from policy makers and planners. Both forms of impact assessment focus on human-initiated activity that involves some form of planning. Whereas large projects can cause cumulative effects, a large number of insults to the environment are not planned or consciously considered or anticipated.

The large number of unplanned, mostly small human activities that cause cumulative effects has been termed the "tyranny of small decisions" by Alfred Knopf. Whenever numerous, small decisions affecting the environment are made independently, their incremental consequences are rarely addressed or recognized as being caused by discrete decisions or activities. As a result, long-term, large-scale environmental perturbation has not been examined by most traditional approaches.

Most environmental impact assessments are narrowly focused upon single-proponent, single-development assessments. These "proponent-driven" assessments (Figure 9.1) rarely treat cumulative effects adequately. No effective assessment and management approach exists for regional patterns of environmental deterioration that may result from small incremental actions having no identifiable proponent or for cases in which so many proponents and human activities occur that no obvious procedures are available to assess their combined effects on the environment. These ecosystem-driven assessments (see Fig. 9.1) beg a question: Who assumes responsibility in such instances? Cumulative effects have ramifications beyond environmental deterioration per se. Broad-scale loss of environmental quality implies severe long-term economic losses and a restructured set of development opportunities, or lack of them.

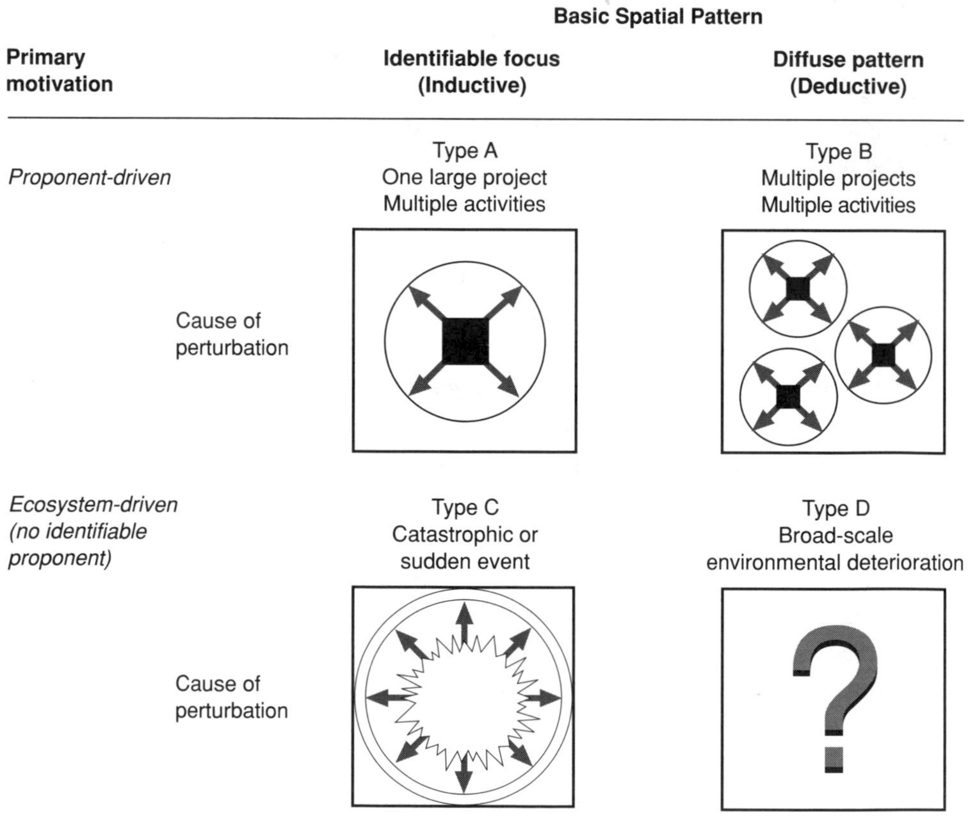

FIGURE 9.1. *Basic characteristics of cumulative effects (periodic, synergistic, or combined).*

In 1987, the World Commission on Environment and Development (Brundtland Commission) published its "Our Common Future" report and created the principles of sustainable development based on the ramifications of cumulative effects on a global scale. Although the commission gave a very convincing rationale for action, it failed to provide a blueprint for such action. We still lack such a blueprint, and the primary difficulty in devising it centers on the problem of how to assess, manage, and prevent cumulative effects. At a fundamental level, sustainable development cannot be achieved without resolving the dilemma of cumulative effects.

Cumulative effects can arise from multiple human activities in a given region or from multiple perturbations to the environment from a single, repeated activity. They can also follow accidental events and catastrophes (such as Chernobyl) that impact regional areas. Cumulative effects can be characterized as occurring over spatially extended areas or regions greater than the size of the local ecosystem, up to and including the total biosphere. Examples of these effects include the deterioration of large water bodies, long-range transport of atmospheric pollutants, climate change, large water diversion projects, groundwater contamination from toxic chemicals, habitat fragmentation, and loss of biodiversity.

In this chapter, we have adopted a working definition of cumulative effects based on the CEARC/U.S. National Research Council's definition (1986):

1. Cumulative effects happen over a period of time when the same type of perturbation occurs with high frequency such that the separate perturbations are not damped out by the ecosystem (time crowding).

2. Cumulative effects happen in space when the same perturbation occurs in locations so close together that effects overlap spatially (space crowding).

3. Cumulative effects occur from different types of perturbations (possibly from separate development, activities, and other sources) that affect similar environmental components if the spatial-temporal scales of the perturbations overlap sufficiently. This situation is referred to as "combined effects" when both periodic and synergistic inputs are coincident.

Cumulative effects are especially important because of their inherent multiplicative and nonlinear nature; combined effects represent a general example of this phenomenon. Many ecological relationships are essentially nonlinear, and these nonlinearities become magnified under many cumulative effects scenarios. Clearly, if cumulative effects are not addressed appropriately, large prediction errors can arise. Although factor interaction is well known in toxicological studies at the individual level, it is less well understood at the population and ecosystem levels. Nonlinearities also complicate the concept of thresholds because discontinuities may appear in the behavior of cumulatively impacted regional ecosystems.

Many diverse and remote factors are now recognized as causing cumulative effects. For example, certain economic sectors may be directed, supported, or controlled by governments, regulatory requirements, or jurisdictional factors in ways that synergistically produce long-term environmental degradation. The agricultural industry provides a good example of this case. Agricultural policies have called for enhanced production at virtually any environmental cost, resulting in an uncontrolled expansion of the agricultural base. These policies have caused additional and significant fragmentation of wetland habitats, loss of biodiversity, serious soil erosion, salination, deterioration of soil quality, deterioration of aquatic systems through chemical contamination in run-off and siltation, and contamination of aquifers over large regional areas.

In addition to the increased complexity of larger time and space scales and the development of a more intricate causality, a more complex level of prediction is required for cumulative effects assessment and management (CEAM) than for traditional impact assessment. In the latter case, we wish to predict and compare systems structure and behavior both with and without a particular development. With CEAM, we must predict not only the future behavior of systems with larger space and time scales, but also the cumulative effects from developments and human activities not yet proposed, but probable for a given region.

Resolving cumulative effects is not simply a problem of developing better analytical tools and improved understanding of spatially extended and perturbed regional ecosystems. Equally important are the plethora of jurisdictional and institutional barriers that impede the identification and management of cumulative effects. Hence, successful CEAM is significantly and simultaneously influenced by decisions made in scientific, social, and jurisdictional areas (Lane et al 1988).

UNDERSTANDING CAUSALITY AND FEEDBACK IN ASSESSING CUMULATIVE EFFECTS

In a simplistic form, large-scale environmental deterioration can be diagrammed as shown in Figure 9.2. In this type of cumulative effect, the causality goes both backward in time to the causes and forward in time to the future state of the environment. Reasoning from the general to the specific constitutes deductive reasoning (to reason backward in time from the poor environmental state to determine its causes); inductive reasoning is used to

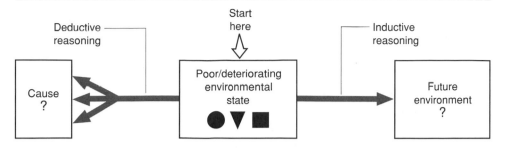

(?) Uncertainty
(■) Feedback among human activities
(●) Feedback among environmental changes
(▼) Feedback among environmental components as risk (ECARs)

FIGURE 9.2. *Causality involved in understanding how to manage perturbed ecosystems.*

reason forward in time (moving from specific to general) to predict how the present environmental state will change. Both inductive and deductive thinking are applied to understand the natural world; neither mode is superior to the other. Scientist and assessors routinely use both. Causality in nature and in human-impacted ecosystems, however, is never this straightforward or simple.

The essence of understanding cumulative effects is the need to understand the causality of complexity interconnected pathways and their feedbacks. Feedback occurs where a variable affects itself through intervening variables. Feedback may also be envisioned as a closed loop where one variable changes a second variable that, in turn, changes a third variable and perhaps a fourth, until the original variable is again affected. A house thermostat represents an everyday example of a negative feedback system that is designed to achieve homeostasis—that is, a room temperature within a narrow range. If the room becomes too hot, the thermostat tells the furnace to make less heat; if the room becomes too cool, it tells the furnace to make more heat. Thus, an increase in heat leads to less heat, and vice versa. This negative feedback is basically stabilizing. Positive feedback occurs when more of one variable leads to more of a second variable and less leads to less. An ever-increasing variable cannot be sustained indefinitely—for example, the human population cannot grow forever and still flourish. Conversely, if a variable is decreasing, it will go to zero and cease to be part of the system. In both instances, positive feedback is considered to have a destabilizing effect.

Feedback has always played a central role in understanding many aspects of nature, such as human physiology, whereby life is played out within rather narrow, homeostatically controlled ranges. In addition, many engineered systems (such as the thermostat and its furnace) work by designed feedback.

Newer concepts of global change (such as the Gaia concept) rely on the notion of feedback, with feedback relationships being seen as keeping the planet suitable for life (Lovelock 1991). To solve cumulative effects problems, feedback and interconnection must be addressed directly; straight-line thinking leads only to more environmental deterioration.

Figure 9.3 illustrates various locations of feedback in an environmental system. Much of our failure to manage ecological systems successfully has occurred because we have not recognized interconnection and especially feed-

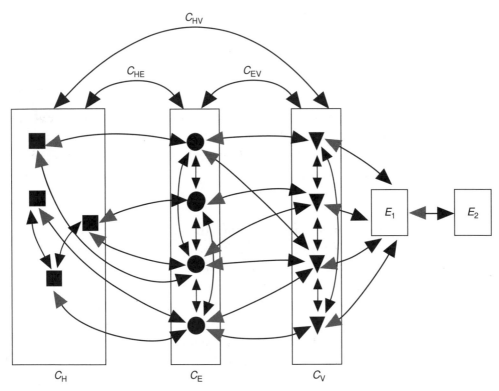

Feedbacks
C_H Feedback among human activity (■)
C_E Feedback among environmental changes (●)
C_V Feedback among ECARs (▼)
C_{HE} Feedback among human activities and environmental changes
C_{EV} Feedback among environmental changes and ECARs
C_{HV} Feedback among human activities and ECARs

State of the environment
E_1 State of the present environment
E_2 State of the environment at time 2

FIGURE 9.3. *Summary of causality and feedback operating to produce cumulative effects (CEs). Black arrowheads indicate direction of inductive reasoning and gray arrowheads show the direction of deductive reasoning.*

back. In addition to the feedback within human activities, environmental changes, or the environmental component at risk (ECAR), these components are also characterized by feedback relationships. Environmental managers, for example, might regulate a particular type of human activity or development as soon as environmental change is observed and before the ECARs deteriorate. Likewise, environmental changes can prompt the ECAR to alter its behavior or physiological patterns, which in turn leads to subsequent ecological changes. An ECAR is most noticeably affected, however, when environmental managers influence human activities so as to lessen these negative consequences. These feedback links, which may be either one- or two-way, represent an important class of control mechanisms and options that might be available to the environmental manager.

The concept of feedback—and hence interconnection—at extended spatial scales is the quintessential concept in cumulative effects assessment. The essence of conducting good cumulative effects assessments involves the identification of the locations and strengths of the key feedback relationships at the appropriate scale. This task is not an easy one, and no precise prescription can be written that will guarantee success for every CEAM problem. Much of the complexity in cumulative effects, however, reflects the existence of these extended feedback relationships. Cumulative effects themselves are not fundamentally different from those effects that occur in local ecosystems: Reproductive capacity may be diminished, feeding behavior may be altered, individuals may die, a lake may become eutrophic, profit may decrease, a way of life might change irreversibly, or a social value may be lost. Instead, the patterns of the changes that occur on a regional scale over longer periods of time differ. In addition, as many of these impacts are cumulative in a multiplicative way, irreversible or sudden environmental damage may take place before the environmental managers can respond appropriately. This problem is coupled with the fact that enough small local changes, which are usually ignored, may actually form part of a severe cumulative effects problem with regional consequences.

DIAGNOSTIC APPROACHES

Cumulative effects assessment (CEA) shares some steps, at least conceptually, with traditional environmental impact assessment. A framework for a bottom-up process is given below for predicting forward into time or inductively (see Fig. 9.2) for cumulative effects types A and B:

1. Scope out the situation (define questions, issues, potential detail of analysis, CEA goals, and logistical support).

2. Develop bounding (define the universe of human activity, potential environmental change, ECARs, and institutional–jurisdictional boundaries).

3. Establish a common basis of understanding as to the type of cumulative effects.

4. List human activities, environmental changes, and ECARs.

5. Prepare a forward CEA diagram (solid arrows; see Fig. 9.2).

6. Decide on the location and amount of feedback (C_M, C_E, C_O, C_{HE}, C_{HO}, C_{EO}).

7. Determine whether the CEA problem is one of space crowding (synergistic), time crowding (periodic), or combined.

8. Select the analytical tools and perform the analysis to determine the predicted state of the environment.

9. Diagram predicted qualitative states.

10. Decide whether the future states are acceptable, have the potential to be mitigated, or are not acceptable and must find ways to alter causality in CEA diagram.

11. Explore management options and the design strategy, and make recommendations.

12. Repeat steps 4–12 if additional scenarios of potential human activities are hypothesized to occur in the future and if decisions need to be made concerning equitable usage of the regional ecosystem.

A similar top-down process, going backward into time for deductive reasoning, is available for cumulative effects types C and D (see Fig. 9.2). Steps 5–9 are unique to CEA and mainly relate to the understanding of the main feedback relationships in the system under study.

In the following discussion, we present a major tool—loop analysis—for elucidating feedback in environmental systems. First, we show how this technique has been used to determine ecological skeletons of marine plankton communities in the western Atlantic. Second, we illustrate how loop analysis was used in a type A cumulative effects problem involving the construction of a large bridge with the potential to alter climatic variables over a large marine environment and harm several important commercial fisheries. This environment, the Northumberland Strait, lies between Prince Edward Island and New Brunswick, two provinces in eastern Canada. The main concern was that the bridge would trap ice normally moving through the strait in the spring, thereby lowering the water temperature and changing other physical variables so as to negatively impact the fisheries there.

Loop Analysis Methodology

We know from network theory that even a single feedback loop strategically positioned in a network of cause and effect can lead to unexpected and even counterintuitive results. Most environmental managers have an intuitive grasp of the counterintuitive aspects of ecological systems. For example, Levins has given the example of pesticides that kill insects in laboratory bottles but fail to control them in nature. Why does this failure occur? It is not because we have observed an incorrect physiological response in the laboratory, but rather because the diverse feedbacks in ecological systems can permit an increase in pests where the physiology predicts a decrease. This situation could happen in many ways. For example, the pesticide may kill predators that grow more slowly than the prey pest of interest. Freed from the threat of predation, the pest species population may increase even though some individuals were killed by the pesticide. Likewise, when nutrients enter a lake, large blue-green algal blooms may occur; subsequently, lowered concentrations of nutrients may be observed. In this case, the turnover rate of the nutrient pool increases greatly, but its concentration observed at a single time decreases. None of these ecological observations negates the physiology upon which all of them are based. In a complex system with feedback, however, many physiological truths become ecological myths (Lane and Levins 1977).

These examples illustrate the importance of feedback in environmental systems. They also demonstrate the importance of choosing a suitable level of organization and scale to study an environmental problem so as to predict cumulative effects. An unsuitable choice leads to great difficulty in characterizing the minimal causality needed for understanding and predictive accuracy. Just as the physiological level is not appropriate to understand a local ecosystem, the local ecosystem is not adequate to understand events at a regional level. Physiological information (in the first case) and local ecosystem understanding (in the second case) may still be not only useful in such case, but often essential. Nevertheless, information at lower levels of organization and scale is often not sufficient to explain behavior at higher levels. If one knew everything about chemistry and physics, it would still be impossible to write the equation for life because the essence of organization of living systems is not wholly contained within the sciences of chemistry and physics. Fundamentally different types of system behavior that are not totally subsumed and apparent at lower levels may eventually emerge at higher levels in a hierarchy.

Central to understanding cumulative effects and the complex causality is our ability to diagram the important causal relationships and understand how they interrelate to cause system change. Qualitative network analyses, such as loop analysis, show the most promise in achieving this end (Levins

1973; Lane 1986a, 1986b, 1986c). This technique can be combined with routine field data and ad hoc expertise to provide direct calculations of feedback and system stability. Loop analysis can also predict qualitative changes in all of the network variables for a set of network stresses. Because the variables are not expressed in quantitative units, such as grams of carbon or kilocalories of energy, disparate variables can be placed together. For example, a fish species, the fisheries manager, market conditions, and fishermen can be incorporated into the same network. Their dynamics affect one another in the real world, and the loop models can capture those relationships. Once the qualitative network is deemed to represent the real world to a reasonable extent, quantitative techniques can be applied to measure the consequences of the system predictions concerning key variables of interest. Thus, the loop models serve as the skeletons upon which data are organized. The process of determining what is most important and measuring it first in a qualitative sense saves a substantial amount of unnecessary quantification and data collection that will not lead to enhanced understanding of the ecosystem under study.

Loop analysis is essentially a tool that enables qualitative understanding of the underlying system causality. It can be regarded as a form of tomography for evaluating ecological skeletons. If we cannot understand the structure of cause-and-effect dynamics at the regional level, we have no more hope of managing cumulative effects than a phrenologist who interprets bumps on a skull from surface observation and touch.

As noted, the modeling efforts enhance the data collection program, and vice versa. Once the basic understanding emerges, we can study the range of management options systematically and determine which set will provide the most environmental improvement and protection in the future at the smallest cost.

Loop analysis employs signed digraphs to represent networks of interacting variables. The technique allows deduction, from routine monitoring data, of the important variables and interactions in a complex system such as a coastal marine community. In addition, loop analysis provides a methodology for analyzing systems based on qualitative representations of variable interactions. These signs are equivalent to the signs of the first partial derivations of the coupled differential equations describing the system. The models derived from routine sampling data indicate the dominant variables and interactions, and the predominant driving forces, or parameter inputs, for each sampling period. Once the network is constructed, the effects of a parameter input can be assessed in terms of directed changes in standing crops and turnover rates of all variables.

Patterns of correlation can also be predicted and the most sensitive components for environmental impacts can be identified. In ecological research, the relationship between correlation and causality has traditionally been an uneasy one. Although significant correlations may be found between pairs of ecosystem variables, ecologists can at best infer only causal mechanisms and, often unknowingly, their inferences support nonsense correlations. As ecosystems undergo parameter input, variables change and produce numerous types of correlation patterns. It has proved difficult to identify the causal pathways between the parameter inputs and the observed changes—especially those that are not directly related. Loop analysis helps identify such pathways. It follows the same methods derived from dynamic systems theory but uses only the signs of the first partials. Thus, this technique provides an advantage in that it relies only on knowledge of the system structure, rather than on detailed knowledge of the functional relationships between system components and quantitative measurement of all variables and constants in the equations.

The graph approach to network analysis came from engineering (Mason 1952) with varying degrees of application in ecology (Levins 1973, 1975). Levins (1973, 1975) developed the theory of loop analysis for physiological and ecological systems. Berryman (1981) and Hutchinson (1975) have given introductions to the technique. To date, the methodology has been applied to several systems, including human physiology and disease (Levins 1973), community evolution and population genetics (Levins 1975), species interactions (Boucher et al 1982; Henry 1980), freshwater plankton communities (Briand and McCauley 1978; Lane and Levins 1977), and marine mesocosms and environments (Lane 1981, 1982, 1984, 1985, 1986a, 1986b; Lane and Collins 1985; Wright and Lane 1986; Stokoe et al 1989). Lane (1986a, 1986b) has described some theoretical developments. Calculation methodology is not detailed in this chapter.

By comparing the theoretical predictions with the field data, it becomes possible to construct a restricted set of loop diagrams that represent the system under study. Once the correct representation has been determined, the pathways producing the directed changes can be extracted, examined, and subjected to a more quantitative form of analysis using standard modeling techniques.

Core Loop Models from Data Sets

To date, loop analysis has been routinely applied to large data sets describing marine plankton communities (Lane 1986a, 1986b). These data sets are similar to those collected in biological monitoring programs for the purposes of conducting an environmental impact assessment. In fitting loop models to data, one of the first procedures involves grouping the raw variables into loop variables. Infrequently sampled species are deleted because too little information is available to characterize them. The data frame involving changes in abundances over time is then transformed into a qualitative data matrix by determining percent relative change of each variable from one sampling period to the next (Lane 1986a, 1986b). Next, the loop models are fitted by hand and then checked by computer.

Core Models of Marine Communities. A set of loop models can be summarized by using variables and links that indicate the dominant or core network or ecological skeleton of the ecosystem. The core diagram is a network formed from the most prevalent variables and linkages in a set of individual models. Although a loop diagram that illustrates changes between two dates might appear to have missing links, the core structure of an annual cycle (a set of 10

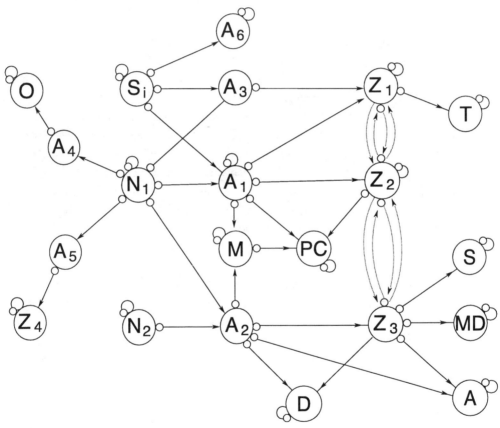

FIGURE 9.4. *Composite marine core loop model for pelagic zone of western Atlantic. Si = silica; N_1 = inorganic N/P ratio; N_2 = organic nutrient pool; A_1 = diatoms; A_2 = dinoflagellates; A_3 = luxury-consuming; A_4 = microflagellates; A_5 = monads and miscellaneous algae; A_6 = silicaflagellates; Z_1 = adult copepods, group 1; Z_2 = immature copepods; Z_3 = adult copepods, group 2; Z_4 = cladocerans; 0 = tunicates (Oikopleura spp.); M = mollusk larvae; PC = polychaete and barnacle immatures; D = decapod larvae; T = ctenophores; S = chaetonaths (Sagitta spp.); MD = hydromedusae; and A = anemone larvae. Links from Z_2 to adult copepods (Z_1, Z_3) are given in dotted lines because they are usually connected by one-way relationships that reverse sign and direction frequently.*

to 12 diagrams) represents the bulk of important relationships among the individual diagrams to be included in the core structure.

A composite core model of several field communities has been developed that is robust and easily transferable among ecosystems in the western Atlantic (Figure 9.4). Approximately 1000 individual models of marine systems have been constructed to date. Environments analyzed to date include St. Margaret's Bay and Bedford Basin, Nova Scotia; Narragansett Bay, Rhode Island; Delaware Bay; and Long Island Sound. These models take into account more than 1000 species. Few other techniques of ecosystem analysis have the power to treat species individually with this degree of feedback complexity. The composite core model includes approximately 22 variables, 8000 causal pathways, and more than 1 million feedback terms.

Loop analysis has also been used successfully with laboratory data sets both for the Dalhousie Aquatron tower tank and the Marine Ecosystems Research Laboratory (MERL) mesocosms at the University of Rhode Island. At the ecosystem level, researchers have encountered considerable difficulty in comparing field and laboratory results. Because few rigorous measures of community structure have been determined, it can be difficult to predict whether perturbations identified in the laboratory will have the same effect on a field community. Changes in the abundance of an individual variable can have four forms (+, −, 0, or absent); for an ecosystem of 20 components, four sets of directed changes, each containing 4^{20} possibilities, exist. No ecological theory has been developed to predict when a particular set of directed changes or two different sets will actually represent similar or equivalent networks. In contrast, loop analysis accomplishes this goal easily. By comparing the "ecological skeletons" of the field and laboratory communities, one can evaluate their degree of similarity, especially in response to a particular pollutant or perturbation. A core model formed from 20 individual Dalhousie tower tank networks produced only two differences in links and one difference in a variable compared with loop models for nearby Bedford Basin in Nova Scotia. Thus, the analysis provides a powerful tool for questions of field–laboratory correspondence. Results to date have indicated that the tower tank will support a community of sufficient complexity to mimic nature effectively.

Cumulative Effects Assessment for the Northumberland Strait. The physical variables that were quantified included total and monthly flows for the St. Lawrence River, monthly air temperatures, monthly surface water temperatures (for April to November only), freezing-degree days, first day of the year the strait was free of ice, an ice severity index (calculated as an average thickness over the strait measured on a chosen date in March), monthly average wind speed, and wind direction (measured as degrees from true north). The biological variables included the total yearly commercial landings for five species of fish in the strait. Climatic data were continuous for the period 1969 to 1987 and commercial fisheries data from 1960 to 1987.

Fish Landings.   Four species of commercially exploited fish have been chosen to be the key species of interest in the loop analysis of the marine food web of the Northumberland Strait. These species are as follows: American lobster, *Homarus americanus;* sea scallop, *Placopecten magellanicus;* winter flounder, *Pseudopleuronectes americanus;* and American plaice, *Hippoglossoides plates-soides.*

Yearly landings of four major species are plotted on Figure 9.5. Patterns for individual species are not similar over time. In the late 1960s, scallops had the largest landings, followed by lobster, hake, and plaice, with flounder being much lower. In the early 1970s, the lobster and scallop landings decreased and the other species' levels increased so that they were all roughly the same (flounder was still the lowest). In the late 1970s, lobster landings began to increase and became dominant. In the 1980s, lobster landings continued to

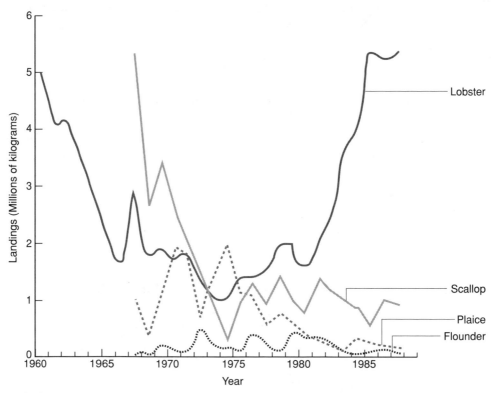

FIGURE 9.5. *Landings from Northumberland Strait (1960–1987).*

rise while those of other species slowly decreased. Scallops currently have the second highest landings.

Loop analysis essentially provides the "ecological skeleton" of an ecosystem by producing the constrained set of variables, links, and parameter inputs (driving forces to the networks). This technique greatly reduces the amount of quantification needed as only some 15 to 25 of all possible links appear to be operative in these networks. Successful completion of the analysis also verifies that variable aggregation has been correct.

Lobster landings decreased by half from 1960 to 1966. They showed a small peak in 1967, but remained at the lower levels from 1968 to 1977. After that year, landings increased steadily (except for a small dip in 1980), surpassing their 1960 levels.

For scallops and the finfish species, the data do not begin until 1967. Scallop landing levels in 1967 declined very steeply, with a small recovery in 1969 and a very low level in 1974. They then climbed for two years and have remained more or less constant ever since, albeit at a much lower level than in the late 1960s.

Winter flounder landings were very low in 1967 and 1968 but increased the next year. Landings remained fairly even, with significant peaks in 1972 and 1976, before they dropped to zero in 1984. In 1985, landings returned to levels that were previously considered "average" and have remained fairly steady since then. Plaice landings declined in 1968 from their 1967 levels, then rose sharply, peaking for three years before dropping again in 1972. They recovered somewhat until 1974, when they began to decrease fairly steadily. Variations in landings, of course, may reflect many factors other than the species populations, such as economic incentives (that is, selling prices).

A literature search was conducted for each of these species to determine life histories and predator–prey interactions as they may exist in the Northumberland Strait. Using this information, a food web was constructed (Figure 9.6). Although statistical analysis was conducted on white hake, *Urophycis tenuis*, in previous sections, it was omitted from this model because of the lack of reliable information for this species. The main predator–prey relationships are readily apparent in Figure 9.6.

For several variables, individual species have been grouped into a single variable. This convention eliminates the need to create an individual variable for each species and significantly reduces the number of pathways within the model. For example, the variable "small invertebrate" includes such species as juvenile scallops, mussels, clams, gastropod mollusks, small shrimp, isopods, and amphipods. The variable "crabs, echinoderms" includes decapod crabs, brittle stars, sand dollars, sea urchins, and sea stars. The food sources for each species within each variable are similar, but not identical, and are predominantly detrital. Juvenile lobsters have a negative effect on small invertebrates.

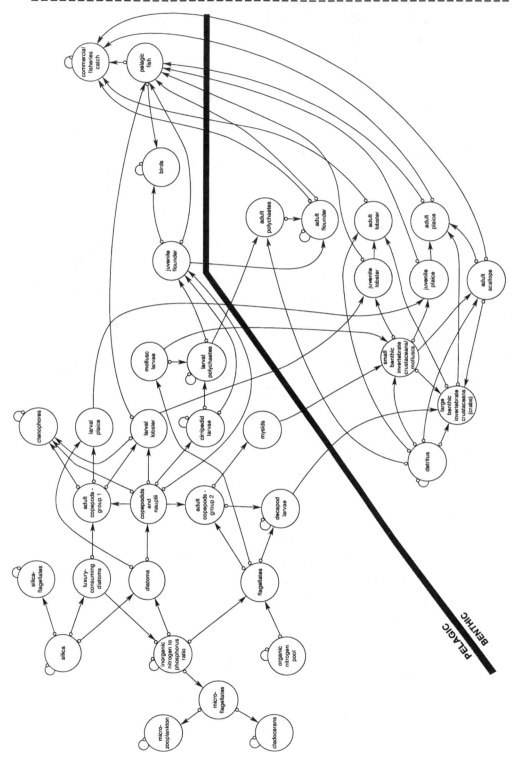

FIGURE 9.6. *Hypothetical 35-variable food web with coupled benthic–pelagic communities in the Northumberland Strait.*

According to the literature, juvenile lobsters will feed on gastropod mollusks and small mussels (Cobb 1976; Weiss 1970), two of the main species included in the "small invertebrate" variable.

Reflecting the nature of the life histories of the key species, the food web has been divided spatially into pelagic and benthic ecosystems. All but winter flounder have pelagic larvae, while the juveniles and adults of these species are predominantly benthic. Winter flounder eggs are demersal and adhesive (Buckley 1989); although the larvae remain mainly on the bottom, they will engage in upward swimming. Larvae of other flatfish are more pelagic, spending all of the larval stage within the water column. Juvenile flounder are benthic and seldom lose contact with the substrate (Buckley 1989); most juveniles tend to remain in or near shallow waters for the first two years, however, and move seaward as they grow larger (Topp 1967). For this reason, juvenile flounder have been incorporated into the pelagic system of this food web, assuming that, as cited in the literature, they feed directly on pelagic species such as larval polycheates, cirriped larvae, and nauplii (Laurence 1977; Pearcy 1962).

Larvae of the other three species are pelagic but settle to the benthos as juveniles. Lobster larvae settle late in stage IV and burrow into the substrate. While sheltered in their burrow, they molt into juveniles (Cooper and Uzmann 1980). Scallop larvae settle out as spat and attach themselves to the underside of shell fragments and other solid materials on the bottom (Culliney 1974). Once attached, they begin their metamorphosis into adults; when they reach a length of 10 mm or more, they detach from the epibenthic substrate and settle on the bottom. Scallops, which remain relatively active until they are approximately 8 cm long, will swim in response to predation and dredging (Caddy 1968). Adult scallops remain recessed in the sediment and rarely move unless they are dislodged by some physical process (Mullen and Moring 1986). Juvenile and adult plaice remain in the benthos but will frequently move off the bottom in pursuit of prey species (Pitt 1984).

Loop Analysis Models of Key Relationships. The food web depicted in Figure 9.6 is a hypothetical description of the key interactions in the pelagic and benthic ecosystems of the Northumberland Strait developed from the literature on individual predator–prey relationships and previous marine loop models (Lane 1986a, 1986b; Lane and Collins 1985). If an appropriate database were available, such as that Lane (1986a, 1986b) used for Delaware Bay, one could determine a core loop structure for the Northumberland Strait, compute several measures of overall community structure and stability for this community, and quantify the pathways of effects and changes in the standing crops of these variables at the end of pathways. Lane, writing in Stokoe et al (1989), illustrated how loop models can be represented as sets of differential equations that will give quantitative predictions. The model in Figure 9.6 has 35 variables, which means that a large universe of possible qualitative structures for the core model exists—more than 10^{26}, at least. It would not be realistic to attempt to delimit this universe without additional data fitting.

Because the set of field data has been restricted to the landing values for the four commercial fish species and physical variables as a time series over several seasons and years, it is necessary to use a modified approach based on examining the statistical relationships summarized on subportions of the larger model. It is also necessary to assume that the yearly data on landings of fish are directly related to the abundances of the key commercial fish species living in the Northumberland Strait, as no other field abundance data were readily available for this study. Data sets are never perfect, and ecosystems—especially regional ecosystems—will probably remain "partially specified" in most cases. In the following analysis, causal interpretations are offered for some of the results observed for pairs of fish species.

Combined Fisheries Food Webs Illustrating Benthic–Pelagic Coupling. Figure 9.7 illustrates the combined four species fisheries for both the benthic and pelagic zones of the Northumberland Strait. The two models are identical except for the predation on adult lobsters (A) and adult scallops (B). Because of the complexity of these models, they are portrayed somewhat differently than the loop models previously given in this section. This type of representation also emphasizes the different types of time lags that operate in these food webs. The concentric circles represent linkages between pairs of successive life stages in the four major commercial fish species and for one important food source, adult and larval polychaetes. Polychaetes appear to play a major role in the benthic–pelagic coupling. Most of these links span both the benthic and pelagic zones. Links between pairs of variables, positioned as spokes in a wheel, that lie wholly within the pelagic or benthic zone signify predator–prey relationships.

The question is then asked, how strong is the benthic–pelagic coupling and what role does it play in determining time lag effects between the physical environment and the commercial fishery? To answer this question, five types of relationships were altered as shown in the numbered boxes. First, all links from the benthic zone to the pelagic zone were severed (box 1). Given the life history information, the timing of these links could be affected by events occurring at the end of the ice season or immediately thereafter. Second, all links from the pelagic zone to benthic zone were severed (box 2). Most of these links would involve the movement of pelagic immature stages to the benthos at the end of the summer and in early autumn before advent of the ice season. Third, links between larval and adult polychaetes were deleted (boxes 3 and 4). Fourth, one-way links from polychaete larvae to adults were

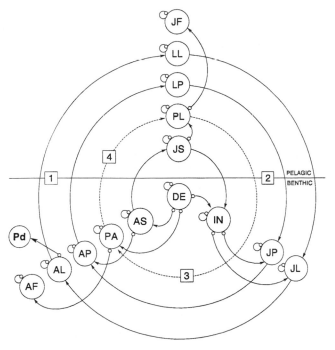

(A) LOBSTER

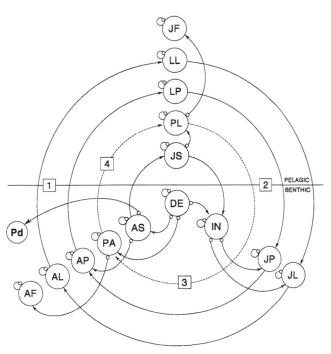

(B) SCALLOP

FIGURE 9.7. *Combined life-stage model of two fish species in the Northumberland Strait. DE = detritus; IN = small invertebrates; PA = polychaete adults; PL = polychaete larvae; AS = adult scallop; AP = adult plaice; AL = adult lobster; AF = adult flounder; Pd = predator; JF = juvenile flounder; JP = juvenile plaice; JL = juvenile lobster; JS = juvenile scallop; LP = larval plaice; LL = larval lobster.*

omitted (box 3). Fifth, one-way links from polychaete adults to larvae were deleted (box 4). Using these combinations, 12 models were developed:

- Lobster fishery: boxes 1 + 3; 1 + 4; 1 + 3 + 4; 2 + 3; 2 + 4; 2 + 3 + 4

- Scallop fishery: boxes 1 + 3; 1 + 4; 1 + 3 + 4; 2 + 3; 2 + 4; 2 + 3 + 4

Properties of the 12 loop models are summarized in Table 9.1.

In the loop model results, predation relationships appeared to be important for both the lobster and scallop fisheries; in contrast, the plaice and flounder fisheries were driven by parameter inputs lower in the food web. The scallop was driven more by predation in the benthic zone than was the lobster. The lobster and scallop fisheries were examined in more detail to discern the potential for benthic–pelagic coupling given that adults of each species are subject to predation (from humans). In the lobster models, the greatest fit occurred during the lobster life stages (LL and JL). As no predator was represented in the pelagic zone on LL or in the benthic zone on JL, we cannot discriminate these causal relationships from effects emanating solely from the immature forms (as contrasted with predators of the immature forms).

TABLE 9.1   Community measures for combined food web of coupled benthic–pelagic fisheries in the Northumberland Strait

Model	Number	Box Alterations	Nodes	Links	F_n	Loops	Paths	C
Lobster	(1)	1 + 3	15	40	−33	25	243	12%
	(3)	1 + 4	15	41	−33	25	279	13%
	(2)	1 + 3 + 4	15	41	−37	28	276	13%
	(4)	2 + 3	15	40	−78	25	222	12%
	(5)	2 + 4	15	41	−78	25	255	13%
	(6)	2 + 3 + 4	15	41	−90	28	343	13%
Scallop	(7)	1 + 3	15	40	−33	25	256	12%
	(9)	1 + 4	15	41	−36	28	390	13%
	(8)	1 + 3 + 4	15	41	−33	25	295	13%
	(10)	2 + 3	15	40	−78	25	236	13%
	(11)	2 + 4	15	41	−87	28	362	13%
	(12)	2 + 3 + 4	15	41	−78	25	272	13%

F_n = total feedback (n variables); C = connectance.

In models with boxes 1 + 3, 1 + 4, and 1 + 5 (with the benthic-to-pelagic zone links severed), parameter inputs were similar regardless of the configuration of the adult and larval polychaetes. These results indicate that, at the end of ice season, the effects on lobster travel through its immature stages, despite the lack of direct links from the benthic adults to the pelagic larvae. Thus, lobster abundances could be influenced by long time lags from events occurring in the pelagic zone. When the links were severed from the pelagic to benthic zone representing the mid- to end of summer conditions, less overall fit with the parameter inputs was observed, especially when the link between polychaete adults and polychaete larvae was severed. A much better fit arose when the polychaete larvae to polychaete adult link was maintained—a relationship indicative of the life history events occurring at this season. Thus, the lobster fishery through its pelagic life stages could be expected to exhibit time lags related to the events occurring at ice-out.

For the scallop fishery, a generally poor fit arose when either the benthic-to-pelagic links or the pelagic-to-benthic links were severed. Thus, this species appears to be more dependent on benthic–pelagic coupling than is the lobster. The scallop is also less dependent on events occurring at ice-out than is the lobster.

As is clear from these models, several causal pathways exist by which ice-climate mechanisms could interact with both the benthic and pelagic food webs and their coupling. Each of the four fish species of interest exhibited different potential responses to environment vis-à-vis the position of their life stages in relation to their food resources and predators. Lobster appeared to be the species most apt to be influenced by events concurrent with ice-out, whereas scallops were most dependent on benthic–pelagic coupling.

Summary of Results

First, a large benthic–pelagic food web was constructed using the available literature and previously constructed loop models for the pelagic zone. It represented the first loop model of its kind based on the relationships of life stages of key commercial fish species in the extended spatial-temporal context of coupled marine food webs and the cumulative effects they can potentially experience from large development projects. Given the literature on the feeding and temperature relationship of these species, it was possible to hypothesize how climatic-ice parameters might enter the food web at particular nodes.

Second, three types of preliminary models were constructed as subcomponents of the larger model. The first set of models identified four categories of parameter inputs and how they would relate to a combined model of the four key species. This set of models represented immediate effects (that is, very short lags). The presence of a particular fishery did not greatly change the locations and signs of parameter inputs. Plaice and flounder reacted similarly and most negatively to likely perturbations that might be caused by a bridge. Scallops also reacted negatively for three categories but favorably to negative inputs from the fishery. Lobster was the most sensitive, reacting positively to all models with the exception of negative category I inputs.

The second set of models delineated how differences in life stages would interact to produce different responses to parameter inputs. These inputs represent mostly short lag phases. Plaice and flounder reacted similarly, being less responsive to predation events and more responsive to food resources (lagged variables); this reaction rendered them only slightly at risk to bridge-related effects. Lobsters responded differently based on the inputs that were entered as food, pelagic larvae, or fishery variables (immediate and short-term lags). Scallop was most sensitive to benthic inputs (lagged variables) and did not respond to pelagic events, freeing them from immediate effects of a delayed ice-out.

In a third set of models, the role of benthic–pelagic coupling was studied for the four fish species. These models represented moderate to long lags. Lobster appeared to be relatively insensitive to this coupling but was influenced by the physical environment that impinged upon the benthic juvenile stages at the ice-out part of the season. A delay of ice-out caused by construction of the bridge would impact these animals and alter the future abundance of adults, with the lag being equal to the recruitment age. In contrast, the scallop was more influenced by the total structure of interconnection between the benthic and pelagic zones; hence, it remained unaffected by delayed ice-out. Plaice and flounder had similar responses.

Third, several causal pathways exist by which ice-climate mechanisms could interact with both the benthic and pelagic food webs and their coupling in a regional marine ecosystem. Each of the fish species exhibited different potential responses to the physical environment vis-à-vis the position of their life stages in relation to their food resources and predators. Lobster appeared to be the species most apt to be influenced by events concurrent with ice-out whereas scallop was most dependent on benthic–pelagic coupling and least influenced by events concurrent with ice-out.

Fourth, cumulative effects assessment centers on understanding the complexity of interconnections among environmental variables and parameters over regional or extended time and space scales. In particular, understanding feedback is essential to this type of assessment. In the example presented in this chapter, a large bridge construction project costing more than $1 billion was projected to affect a large marine environment, inspiring multiple interactions that would involve a variety of climatic variables and four key commercial fisheries interconnected with three-dimensional benthic–pelagic coupling. The level of complexity represented in the loop models is probably appropriate to that needed to address cumulative impacts successfully. Too often, more simplistic approaches have failed to include causal relationships or predict cumulative impact accurately. Loop analysis is perhaps one of the best methodologies currently available for understanding and measuring feedback in impacted environments, both local and regional areas.

ACKNOWLEDGMENTS

The author gratefully acknowledges research funding from the Canadian National Science and Engineering Research Council (NRSERC) to Dalhousie University, from the National Oceanic and Atmospheric Administration (NOAA), U.S. Department of Commerce to Harvard University, and by a contract from the Northumberland Strait Crossing Project, Public Works Canada.

REFERENCES

Berryman AA. Population systems: a general introduction. New York: Plenum Press, 1981.

Boucher DH, James S, Keeler KH. The ecology of mutualism. Ann Rev Ecol Syst 1982;13:315–347.

Briand F, McCauley E. Cybernetic mechanisms in lake plankton systems: how to control undesirable algae. Nature (London) 1978;273:228–230.

Buckley J. Species profiles: life histories and environmental requirements of coastal fishes and invertebrates (North Atlantic)—winter flounder. U.S. Fish Wildlife Serv Biol Rept 1989;82(11.87). U.S. Army Corps of Engineers, TR EL-82-4.

Caddy JF. Underwater observations on scallop (*Placopecten magellanicus*) behaviour and drug efficiency. J Fish Res Board Can 1968;25:2123–2141.

Canadian Environmental Assessment Research Council (CEARC) and U.S. National Research Council. Proceedings of a workshop on cumulative effects: a binational perspective. Hull, Quebec: CEARC, 1986.

Cobb SJ. The American lobster: the biology of *Homarus americanus*. Rhode Island University Mar Tech Rept 49, 1976.

Cooper RA, Uzmann JR. Ecology of juvenile and adult *Homarus*. In: Cobb JS, Phillips BF, eds. The biology and management of lobsters, vol. II. New York: Academic Press, 1989:97–142.

Culliney JL. Larval development of the giant scallop *Placopecten magellanicus*. Biol Bull (Woods Hole) 1974;147:321–332.

Henry C. Competitive interactions studied by the mathematical method of loop analysis. CR Seances Acad Sci Ser D 1980;290:787–790.

Hutchinson GE. Variations on a theme by Robert MacArthur. In: Cody ML, Diamond JM, eds. Ecology and evolution of communities. Cambridge, MA: Belknap Press of Harvard University Press, 1975:492–521.

Lane PA. Symmetry, change, perturbation and observing mode in natural communities. Ecology 1986a;67:223–239.

———. Preparing marine plankton data sets for loop analysis. Ecology. Supplementary Publication Source Document No. 825B, 1986b.

———. A foodweb approach to mutualism in lake communities—distinguishing direct, indirect and community effects. In: Boucher D, ed. Mutualism: new ideas about nature. Montreal: Universite du Quebec, 1985:344–374.

———. Notion of core structure in marine plankton communities. Conference on ecosystem theory in relation to biological oceanography. Sponsored by SCOR/UNESCO Working Group no. 73, National Science Foundation, and U.S. Office of Naval Research. Quebec City: Universite Laval, 1984.

———. Using qualitative analysis to understand perturbations in marine ecosystems in the field and laboratory. In: Archibald P, ed. Environmental biology state-of-the-art seminar. EPA-600/9-82-007. Washington, DC: Office of Exploratory Research, U.S. Environmental Protection Agency, 1982:94–122.

———. Analyzing perturbations to marine ecosystems using loop analysis and time averaging. In: Lane P, Morison R, eds. Harvard University Technical Report. Cambridge, MA: Harvard Marine Ecosystems Research Group, 1981.

Lane PA, Collins TM. Foodweb models of a marine plankton community: an experimental approach. J Exp Mar Biol Ecol 1985;94:41–70.

Lane PA, Levins R. Dynamics of aquatic systems. II. The effects of nutrient enrichment on model plankton communities. Limnol Oceanogr 1977;21:454–471.

Laurence GC. A bioenergetic model for the analyses of feeding and survival potential of winter flounder (*Pseudopleuronectes americanus*) larvae during the period from hatching to metamorphosis. US Natl Mar Fish Serv Fish Bull 1977; 75:529–549.

Levins R. Evolution in communities near equilibrium. In: Cody ML, Diamond JM, eds. Ecology and evolution of communities. Cambridge, MA: Belknap Press of Harvard University Press, 1975:16–50.

———. The qualitative analysis of partially specified systems. Ann NY Acad Sci 1973;231:123–138.

Lovelock J. Healing Gaia: practical medicine for the planet. New York: Harmony Books, 1991.

Mason SJ. Some properties of signal flow graphs. Proc IRE 1952;41:1144–1156.

Mullen DM, Mooring JR. Species profiles: life histories and environmental requirements of coastal fishes and invertebrates (North Atlantic)—sea scallop. US Fish Wildlife Serv Biol Rep 1986:82(11.67) U.S. Army Corps of Engineers. TR EL-82-4.

Pearcy WG. Ecology of an estuarine population of winter flounder *Pseudopleuronectes americanus*. Bull Bingham, Oceanogr Collect Yale Univ 1962:18(1).

Pitt TK. Underwater world fact sheet. American Plaice. Department of Fisheries and Oceans, Canada. DFO/1360, 1984.

Stokoe PK, Lane PA, Cote RP, Wright JA. Marine ecosystem models and modelling approaches; evaluation of holistic marine ecosystem modelling as a potential tool in environmental impact assessment. Biological Sciences Branch, Scotia-Fundy

Region, Department of Fisheries and Oceans, Bedford Institute of Oceanography, Dartmouth, Nova Scotia, 1989.

Topp RW. An estimate of fecundity of the winter flounder, *Pseudopleuronectes americanus*. J Fish Res Board Can 1967;25:1299–1302.

Weiss HM. The diet of feeding behaviour of the lobster, *Homarus americanus*, in Long Island Sound. Ph.D. thesis, Univ. of Connecticut, Storrs, 1970.

World Commission on Environment and Development (Brundtland Commission). Our common future. Oxford, UK: Oxford University Press, 1987.

Wright JA, Lane PA. Theory of loop analysis. Ecology. Supplementary Publications Source Document no. 825A. 1986.

CHAPTER 10

Integrating Health Surveillance and Environmental Monitoring

Paul R. Epstein

EMERGING INFECTIOUS DISEASE AND GLOBAL CHANGE

Widespread social inequities, ecological dysfunction, and climate instability are all contributing to the emergence, resurgence, and redistribution of infectious disease on a global scale. Changes in these three components of global systems are transforming biodiversity—disrupting the species compositions and relative abundances essential for the biological control of pests and pathogens, thereby creating conditions conducive to the spread of such disease. Today, for example, urban dengue and rural malaria are in resurgence, and morbilliviruses (measles and distemper viruses) have begun to affect terrestrial mammals (such as Australian horses and Serengeti lions) as well as aquatic mammals (such as seals and dolphins). An iridovirus may be involved in the disappearance of frogs from six continents. Plant diseases have also begun a resurgence in Latin America, threatening trade and food security. These evolving epidemics, which affect a wide taxonomic range, may reflect the existence of altered agents, weakened hosts, or environmental change, or contributions of all three factors.

Climate Change and Disease

Heatwaves are not healthy for humans and other animals. In the summer of 1995, for instance, excess deaths from heatwaves occurred from the United States to India. Subtropical cities are generally less affected by this type of environmental change because of acclimatization. Recently developed climate change scenarios predict the emergence of more prolonged and intense heatwaves in the future.

154

The direct impact of heatwaves is not the only health consequence of climate change. For infectious diseases, climate circumscribes the range of vector-borne diseases (carried by mosquitoes and rodents, for example), and weather affects the timing of outbreaks. An integrated assessment of climatic, ecological, and social factors is necessary to understand the current emergence of infectious disease as a global problem. In particular, mosquitoes are highly sensitive to climate factors and change, as paleoclimatic data demonstrate. Models of global warming project that increased temperatures at high altitudes and temperate latitudes will favor the spread of mosquito-borne diseases. With misquitoes having clear thresholds for survival (for example, 10°C winter isotherm for *Aedes aegypti* and dengue fever, and a 16°C winter isotherm for *Anopheline* spp. and falciparum malaria transmission), shifts in the latitude and altitude of isotherms may extend the area of disease transmission (in the case of malaria, dengue fever, encephalitis, and yellow fever, for example). The transmission may also become more seasonal in areas now on the margins of regions with bioclimatic (temperature and moisture) conditions conducive for vector survival. While 42% of the globe presently has conditions that can sustain malaria transmission, for instance, some scenarios suggest that this figure could rise to 60% of the globe.

The consensus of the Intergovernmental Panel on Climate Change (IPCC) Working Group I (WG I) is that altered in-climate variability suggests that overall climate is changing; the changes observed are most likely due to human activities. According to WG II, the health impacts from this trend will continue to be overwhelmingly negative. Indeed, current changes in the distribution of emerging infectious diseases (EIDs) are consistent with those projected under climate change scenarios. Montane regions, where isotherms may first noticeably shift, are sentinel areas to examine in this light.

Montane Regions

Vectors and the diseases they carry (for example, malaria and dengue) are currently being reported at higher altitudes in Africa, Asia, and Latin America than at any time during this century. Highland malaria is becoming a major problem for urban centers in central Africa, and dengue fever has blanketed the Americas, crossing mountain ranges that previously posed obstacles to its spread. In addition, upward plant migration has been documented on 26 Alpine peaks and observed in Alaska, the U.S. Sierra Nevada, and New Zealand. These biological findings are occurring in tandem with widespread physical phenomena. For instance, montane glaciers are retreating in Argentina, Peru, Alaska, Iceland, Norway, the Alps, Kenya, the Himalayas, Indonesia, and New Zealand.

Temperature measurements support these statements. Henry Diaz and Nicholas Graham, using weather-station measurements, have detected a

160-m rise in the freezing level in montane areas since 1970. Considering the adiabatic drop of approximately 1°C per 100 m of elevation, this shift represents a warming of 1.5°C. Joel Susskind of NASA, examining MSU and HIRS2 satellite data from the 1980s, has found preliminary evidence suggesting that mid-troposphere temperatures during El Nino years may increase disproportionately with respect to temperatures on the Earth's surface. The most recent data and models (IPCC 1995, which combines aerosolized sulfates and greenhouse gases) confirm these findings on warming in the mid-troposphere (in the southern hemisphere), indicating that the warming there is even greater than that evident on the Earth's surface.

Marine Ecosystems

Each year 30,000 to 60,000 U.S. residents become ill from eating contaminated seafood. New diseases are emerging from marine systems that threaten human health, food safety, and food security. In the marine environment, ecologists portray complex microecosystems involving algae, bacteria, and viruses associated with the emergence of new infections and biotoxic events. These events include the growth of vibrio (*Vibrio cholerae*) (Colwell et al 1980; Byrd et al 1991; Epstein et al 1993) and *V. vulnificus* (CDC 1993a); viral contamination of oysters in the Gulf of Mexico (CDC 1993b); domoic acid–induced amnesic shellfish poisoning on both coasts of North America (Todd 1993); the emergence of *Pfeisteria piscicida* (a highly toxic dinoflagellate) as a cause of large fish mortalities on the East Coast (Burkholder et al 1993); multiple marine mammal die-offs; a green turtle fibropapilloma epizootic (Willams et al 1994); and a brown tide off Long Island that first crippled the scallop industry and then turned toxic, with a viral agent suspected.

Biodiversity in marine ecosystems is important for controlling diseases in marine flora and fauna, and those species dependent upon them (such as shorebirds and humans). For example, among the factors influencing the proliferation and persistence of coastal phytoplankton populations are eutrophication, warm water events, and diminished grazing due to overfishing. The widespread epidemic of harmful algal blooms—the increase in their extent, duration, and appearance of novel toxic species—has multiple health effects, including the following:

- An increase in shellfish and finfish poisonings

- Anoxia in seagrass beds (shellfish nurseries)

- Increased health hazards related to in-land and seashore swimming (gastroenteritis, and ear and eye infections)

The "Taura syndrome," which has plagued aqua/mariculture in Ecuador, Panama, Honduras, Venezuela, China, and Japan, serves to illustrate how such epidemics arise. These marine monocultures are confined in coastal ponds carved from surrounding mangroves and vegetation. They are also contaminated with pesticides (so as to prevent banana pest infestations, for example) and flooded with broad-spectrum antibiotics (such as chloramphenicol). Shrimp in such farms have become prey to opportunistic infections including several viruses (one baculo- or insect-borne virus) and vibrios; oysters have been infected with rickettsia. Such semi-enclosed systems are vulnerable and "immunocompromised"—devoid of refugia and in- and out-migration; they may not be a sustainable answer to the worldwide decline in ocean fisheries.

Global Change, Biodiversity, and Marine-Related Disease. Changes in climate dynamics and variability (CLIVAR 1992) can have profound effects on the distribution of flora and fauna that were previously acclimatized to particular regions (Bakun 1993; Barry et al 1993; Tester et al 1991). Ocean warming may alter functional relationships within the marine food web, beginning with plankton biomass (Roemmich and McGowan 1995). Paralytic, diarrheal, neurologic, and amnesic shellfish poisoning as well as ciguatera, pufferfish, and scromboid fish poisoning all related to algal biotoxins (Epstein 1993b)—appear to be spreading in an apparent "global epidemic" of coastal algal blooms (Anderson 1992; Smayda and Shimizu 1993), with direct consequences for human health and nutrition.

A species with yearly rhythms may be able to tolerate specific temperatures at one time during the year, but it—or its predators or food sources—may be incapable of withstanding the same temperatures during another period. For example, in a warm El Nino–Southern Oscillation (ENSO) event, salmon move north along the North American Pacific coast. The northward shift of California mackerel—consumers of salmon exiting Northwest rivers—actually decrease salmon stocks two years later, however. Thus, the timing, intensity, duration, and frequency of anomalies are important.

Scientists have yet to evaluate the cumulative ecological impacts of the prolonged El Nino (five years, eight months) that ended in August 1995 and transitioned into a cold ENSO ("La Nina") event during 1995–1996 (Trenberth and Hoar 1996). Temperature thresholds, however, are known to be important for both humans and mosquito activity. Surpassing these thresholds affects the health of coral reefs and other marine life; it is also important for human nutrition. In 1995, warming in the Caribbean led to coral bleaching for the first time in Belize, as sea surface temperatures (SSTs) surpassed the 29°C threshold (Hayes and Goreau 1996). A 1996 report (Kushmaro et al 1996) implicates a *Vibrio* bacteria in coral bleaching, with warm SSTs a contributing factor; this development may have increased bacterial growth and lowered the resistance of the coral. The die-off of manatees (and cormorants) from a viral pneumonia—which opportunistically invaded hosts that were weakened by a combination of synthetic toxins (PCBs, dioxins), biotoxins (red tides), and warm sea surfaces—may be another example of the cumulative impacts of the "longest [El Nino period] on record" (Trenberth and Hoar 1996). During the cold phase occurring over the 1995–1996 period, many regions of the globe experienced intense rains and flooding, following prolonged drought. Conditions in Colombia and Central America, for example, were optimal for insect breeding (dengue fever and Venezuelan equine encephalitis) and allowed rodents carrying leptospirosis to escape the flooded areas (Epstein et al 1995). Clearly, the impacts of climate change (warming trends and patterns of variability) on biodiversity (Peters and Lovejoy 1992) and biological interactions deserve greater attention.

Mass and chronic mortalities of marine biota result from alterations of the food web and cause substantial economic losses. Changes in climate (temperature and precipitation means and the frequency of anomalies), for their part, can affect the ranges and dynamics of toxic phytoplanktons, as well as the health of coral reefs, seagrasses, shellfish and finfish, sea mammals, sea birds, sea turtles, invertebrates, and humans. From a broad health perspective, we cannot afford to maintain an anthropocentric outlook.

Climate Variability and Epidemics

Two primary aspects of climate change exist: the global warming trend, and the variability about the means. Since the late 1800s, a global warming pattern has been documented. The Intergovernmental Panel on Climate Change (IPCC, which gathered together 2500 of the world's leading scientists) predicts that more intense heatwaves and extreme precipitation will accompany that trend in the future. Data from the National Climatic Data Center—the United States' main repository for meteorological data—indicate that extreme weather events have increased in the United States, northern Australia, and the former Soviet Union since the 1970s.

Extreme events such as droughts, floods, storms, and fires can have devastating consequences for health. Heatwaves and winter storms both usher in cardiac deaths. Floods foster fungi growth and promote insect breeding. Prolonged droughts punctuated by heavy rains favor insect and rodent growth. They also reduce the populations of predators such as owls, snakes, birds, and reptiles that traditionally control the opportunistic rodents and bugs. In contrast, torrential rains provide the food and breeding sites for these disease carriers.

Increased climate variability and extreme events (most associated with El Nino/La Nina events)—in particular prolonged droughts punctuated by torrential rains—have immediate health consequences (such as malaria outbreaks and harmful algal blooms in Asia, and the development of water-borne diseases such as typhoid, hepatitis A, and bacillary dysentery in Peru and

Chile, and cholera in Bangladesh). Moreover, prolonged droughts punctuated by torrential rains have long-term effects on predator populations that limit the growth of pests and pathogens.

In Colombia, a June 1995 heatwave followed by the heaviest August rainfall in 50 years precipitated a cluster of diseases involving mosquitoes (dengue and Venezuelan equine encephalitis), rodents (leptospirosis), and toxic algae (killing 350 tons of fish). In the United States, Latin America, southern Africa, India, and Europe, rodents are emerging as bothersome crop pests and/or carriers of disease. Deforestation, the decline of predators, and war conditions create conducive situations, and climate variability provides the trigger, for population explosions.

Minimum Temperatures. Another aspect of climate change is the disproportionate rise in minimum (night-time and winter) temperatures, reported by Karl and colleagues in 1993. Elevated temperature minimums (TMINs) can allow insects to overwinter and survive. New Orleans, for example, went five years without frost (1990–1995, during the prolonged El Nino) and subsequently experienced an explosion in its mosquito, termite, and cockroach populations. The same considerations apply to agricultural pests that are stenotherms—requiring specific temperature thresholds for survival. Finally, both farm animals and humans are adversely affected when night-time temperatures offer no relief during heatwaves (as occurred in Chicago, which had more than 500 excess deaths, and other U.S. cities in the summer of 1995).

Biodiversity and Emerging Infectious Diseases

The role of biodiversity in controlling pests, pathogens, and human parasites has received little attention to date. Nevertheless, this protective "function" of the diversity of defensive responses, acting at all scales, is essential for the preservation of the health of living systems. How will losses of species that are key to ecosystem-level processes (Likens 1992), such as large predators, affect natural biological controls? The role played by declining biodiversity in the current trend of emergence, resurgence, and redistribution of infectious diseases may have far-reaching consequences for the health of animal and plant life alike.

Mechanisms that regulate harmful mutations on the cellular level are mirrored by interspecific actions existing on the macro level that prevent opportunistic species from overexploiting and dominating all but the most degraded natural systems. Although each individual organism works to maximize its own reproductive success over its lifetime, the numerous interactions between exploiters and victims (that is, between predator and prey, parasite and host, herbivore and plant, and so on), the diversity of which characterizes the biodiversity of a region, also often act to regulate the abundance of any individual species and preserve the health and dynamic integrity of the ecosystem as a whole.

Predator–prey relationships are central to biological control. Owls, coyotes, and snakes all help to regulate populations of rodents—opportunistic species that are involved in the transmission of Lyme disease, hantaviruses, arenaviruses, leptospirosis, and human plague. Freshwater fish, reptiles, and bats help to limit the abundance of mosquitoes, some of which carry malaria, yellow fever, dengue fever, and many encephalitides. Finfish, shellfish, and sea mammals affect the dynamics of marine algal populations—some being toxic, others being anoxic, and still others acting as transporters for cholera bacteria.

Population explosions of nuisance organisms and disease carriers may be viewed as signs of ecosystem disturbance, reduced resilience, and lowered resistance. Rodents, insects, and algae represent key biological indicators that respond rapidly to environmental change. The current rate of extinctions assumes additional significance in this respect: Periods of mass extinctions— punctuations in evolutionary equilibrium—are typically followed by the emergence of new species. Will a collapse initially favor opportunistic species?

Evolutionary Biology: The Environment and Disease Emergence

When rodents are deprived of an essential nutrient (such as selenium), viruses that are otherwise benign may mutate at an accelerated rate to become virulent (causing a heart ailment). The virulent viruses can then infect healthy mice. An experiment demonstrating this effect (Beck et al 1995) showed that the environment influences pathogen evolution at the most fundamental level; it also indicated that virulent agents that evolve in malnourished, vulnerable populations can affect the previously healthy. Thus, what happens in poor communities and in poor nations affects everyone.

Ecology, Pests, and Disease: Terrestrial Ecosystems

Competitive *and* cooperative interactions act within nuclei, cells, and organisms and at community levels to regulate anomalies; harmful mutations, cancerous cells, and opportunistic species. Fitness for survival takes into account how species co-evolve and fit together in complex networks to create and sustain well-functioning, self-regulating systems.

Explicit regulatory systems have co-evolved to control mutations: for example, p53 gene products (exonucleases) "proofread" DNA errors by excising mismatched base-pairs (Holland 1993; Modrich 1994; Sancar 1994; Hanwalt 1994), and peroxidases detoxify harmful oxygen-derived free radicals in the cytoplasm (de Duve 1996). In organisms, cells, cytokines, and antibodies tender a multifarious orchestra of defensive responses against parasitic invasion. In nature, birds, pheromones, and parasitic wasps limit insect herbivore populations (Blaustein and Wake 1995). Co-evolved control systems can involve multiple species and a variety of mechanisms. Just as cytokines summon immune cells at the organismic level, pheromones (such as terpenes)—which are released when herbivores ruminate on leaves (not from each alone)—attract parasitic wasps that infest and reduce the plant pests (Tumlinson et al 1993). Thus, three species (plant, herbivore, and wasp) have co-evolved in a regulatory system that offers resistance to pest overgrowth.

Perturbations can act at several points in regulatory dynamics. On the genetic level, toxins, ultraviolet-B and X-radiation harm both informational and structural genes; the latter development removes the system for regulating mutations for the remainder of the cell's future. In nature, excessive pesticides select for hardy, resistant insects *and* harm predators (birds, ladybugs, and lacewings) that control insect overgrowth. On the macro, phenotypic level of communities, cooperative and regulatory mechanisms also act. For example, a colony of fire ants behaves and develops in relation to environmental constraints and competitors (Gordon 1995; Holldobler and Wilson 1994), as does a stand of trees.

Redundancy, often mediated by spatial heterogeneity, allows large levels of species diversity to accumulate; thus, complex systems tend to be more stable than simpler, more degraded ones. A variety of organisms perform similar tasks (predation, protein breakdown, nutrient recycling, and so on), and the often intense competition between species performing similar functions means that the independent decline of one species usually leads to an increase in the abundance of another (Tilman and Downing 1994; Tilman et al 1994; Dobson 1996).

At each level, a diversity of responses is key to the resilience and resistance of the system. This proliferation reduces its vulnerability to colonization, invasion, and dissemination of infections and nuisance organisms. Stresses acting at multiple points, however, can destabilize ecosystem function. Multiply stressed systems—prone to oversimplification (monocultures) and habitat fragmentation, overuse of pesticides, and a changing climate, for example—can experience a disruption of synchronies, leading to collapse of their defenses and an invasion of opportunistic infections, much as one sees with HIV/AIDS patients.

r-Selected and K-Selected Species. According to MacArthur and Wilson (1967), the most desirable attributes for a successful invader and good colonizer are good dispersal ability, high reproductive rates, and hermaphroditism or asexual reproduction. The first attribute enables species to cross barriers to dispersal easily. The second allows invaders to build up a population rapidly after introduction. The third attribute allows the colonizer species to found a new population from a few propagules. Most helminths, arthropods, protozoa, bacteria, and viruses excel at these "r-selected" attributes.

Most parasites have much higher rates of reproduction than free-living species. Data comparing the fecundity of free-living and parasitic helminths suggest that parasitic species produce two to three orders of magnitude more eggs than free-living species of similar size. Part of the massive increase in reproduction reflects increases in the number of reproductive stages needed to ensure that at least one parasite infective stage locates a suitable host; part of it must also reflect the huge energetic savings achieved by living a parasitic lifestyle (Calow 1979, 1983).

Species exhibiting K-selection are larger, reproduce later in life, and develop more slowly; they are, however, superior competitors in a stable environment (Ehrlich 1986). r-Strategists would proliferate exponentially if not kept in check by the K-strategists through predator–prey relationships and competition—the systems of biological control or the "immune," regulatory systems of the environment. As "specialists," dependent on localized niches and often more limited diets, predator K-strategists are ultimately more fragile than their opportunistic prey (Davis and Zabinski 1990).

Weeds, rodents, insects, and microorganisms are all opportunistic species; they grow rapidly, have a small body size, reproduce in huge broods, and spread through good dispersal mechanisms (thereby giving them high r, the intrinsic rate of increase). Small mammals, of course, consume insects; thus, the interplay and emergence of pests is complex. Extreme weather events (floods and droughts) can be conducive to insect explosions and rodents fleeing inundated burrows (Epstein et al 1995). Overall, opportunists and r-strategists are good colonizers and achieve dominance in disturbed environments (and in weakened hosts).

The dominance of scavengers and "generalists" (such as crows) over "specialists" (losing circumscribed niches) in rural and urban settings may influence the spread of disease. The transmission of eastern equine encephalitis, a viral disease present in Massachusetts, involves starlings and robins ("generalists") in the bird–mosquito–human–equine life cycle (A. Spielman, personal communication, 1995). Hardy generalists might be better equipped to acquire, maintain, and circulate pathogens than those with isolated ranges and restricted diets.

The Volterra predator–prey relationship (described by the Lotka-Volterra equation; Ehrlich 1986) is an ecological principle fundamental to understanding disease emergence and resurgence. When both predator and prey populations are reduced because of habitat fragmentation, pesticides, or climate extremes, the prey—which reproduces and evolves more rapidly, giving it greater resistance—can rebound with punishing ferocity (as Rachel Carson eloquently depicted).

Models of population dynamics of predators and prey that exclude refuges, camouflages, and migration from and to other systems are not sustainable. Predator *and* prey are eliminated. Habitat reserves are thus necessary to preserve refugia for raptors (like owls) that control rodent populations elsewhere. Corridors can be crucial in maintaining metapopulations of predators and competitors (Levins 1970) so as to limit the opportunistic prey populations.

So-called redundant species may provide insurance against invasion in the face of reduced populations of one or more predator, or with severe environmental stresses. If coyote populations decline, for example, their competitors—snakes and owls—may compensate in regulating rodent populations.

CUMULATIVE ECOLOGICAL IMPACTS OF GLOBAL CHANGE

Global change may be defined as: (1) climate change, (2) altered stratospheric ozone, and (3) those changes that occur on such a widespread basis that they influence global systems (for example, deforestation, coastal pollution, and urbanization). Practices by one species (humans) are rapidly altering the global environment, reducing the diversity of species and enhancing susceptibility to disease. These practices include the following:

- Habitat loss and fragmentation

- Monocultures (marine and terrestrial)

- Excessive use of synthetic and radioactive chemicals

- Ultraviolet-B penetration

- Climate change and instability

All of these factors harm defense systems by reducing more predators than prey (and more specialists than generalists)—thereby allowing opportunistic pests and pathogens to prosper unchecked. Habitat fragmentation with "edge effects" and excessive use of pesticides harms predators and favors pests. Monocultures diminish genetic and species diversity, increasing their vulnerability to infections and invasions.

Significant synergies arise among these factors. Deforestation and climate extremes disrupt habitats and favor pests; ultraviolet penetration, which is increased by acid precipitation and warm aquatic temperatures, harms aquatic life. Climate change and increased variability (more extreme events) affect the overall synchronies among species, altering the populations of predators, competitors, and cooperators relative to pathogens, parasites, and pests.

Synergies and Pests

Synergistic effects among environmental stresses can play important roles in precipitating population explosions of pests and pathogens. Rodents—acting as agricultural pests and transporters of pathogens—represent an emerging problem in the United States, Latin America, Africa, Europe, Asia, and Australia. The hantavirus story in the United States illustrates this point. Prolonged drought in the U.S. Southwest reduced rodent predator populations (owls, coyotes, and snakes), while heavy rains in 1993 precipitated a tenfold increase in rodent populations. Consequently, a "new" disease (hantavirus pulmonary syndrome) emerged that was spread in rodent feces and urine. A study of Canadian snowshoe hares depicts a similar type of synergy: Reduced predation led to twofold increases; new food sources enabled populations to increase threefold; and the combination of both developments engendered a tenfold population explosion.

Multiple negative synergies are possible between land clearing and other stresses. In many Latin American nations, for example, land clearing has led to a shift from forest to field mice (*Calomys* spp.). The overuse of pesticides and herbicides also reduces predators in many cases, whether directly through poisoning or indirectly by removing tall grasses where large predators take refuge. This combination of events produced the sudden appearance of Machupo virus (Bolivian hemorrhagic fever) in 1962 and the emergence of Junin (Argentinean hemorrhagic fever) in 1953. Similar mechanisms have been involved with the explosion of rodent populations related to Guaranito (Venezuela) and Sabia (Brazil) arenavirus emergence. In Bolivia, cats were killed from the heavy DDT used to control malaria. When cats were reintroduced, the epidemic abated.

In southern Africa, rodent populations exploded in 1994 following years of prolonged drought that had been alleviated by deluges in 1993. (Since the 1970s, rodent populations have typically surged after El Nino years.) The maize crop in Zimbabwe was crippled, and plague broke out in Zimbabwe, Malawi, and Mozambique. In Kruger Park in South Africa, a rodent-borne virus caused the deaths of 81 elephants (harming tourism).

Hantaviruses have resurged in several European nations as well, particularly in former Yugoslavia. Plague resurged in India in 1994, following a blistering summer (124°F) and unusually heavy monsoons. Rodents became a pest in Australia after five drought years associated with the prolonged El Nino of the first half of the 1990s.

Insect populations may also respond to multiple insults. In the United States, excessive use of pesticides intended to control boll weevil populations in the South has reduced populations of the wasps and spiders that control other pests; the explosion of other pests has caused a backlash among farmers in Texas and Alabama. These stories echo the message of *Silent Spring*, in which pesticide-resistant insects were no longer met by the chorus of birds to consume them.

Do current land-use practices, overuse of chemicals, and habitat fragmentation increase the chances for such "nasty synergies"? A disturbance in one factor can prove destabilizing; multiple perturbations can affect the resistance and the resilience of a system.

Ocean Warming

In the marine environment, deep-ocean warming has been reported in the tropical Pacific, Atlantic, and Indian Oceans. Water evaporating from warmer seas reinforces the greenhouse effect, and warm seas can fuel hurricanes. Ocean warming may harm marine zooplankton as well, and it has been associated with a shift in marine flora and fauna occurring along the California coast (since the 1930s). In the presence of sufficient nutrients (amply supplied, in rivers and coastal areas throughout the world, by sewage, fertilizers, and acid precipitation), this trend may contribute to the proliferation of coastal algal bloom (warming enhances photosynthesis and metabolism). Warming and stratification in the water column favor the existence of anaerobic and toxic cyanobacteria and dinoflagellates, which are the smaller phytoplankton containing the more toxic species. "Red tides" (harmful algal blooms) of increasing extent, duration, and intensity that involve novel species are being reported throughout the globe. Indeed, "the worldwide increase in coastal algal blooms may be one of the first biological signs of global change" (T. Smayda, personal communication, 1995). Finally, plankton and seaweeds form a reservoir for cholera and other bacteria that cause diarrhea, the world's leading cause of mortality.

From 1990 to 1995, the Pacific warm pool persisted in the El Nino phase. It shifted to the cold La Nina phase in August 1995. Since 1877, no El Nino had endured for more than three years until that period. Both anomalous phases (El Nino/La Nina) bring climate extremes to many regions across the globe.

Decadal Variability. The cumulative ecological impacts of the prolonged El Nino (five years, eight months) that ended in August 1995 have yet to be evaluated. In 1995, warming in the Caribbean led to coral bleaching for the first time in Belize as sea surface temperatures surpassed the 29°C threshold. The warm Caribbean provided moisture for the northeast storms of 1995–1996. In addition, a widespread die-off of sea urchins occurred in the Caribbean in 1995.

The die-off of manatees, cormorants, and other shore birds from a red tide (*Gymnodinium breve*) along Florida's coast may represent another example of the cumulative impacts of the "longest [El Nino period] on record" (Trenberth and Hoar 1996).

Discontinuities

Ecosystems exist in steady states that are interrupted by collapses and explosions of new species. Six major evolutionary upheavals with ecological and climate change have been associated with extinctions of some species, followed by the emergence of new ones. In the collapse and transition phase, pioneering species, pests, and pathogens may be among the first to emerge, with experimentation and specialization increasing once the system reaches relative stability.

The Earth's climate may also tend toward certain steady states. Three such states have been the most common: small, medium, and large polar ice caps. The Holocene, which has lasted for the past 10,000 years and involves medium-size caps, has been associated with agriculture and modern civilization. Instability around the present attractor—a global temperature of 15°C—may be manifested in greater compensatory overshoots (that is, greater extreme events). The one- to two-decade lag time of ocean warming and subsequent release of stored heat (for example, during El Nino phases), followed by deep cold upwellings (La Nina phases), may help to explain the discontinuities that have accompanied the gradual global warming during the past century and a half.

Are overshoots and extreme weather events characteristic of an unstable global climate system? Was the prolongation of the seasons beginning in 1940 indicative of the first minor readjustment of system attractors? Was the shift of sea surface pressures in the mid-1970s another such shift? Does the present instability herald a further "jump" in the climate system state? Has the baseline shifted, and will further perturbations and minor jumps lead to a major jump of the type evidenced in the Greenland Ice Core records of 10,000 years ago? These questions have yet to be answered.

COSTS OF EPIDEMICS

The impacts of disease on humans, agriculture, and livestock are both costly and pervasive, sometimes rippling through societies and economies. A 1991 cholera epidemic, for example, cost Peru more than $1 billion in seafood exports and tourism, and airline and hotel industries lost more than $2 billion from the 1994 Indian plague. Cruise boats are turning away from islands racked by dengue (breakbone) fever, threatening a $12 billion tourist industry that supports 500,000 workers. The global resurgence of malaria, dengue fever, and cholera, along with the emergence of relatively new diseases like ebola and mad cow disease, can affect trade, tourism, politics, and development.

Additionally, escalating extreme weather events have high costs to the insurance industry. This trend suggests that we can no longer insure our future—a development that offers perhaps the best integral of the nonsustainability of current practices.

INTEGRATED ASSESSMENT AND MONITORING

Integrated Ecological Risk Assessment

A new framework for scientific work and policy development must involve the integration of social, ecological, and climatological factors. Figure 10.1 provides a conceptual framework for such an integrated approach. The risks of inputs to ecosystems (for example, chemicals) and activities must be assessed in terms of both direct and indirect outcomes, and then integrated with background changes. Dioxin, pesticides, and PCBs, for example, may—through impacts on bird and inland fish populations (insect larvae consumers)—increase the risks for arboviral infections in three ways:

- By harming avian immune systems and increasing viral burden

- By acting as hormone imitators to reduce male/female ratios

- By reducing larval-consuming fish in ponds

Loss of habitat and crowding of bird reservoirs may increase contagion, and conducive bioclimatic conditions may precipitate insect population explosions. Side effects of chemicals may not be appreciated initially, however. Malathion, for example, harms bees, which are important pollinators. Fungicides reduce soil fungi that control bacteria, limit gypsy moth populations, and provide essential underground nourishment for ants, which are the quintessential recyclers.

Outcome measures for global change research may focus on health, water, and food—three basic needs. These outcomes can, in turn, be incorporated into integrated assessments and projects involving oceanic, atmospheric, and terrestrial processes. Geographic information systems (layered data sets) can provide the integrating methodology for *mapping* sets of environmental data, allowing us to understand emerging patterns of associations and possible causes. Satellite imaging integrated into GISs and landscape epidemiology can also guide risk analysis and interventions.

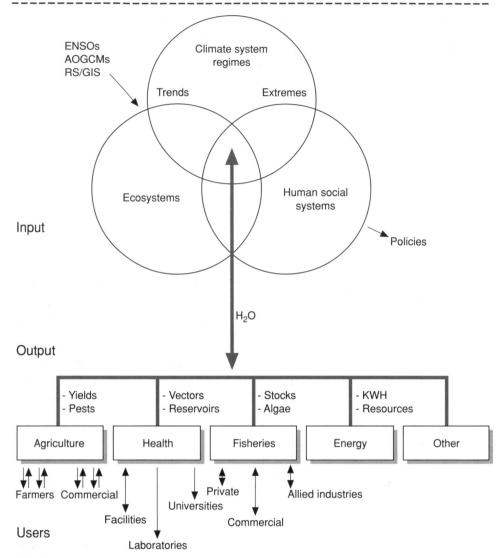

FIGURE 10.1. *A multidisciplinary, multisectoral framework for integrated health surveillance. ENSOs = El Nino/Southern oscillations; AOGCMs = atmospheric–ocean general circulation models; RS/GIS = remote sensing/geographic information systems; KWH = kilowatt-hour.*

Multisectoral application of multidisciplinary understanding can be used for three purposes:

■ To develop early-warning systems

■ To improve ecologically based, environmentally sound interventions

■ To formulate anticipatory policies that reduce overall vulnerability to disease emergence and spread

Biological Indicators for Integrated Monitoring

Rodents and insects (abundance and species composition) are sensitive indicators of rural and urban environmental conditions because, as generalists, they multiply rapidly. In the coastal zone, algal blooms—red, green, blue, brown, and phosphorescent, including some noxious varieties— are functions of local ecology, meaning that sea temperatures, winds, and ocean circulation serve as sensitive indicators. These indicator species are integral to the emergence, resurgence, and rapid redistribution of infectious diseases.

As a consequence, direct links between ecological programs and regional disease surveillance centers might prove beneficial. Monitoring key biological indicators (and plant and animal health) can become essential components of global monitoring systems—such as the Global Terrestrial, Ocean, and Climate Observing Systems (GTOS, GOOS, GCOS)—that are now in their formative stages (described later in this chapter). Other programs plan to monitor ecosystems (the World Hydrological Observing Systems, or WHYCOS) by using a coordinated Integrated Global Observing System (IGOS).

New Methodologies for Surveillance and Integrated Monitoring

The Long-Term Ecological Research site (LTER, funded by the National Science Foundation) in New Mexico has an established collaboration with the Centers for Disease Control and Prevention (CDC) intended to monitor rodent ecology and hantaviruses. In addition, the 17 other U.S. LTERs and international Man and Biosphere (MAB) programs of the United Nations monitor rodent populations and conduct surveillance studies for viruses and bacteria. The groups also monitor insect populations. Potentially, studies intended to determine disease burden and incidence could be conducted in such sentinel sites.

Advances in climate forecasting can prove useful in developing Health Early Warning Systems (HEWS), which provide advanced notice of biocli-matic conditions conducive for disease and pest outbreaks. HEWS can enable timely implementations of environmentally sound interventions, such as community education, vaccination campaigns, water boiling, benign bacterial larvacide applications (*Bacillus thuringiensis*), removal of perido-mestic mosquito breeding sites, and the like. Critical, vulnerable, and partic-ularly sensitive regions (such as mountain ranges) can be targeted for monitoring.

The International Research Institute (IRI) for climate prediction plans to establish regional application centers, which will first focus on hydrology, agriculture, and health. The Global Change System for Analysis, Research and

Training (START) program of the International Geosphere-Biosphere Program (IGBP), World Climate Research Program, and International Human Dimensions Program will involve 13 regional research networks, numerous regional research centers, and affiliated sites.

Large Marine Ecosystem (LME) monitoring of the sustainability and biomass yields of coastal waters is being organized for the Gulf of Guinea, with submitted proposals for the Yellow Sea and the Black Sea—with the intention of eventually expanding coverage to the world's 49 LMEs. These and similar schemes with comprehensively monitored marine ecosystems can serve as coastal components of a GOOS. Monitoring of the health of the living marine components—and of plankton, bivalves, and finfish for biotoxins and vibrios (and coastal nations for shellfish poisoning and cholera)—can form an integral part of these projects. The relative contribution of the chief driving forces for each system (for example, pollution/habitat loss, overfishing, and climate changes) will be evaluated to develop policies for mitigation and prevention.

Such monitoring will be augmented by remote sensing (carried out by NASA and other national space agencies, and Mission to Plant Earth), with data being assimilated into mapped, overlapping geographic information systems. In the United States, the Climate Analysis Center (NOAA) defines four regions for weather surveillance. Coordinating ecological (EPA and National Biological Survey), agricultural (USDA), and health data (CDC) through these centers could provide the basis for integrated regional assessment and monitoring.

The CDC, the World Health Organization, and the Program for Monitoring Emerging Diseases (ProMED) are planning an international consortium of regional centers for enhanced surveillance and response through clinical, laboratory, and epidemiological means. These centers are intended to enable these organizations to respond to the global emergence, resurgence, and redistribution of infectious diseases. Sentinel sites and data gathering will be coordinated through the consortium.

For each region, primary, secondary, and tertiary driving forces of environmental change will differ. Among the principal forces of global change are extraction and exhaustion of nonrenewable resources (including fossil fuels and wildlife species), overexploitation of renewable resources (forestry and fisheries), and the generation of wastes beyond the capacity of biogeochemical systems to recycle them. Forces are sometimes "exported" to other regions, through the import of goods or export of aerosolized, liquid, and solid wastes. In addition, all of the driving forces are influenced by population trends (growth, aggregation, and migration), income gaps, economic and energy policies, technological and behavioral changes, and political will.

A HISTORICAL NOTE ON INFECTIOUS DISEASE PANDEMICS

Is the current global resurgence of infectious disease a new phenomenon? By the 1960s, improvements in hygiene, sanitation, and mosquito control had led most public health authorities to believe that infectious disease would soon be mastered. In the 1970s, public health schools turned their curricular attention to chronic ailments: heart disease, stroke, diabetes, and cancer. But the "epidemiological transition" to diseases of modernity never materialized in most nations. And in the 1980s, the picture shifted dramatically.

According to *The World Health Report 1996: Fighting Disease, Fostering Development* (WHO 1996), at least 30 new infectious diseases—including HIV and ebola—have emerged in the past 20 years. "Drug-resistant strains of bacteria and other microbes are having a deadly impact on the fight against tuberculosis, malaria, cholera, diarrhea and pneumonia . . . which together killed more than 10 million people in 1995." Diseases transmitted person-to-person—like diphtheria and whooping cough—have reappeared where social structures have failed. Diphtheria epidemics that began in the former USSR in 1990, for example, have spread to 15 Eastern European countries. Having essentially disappeared from the Americas, dengue fever has resurged in that continent, infecting more than 200,000 people in 1995. In 1995, the largest epidemic of yellow fever in the Americas since 1950 hit Peru, while other epidemics have struck western Africa. In west Africa, more than 100,000 persons contracted meningitis in 1996, leaving 10,000 dead. Meningitis epidemics are associated with drought in Africa; the 1996 epidemic was the largest ever recorded that accompanied a severe drought.

According to the WHO report, conditions conducive to emerging infectious diseases (EIDs) are now present worldwide. Indeed, hantavirus pulmonary syndrome and Lyme disease were not imported into the United States, and the conditions that enable the spread of tuberculosis are known to exist in that nation's homeless shelters and prisons. In the United States, mortality from infectious diseases rose 58% from 1980 to 1995.

From a long-term historical perspective, pandemics (epidemics involving multiple continents) have often been associated with major historical transitions in the past, and have had profound impacts on civilizations. For example, a pandemic occurred as the Roman Empire was in decline (541 A.D., Plague of Justinian, though the organism remains a source of dispute) and raged for two centuries, claiming more than 40 million lives. As a consequence of this outbreak, urban centers were shattered and populations resettled into rural feudal communities.

Classical plague (due to *Yersinia pestis*) resurged after a 600-year hiatus during the Middle Ages (1346 A.D.), when reurbanization had once again outstripped infrastructure and sanitation. Other factors compounded the situation at the time: human populations had shifted from East to West, and many cats had been killed (in the belief that they were witches). Elevated temperatures accompanying the Medieval Warm Period may also have contributed to the proliferation of infected rats and fleas. In just five years, 25 million people died (30% of Europe's population).

From 1790 to 1850, European cities grew sevenfold under the intense pressure of industrialization. Conditions in these cities (vividly portrayed by Charles Dickens) were breeding grounds for three major infectious diseases: cholera, smallpox, and tuberculosis. Major upheavals occurred throughout Europe, the Sanitary Movement was conceived, and disease epidemiology with modern public health principles was born.

Past epidemics have been generated by major societal conditions. In turn, these breakdowns have helped motivate dramatic changes in social practices and—sometimes—in the social order itself. Have we entered a period of ecological, climatic, and social transition? Will our society react to the symptoms of social and environmental dysfunction—the resurgence of infectious disease affecting multiple ecosystems—and act before we exceed the resilience of our global systems?

CONCLUSIONS

Today, humans are using resources (living and fossilized) and generating wastes at rates that exceed the capacity of biogeochemical systems to adequately recycle them. Policies, programs, projects, and practices affecting forestry, fisheries, petrochemicals, and fossil fuels must all be examined in light of the global resurgence of infectious diseases across a wide taxonomic range. At stake is our common environment and the security of healthy growth for human society, and for other animals, crops, and forests.

Current changes in economics are rapidly changing world demographics. Migrations—both within nations and on an international scale—now surpass the great migrations of the 1800s in terms of size. Population growth (with coastal growth occurring at twice the overall rate) puts enormous pressure on the environment. World Bank figures (GENI coefficients) indicate, however, that the ability of a society to stabilize its population is directly related to the degree of equity and land distribution within it.

Changing course in societal and industrial practices will not be easy or inexpensive. Nevertheless, it is necessary. The costs of global change are mounting, and we can simply no longer afford "business as usual."

International governance is a key concept to sustainable development. The Montreal Protocol for eliminating chlorofluorohydrocarbons offers a prime example of how this type of collaborative effort might be effected;

the United Nations Convention on the Laws of the Seas (UNCLOS) provides another. The behavior of international industries with respect to the commons must be regulated to ensure the future health of all industry.

Carbon is the key element. Along with water, it forms the basis of all organic compounds. Some 3.8 billion years ago blue-green algae—the first photosynthetic organisms—began the transformation of our atmosphere from 95% CO_2 to just 0.03% CO_2 and 20% oxygen. We are now reversing this process by burning wood, coal, oil, and gas—the latter three are the living and fossilized life forms that recast our atmospheric chemistry. We have developed the technology to replicate photosynthesis and to capture the energy of electrons as they are catapulted from one quantum state to the next level.

International disincentives to healthy and environmentally sound growth must be removed and replaced by positive incentives and creation of new markets to encourage fuel efficiency and alternative sources. The manufacture and distribution of new technologies and programs for environmental restoration will provide major stimuli for global and national economies in the next century.

Environmental policies to protect ecosystem health and the health of its components are central, and capping carbon emissions is essential for a healthy future. Thus, strong international climate, forestry, and biodiversity conventions are critical for our health.

REFERENCES

Almendares J, Sierra M, Anderson PK, Epstein PR. Critical regions, a profile of Honduras. Lancet 1993;342:1400–1402.

Anderson DM. The fifth international conference on toxic marine phytoplankton: a personal perspective. Harmful Algae News, suppl to International Marine Science (UNESCO) 1992;62:6–7.

Bakun A. The California Current, Benguela Current, and Southwestern Atlantic Shelf ecosystems: a comparative approach to identifying factors regulating biomass yields. In: Sherman K, Alexander LM, Gold BD, eds. Stress, mitigation, and sustainability of large marine ecosystems. Washington, DC: AAAS Press, 1993:99–221.

Barry JP, Baxter CH, Sagarin RD, Gilman SE. Climate-related, long-term faunal changes in a California rocky intertidal community. Science 1995;267:672–675.

Beck MA, Shi Q, Morris VC, Levander OA. Rapid genomic evolution of a non-virulent Coxsackievirus B3 in selenium-deficient mice results in selection of identical virulent isolates. Nature Med 1995;1:433–436.

Bengtsson L, Schlese U, Roeckner E, Latif M, Barnett TP, Graham N. A two-tiered approach to long-range climate forecasting. Science 1993;261:1026–1029.

Blaustein AW, Wake DB. The puzzle of declining amphibian populations. Scientific American 1995;April:52–57.

Bothwell ML, Sherbot MJ, Pollock CM. Ecosystem response to ultraviolet-B radiation: influence of trophic-level interactions. Science 1994;265:97–100.

Bouma MJ, Sondorp HE, van der Kaay JH. Health and climate change. Lancet 1994;343:302.

————. Climate change and periodic epidemic malaria. Lancet 1994;343:1440.

Burkholder JM, Glasgow HB, Steidinger KA. Unraveling environmental and trophic controls on stage transformations in the complex life cycle of an ichthyotoxic "ambush predator" dinoflagellate. Sixth international conference on toxic marine phytoplankton, Nantes, France, 18–22 Oct 1993 (abstract).

Byrd JJ, Huai-shu XU, Colwell RR. Viable but non-culturable bacteria in drinking water. Appl Environ Microbiol 1991;57:875-878.

Calow P. Costs of reproduction—a physiological approach. Biol Rev 1979;54:23–40.

————. Pattern and paradox in parasite reproduction. Parasitology 1983;86:197–207.

Carreto JI, Benevides HR. World record of PSP in southern Argentina. Harmful Algae News, suppl to International Marine Science (UNESCO) 1993;62:7.

CDC. Addressing emerging infectious disease threats: a prevention strategy for the United States. Atlanta, GA: CDC, 1994.

————. *Vibrio vulnificus* infections associated with raw oyster consumption. MMWR 1993a;42:405–407.

————. Multistate outbreak of viral gastroenteritis related to consumption of oysters—Louisiana, Maryland, Mississippi, and North Carolina, 1993. MMWR 1993b;42:945–948.

CLIVAR. A study of climate variability and predictability, World Climate Research Program. Geneva: WMO, 1992.

Colwell RR, Belas MR, Zachary A. Attachment of microorganisms to surfaces in the aquatic environment. Dev Ind Microbiol 1980;21:169–178.

Davis MB, Zabinski C. Changes in geographical range resulting from greenhouse warming on biodiversity in forests. In: Peters RL, Lovejoy TE, eds. Proceedings of the World Wildlife Fund Conference on Consequences of Greenhouse Effect for Biological Diversity. New Haven, CT: Yale University, 1990.

de Duve C. The birth of complex cells. Scientific American 1996;April:50–58.

Dobson A, Carper R. Biodiversity. Lancet 1993;342:1096–1099.

Dobson AP. Conservation and Biodiversity. New York: Scientific American Books, 1996.

Ehrlich PR. The Machinery of Nature. New York: Simon and Schuster, 1986.

Elias SA. Quarternary insects and their environments. (Based on work of R. Coope and others.) Washington, DC: Smithsonian Institution Press, 1994.

Epstein PR. Emerging diseases and ecosystem instabilities: new threats to public health. Am J Pub Health 1995a;85:168–172 (reader pp. 113–117).

————. Emerging infections and global change: integrating health surveillance and environmental monitoring. Curr Iss Pub Health 1995b;1:224–232.

————. The role of algal blooms in the spread and persistence of human cholera. BioSystems 1993a;31:209–221.

———. The costs of not achieving climate stabilization. Ecol Econ 1993b;8:307–308.

———. Pestilence and poverty—historical transitions and the great pandemics (commentary). Am J Prev Med 1992;8:263–265.

Epstein PR, Chikwenhere GP. Biodiversity questions (letter). Science 1994; 265:1510–1511.

Epstein PR, Ford TE, Colwell RR. Marine ecosystems. Lancet 1993;342:1216–1219.

Garrett L. The coming plague: newly emerging diseases in a world out of balance. New York: Farrar, Strauss and Giroux, 1994.

Gill CA. The role of meteorology and malaria. Indian J Med Res 1920a;8:633–693.

———. The relationship between malaria and rainfall. Indian J Med Res 1920b;37:618–632.

Glantz MH, Katz RW, Nicholls N, eds. Teleconnections linking worldwide climate anomalies. Cambridge, UK: Cambridge University Press, 1991.

Gordon DM. The development of organization in an ant colony. American Scientist 1995;83:50–57.

Haines A, Epstein PR, McMichaels AJ. Global health watch: monitoring impacts of environmental change. Lancet 1993;342:4464–4469.

Hanwalt PC. Transcription-coupled repair and human disease. Science 1994; 266:1957–1958.

Hayes RL, Goreau TJ. Mass coral reef bleaching: climate, temperature, and ultraviolet effects. Abstract. Baltimore, MD: AAAS, 1997 (Feb. 8–13).

Holland J. Replication error, Quasispecies populations, and extreme evolution rates of RNA viruses. In: Morse SS, ed. Emerging viruses. Oxford: Oxford University, 1993:200–218.

Holldobler B, Wilson EO. Journey to the ants: a story of scientific exploration. Cambridge, MA: Belknap Press of Harvard University, 1994.

Houghton JT, Callander BA, Varney SK. Climate change 1992: the supplementary report to the IPCC scientific assessment. Cambridge, UK: Cambridge University Press, 1992.

Karl TR, Jones PD, Knight RW, et al. A new perspective on recent global warming: a symmetric trend of daily maximum and minimum temperature. Bull Am Meteorological Soc 1993;74:1007–1023.

Karl TR, Knight RW, Plummer N. Trends in high-frequency climate variability in the twentieth century. Nature 1995;377:217–220.

Koopman JS, Prevots DR, Marin MAU, et al. Determinants and predictors of dengue infection in Mexico. Am J Epidemiol 1991;133:1168–1178.

Krebs CJ, Boutin S, Boonstra R, et al. Impact of food and predation on the snowshoe hare cycle. Science 1995;269:1112–1115.

Kruess A, Tschamtke T. Habitat fragmentation, species loss and biological control. Science 1994;264:1581–1584.

Kushmaro A, Loya Y, Rosenberg E. Bacterial infection and coral bleaching. Nature 1996;380:396.

Levins R. Extinction. Lectures on Mathematics in the Life Sciences. 1970;2:75–107.

Levins R, Auerbach T, Brinkmann U, et al. The emergence of new diseases. Am Scientist 1994;82:52–60.

Levins R, Epstein PR, Wilson ME, et al. Hantavirus disease emerging. Lancet 1993;342:1292.

Likens GE. The ecosystem approach: in use and abuse. Ecology Institute, Germany, 1992.

Loevinsohn M. Climatic warming and increased malaria incidence in Rwanda. Lancet 1994;343:714–718.

MacArthur RH, Wilson EO. The theory of island biogeography. Princeton, NJ: Princeton University, 1967.

Martens WJM, Rotmans J, Niessen LW. Climate change and malaria risk: an integrated modeling approach. GLOBE report series no. 3, RIVM report no. 461502003. Bilthoven, Netherlands: Global Dynamics and Sustainable Development Programme, 1994.

Matsuoka Y, Kai K. An estimation of climatic change effects on malaria. J Global Environ Eng 1994;1:1–15.

McMichael AJ, Ando M, Carcavallo R, Epstein P, Haines A, Jendritzky G, Kalkstein L, Odongo R, Patz J, Piver W, with Anderson R, Curto de Casas S, Giron IG, Kovats S, Martens WJM, Mills D, Moreno AR, Reisen W, Slooff R, Waltner-Toews D, Woodward A. Human health and climate change. Geneva: WHO/WMO/UNEP, 1996.

Modrich P. Mismatch repair, genetic stability, and cancer. Science 1994; 266:1959–1960.

Morse SS. Factors in the emergence of infectious diseases. Emerging Infect Dis 1995;1:7–15.

Peters RL. Consequences of global warming for biological diversity. In: Wyman RL, ed. Global climate change and life on Earth. New York: Routledge, Chapman and Hall, 1991.

Peters RL, Lovejoy TE. Global warming and biodiversity. New Haven, CT: Yale University, 1992.

Pimm SL, Russell GJ, Gittleman JL, Brooks TM. The future of biodiversity. Science 1995;269:347–350.

Roemmich D, McGowan J. Climatic warning and the decline of zooplankton in the California current. Science 1995;267:324–326.

Root T, Schneider SH. Ecology and climate: research strategies and implications. Science 1995;269:334–341.

Sancar A. Mechanisms of DNA excision repair. Science 1994;266:1954–1956.

Sherman K. Sustainability, biomass yields, and health of coastal ecosystems: an ecological perspective. Marine Ecol Prog Series 1994;112:277–301.

Sherman L, Alexander LM, Gold BD, eds. Large marine ecosystems: stress, mitigation, and sustainability. San Francisco, CA: AAAS Press, 1993:301–319.

Smayda TJ, Shimizu Y. Toxic phytoplankton blooms in the sea. London: Elsevier, 1993.

Stenseth NH. Snowshoe hare populations squeezed from below and above. Science 1995;269:1061–1062 (reader pp. 251–252).

Swerdlow DL, Mintz ED, Rodriquez M, et al. Waterborne transmission of epidemic cholera in Trujillo, Peru: lessons for a continent at risk. Lancet 1992;340:28–33.

Tester PA, Stumpf RP, Vukovich FM, Fowler PK, Turner JT. An expatriate red tide bloom, transport, distribution, and persistence. Limnol Oceanogr 1991; 36:1053–1061.

Thompson LG, Mosley-Thompson E, Davis ME, et al. Late glacial stage and Holocene tropical ice core records from Huascaran, Peru. Science 1995;269:46–50.

Tilman DR, Downing JA. Biodiversity and Stability in Grasslands. Nature 1994; 367:363–365.

Tilman DR, May RM, Lehman CL, Nowak MA. Habitat destruction and the extinction debt. Nature 1994;371:65–66.

Todd ECD. Seafood-associated diseases in Canada. J Assoc Food Drug Officials 1993;56:45–52.

Trenberth KE, Hoar TJ. The 1990–1995 El Nino–Southern Oscillation event: longest on record. Geophys Res Letters 1996;23:57–60.

Tumlinson JH, Lewis WJ, Vet LEM. How parasitic wasps find their hosts. Scientific American 1993;March:100–106.

Valiela I. Marine ecological processes. New York: Springer-Verlag, 1984.

Williams EH, et al. An epizootic of cutaneous fibropapillomas in green turtles *Chelonia mydos* of the Caribbean: part of a panzootic? Journal of Aquatic and Animal Health 1994;6:70–78.

World Health Organization. The world health report 1996: fighting disease, fostering development. Geneva: WHO, 1996.

CHAPTER 11

Qualitative Mathematics for Understanding, Prediction, and Intervention in Complex Ecosystems

Richard Levins

MATHEMATICS plays two very distinct roles in the analysis of complex systems. On the one hand, it is used to obtain numerical results that can predict outcomes and guide interventions. On the other hand, it can provide an understanding of what is otherwise a bewildering web of often nonlinear interactions among variables, external perturbations, and random processes. In this case, its task is to educate the intuition so that the arcane becomes obvious and even trivial.

In the past, mathematical tractability was a major consideration that biased mathematical research in ecology toward questions of the existence and stability of equilibrium solutions, where linear methods were appropriate. Different models satisfied different desiderata, as a trade-off among generality, realism, precision (Levins 1965), and tractability was necessary. A vast literature was produced to answer questions about the coexistence of species, provide explanations for genetic polymorphism, describe the role of random processes in evolution, and determine the stable age structures of populations. All of these operations employed simplifications of the real systems to make them manageable.

Newly available mathematical tools allowed these restrictive assumptions to be relaxed. With the advent of high-speed computers, new possibilities emerged. Tractability was no longer a requirement, the number of variables studied simultaneously could be increased as much as desired, trajectories toward equilibria could be studied as well as the equilibria themselves, and periodic or random environmental perturbations could be introduced at will. Computational methodology became an increasingly prominent part of ecological training, often to the detriment of the biology. Nancy Krieger (1994) has observed the same preponderance of "method over content" in epidemiology.

If a system can be well specified and equations developed to describe the interactions among variables, then in principle we can solve these equations to project the future pathways of change with some accuracy. If the equations cannot be solved analytically, increasingly sophisticated numerical methods using computers can estimate the variables at some future time. Such large-scale computation has its limitations, however:

- It requires exact equations. As the number of variables increases, however, the number of equations grows as well. Each equation may be quite complex and require years of study to develop.

- It requires vast amounts of data. Consequently, a detailed numerical analysis is usually available for only one or a few examples of a formation such as a rainforest or prairie, precluding comparative studies.

- Not all variables can be measured. Those that are less measurable are frequently omitted from the models and at best acknowledged in the text as only additional considerations. In particular, social variables are often omitted in ecosocial systems.

- Numerical methods may provide a good statistical fit of data to equations, but do not enhance our understanding of the processes involved.

The practical successes of numerical methods in applications, as well as the demands of the sponsors of research for quantitative predictions, have made numerical simulation modeling almost synonymous with modeling in general. In reality, it represents merely one possible approach.

Qualitative modeling methods are also available. These models offer several advantages relative to numerical models:

- They pose the question, How much can we get away with not knowing and still understand the system?

- They permit the inclusion of variables that are not readily measurable.

- They allow the use of variables that differ dramatically in physical form, such as rate of profit from fishing and the rate of reproduction of fish, and contagion rates of an epidemic and the caution or panic of the ministry of public health.

- They provide criteria for constructing models of the system, indicating which variables to include and which to treat as external influences.

- They indicate what must be measured by identifying the loci of long and short positive and negative feedbacks and delays.

- They can tell us how dynamics are affected as the reality deviates from the initial simplifying assumptions of the models.

- They are inexpensive, which makes them (in principle) available to non-professional, community, and unsupported parties to a dispute. Their use would signal a continuing democratization of science.

At present, developments in both computation and mathematics and a greater philosophical awareness make it possible to combine analytic and numerical methods and quantitative and qualitative approaches in ways that can partially compensate for each strategy's disadvantages.

SOME METHODS OF QUALITATIVE ANALYSIS

For our purposes, a system is considered to be a network of partly opposing and partly reinforcing processes, observable as changes in their intersections at specified variables. For example, the prevalence of infectious disease is the result of contagion and immigration versus recovery, emigration, and death. The interplay of these processes may lead to increases and decreases in the number of cases that we ultimately designate as "an epidemic." Thus the specification of a system requires the identification of the opposing processes and of the outcomes resulting from them.

Preliminary Description of the System

The system to be studied may be defined as a habitat, geographic area, human community, or network of communities (see chapter 7). Variables are then identified with a view toward evaluating the criteria that motivated the formation of the model. For instance, when biodiversity is a desired goal, we must consider the species, their abundances, and their interactions. If infectious disease is a problem, then the invasion of the habitat by mosquitoes, ticks, rodents, and other reservoirs and vectors represents an important variable.

After the core variables have been identified, however, we must continue to reach out beyond them. Mosquitoes depend on water relationships and vegetation. Contact with humans depends on the mosquitoes' habits as related to people's work and residence patterns, seasonality of employment, and options for protection against mosquitoes. Thus, the analysis must take into account factors affecting rainfall, standing water, irrigation, and deforestation. For some diseases, the status of the immune system is important; this criterion leads us by way of nutrition to the inequality and variability of access to food, land tenure, sex discrimination, the production of subsistence crops, or the dependence on imports and cash income.

As we reach into widening circles of causation, we confront greater complexities. We have more variables to monitor—some of which have not been measured, and others that defy precise measurement. We also gain insight: what would otherwise appear as an arbitrary external perturbation (for example, variable food supplies) is revealed to depend in part on the structure and internal processes of the community itself. Conditions such as the terms of trade and the assortment of products exchanged, which might be taken as givens affecting the rates of pollution or erosion, must then be examined as the outcomes of past events and, in principle, as being changeable. The larger the system, the more the agenda for critical evaluation holds. The choice of system boundaries will, therefore, serve as an object of contention.

Analysis of Fundamental System Characteristics

After the preliminary description of a system is complete, we ask two fundamental questions: Why are things the way they are instead of a little bit different, and why are things the way they are instead of very different?

Homeostasis of the System. The first question deals with the homeostasis of the system—that is, its capacity to adjust to perturbations in ways that maintain it as recognizably the same system. Here the formal tools of mathematical systems theory become appropriate. We can examine a system for local stability properties, resilience (the rate of return after perturbation), resistance to changes in a parameter, and susceptibility to discontinuities, thresholds, catastrophes, or chaos. These properties depend on the qualitative structure of the network of interacting variables.

Among the important descriptors of a system from the dynamic point of view are the redundancy of the set of variables (if they are species, niche overlap expresses this property), self-damping of a variable, the positive and negative feedbacks among variables, long and short pathways in the system, the connectivity of the network, time delays and sinks, the heterogeneity of flows and interactions, and the "shapes" of functional relationships.

Self-damping is the rate at which a variable erases its history and returns toward some equilibrium condition or preserves a "memory" of the past. With strong self-damping, a variable appears relatively more responsive to recent conditions, with less information about the past being preserved. Strong self-damping protects against short-term and random changes in conditions but interferes with the system's ability to adapt to long-term trends.

Both long and short "memory" variables are important in evaluating the state of a system. For instance, in the study of nutrition in human populations, a comparison of weight with height provides a measure of recent acute malnutrition. Body weight responds to the food intake on a time scale of weeks to months. In contrast, height depends on the accumulated growth since birth. Therefore, height for age is an indicator of chronic malnutrition, while a combination of weight for height and height for age gives us the direction and velocity of change (the first derivative of nutrition).

Self-inhibition in a process usually gives rise to self-damping in a mathematical model, so that a verbal description is often sufficient to determine a qualitative model. The two are not identical, however. In cases where doubt exists, we must refer to the mathematical definition of the self-damping of variable x_i, $a_{ii} = \partial(dx/dt)/\partial x$, evaluated at an equilibrium for the system (Box 11.1). This analysis allows us to identify factors that cause self-damping.

Box 11.1   Self-Damping

Simple population reproduction,

$$dx/dt = rx \qquad\qquad 11.1$$

where r includes other variables such as food supply or predators but not x itself, has zero self-damping. The derivative with respect to x is r, but $r = 0$ at any equilibrium in which $x - 0$. For instance, suppose

$$dx/dt = x(aR - m), \qquad\qquad 11.2$$

with resource R, mortality m, and $aR - m$ playing the role of r. Then variable x is not self-damped in this case. If it is at the top of a food chain, then it will serve as a sink.

Self-damping enters an equation in a number of ways. First, a population may be self-inhibiting through crowding, behavioral inhibition of reproduction, or territoriality. The equation is then modified to include x in r. The most familiar way of accomplishing this goal is the logistic model,

$$dx/dt = rx(K - x) \qquad\qquad 11.3$$

At equilibrium, $x^* = K$. The derivative is $r(K - x) - rx$. The first term is zero at equilibrium, so the self-damping equals $-rx$. As $x = K$ at equilibrium, the self-damping is proportional to K.

Another formulation of population growth is

$$dx/dt = rx(K/(a + x) - m) \qquad\qquad 11.4$$

where m is mortality for reasons independent of x. The self-damping is $-rKx/(a + x)^2$, which is again negative but is now strongest at some intermediate population density and diminishes with further increase in the equilibrium. Where population growth of this type operates—for instance, when the response to the food supply or the capacity to eliminate a toxin becomes saturated at high levels—

Box 11.1   *(Continued)*

then a variable may become a sink under extreme conditions. *The effective structure of the network itself can be altered when parameters change.*

Second, population growth causes an accumulation of waste products, reduction in a resource, or increase in natural enemies. In this situation, we have a choice of analytical methods. If these factors are included in the model explicitly with their own equations, then the equation for x can remain simple population growth. If they are not represented in the model as variables, however, then their effects would be included in the equation for x and their own self-damping is passed along to x.

Third, immigration may be used. In the simple growth model, add a migration rate M. Then

$$dx/dt = rx(aR - m) + M \qquad\qquad 11.5$$

Differentiating with respect to x gives $r(aR - m)$. But now at equilibrium

$$rx^*(aR - m) + M = 0 \qquad\qquad 11.6$$

so that

$$r(aR - m) = -M/x^* \qquad\qquad 11.7$$

represents the fraction of the population that consists of immigrants. When several predators prey on a common resource, the impact of each one is proportional to its predation rate but inversely proportional to the immigrant fraction of the predator population. This relationship partially explains why an effective predator may be an ineffective biological control agent. Even a predator that does not affect the average level of prey can alter the synchrony between a prey species and its resource, protecting the crop. Therefore, we should look toward multiple natural enemies in pest control.

Fourth, if a self-damped variable such as a resource or nutrient interacts with a nondamped consumer, we would have a pair of equations

$$dR/dt = A - R(px + c)$$

$$dx/dt = x(pR - m)$$

where A is the inflow rate of R, p is the harvesting rate by x, c is the removal rate of R by all processes other than utilization by x, and m is the mortality of x. Suppose that R reaches an equilibrium rapidly compared with changes in x. Then we can solve the first equation for R,

$$R = A/(px + c)$$

Substituting this value into the equation for x gives

Box 11.1   *(Continued)*

$$dx\big/dt + x\big(pA\big/\big(px+c\big)-m\big)$$

which is self-damped. A variable that is not included explicitly can transfer its self-damping to its consumer.

Fifth, if a variable is not self-reproducing, the equation would look like

$$dx\big/dt = A - Bx \qquad\qquad 11.8$$

where A is the rate of production of x or of its introduction into the system and B is its removal rate. A and B may depend on other variables in the system, in which case the self-damping is equal to B. If x is the prevalence of a disease in a population, then B is the recovery rate plus the death rate (the reciprocal of the duration of the disease). If x is the concentration of a mineral in a lake, A is the input either from external sources or from release by decomposition of organic matter and B is the outflow from the lake plus its uptake by the organisms. In biochemistry, A is a synthesis rate of a compound and B is its rate of removal and transformation into something else by enzymatic action. For instance, Michaelis-Menton kinetics would set $B = v_{max}/(k + x)$. Then

$$dx\big/dt = -v_{max} k\big/\big(k+x\big)^2 \qquad\qquad 11.9$$

Therefore, the self-damping diminishes as x increases and the enzyme surfaces become saturated. In many cases, a control mechanism expressed through self-damping is effective only in some range of the variable; beyond that point, it becomes ineffective. This consideration is important when evaluating ecosystem health, toxicology, public health, and pest management.

If the inflow of a variable is itself variable, we can take average or expected values from Equation 11.1

$$E\big(x\big) = E\big(A\big)\big/r \qquad\qquad 11.10$$

where the operator E means expected value.

The standing stock of x is reduced by increasing r. The turnover of x is increased by r. Thus, increasing r reduces the stock to flow ratio—an important descriptor of a system.

Sometimes autoinhibition is replaced by self-excitation, where $\partial(dx/dt)/\partial x$ is positive. This situation may arise in populations where the finding of a mate is uncertain, so that a population increase from small numbers therefore increases the rate of growth. It may also occur when organisms jointly produce a suitable habitat (such as trees jointly creating the forest environment) or detoxify their surroundings or provide mutual protection, where infection is sensitive to multiple exposures to a microorganism, where a social group can combine for political

---

Box 11.1   *(Continued)*

action, and where the accumulation of wealth facilitates further accumulation by making credit or political influence available.

In assessing ecosystem changes, factors that increase or decrease self-damping should be noted.

Negative feedback is a process in which an initial change in a variable gives rise to events that eventually turn the variable back toward its original value. If the feedback is indirect or includes a delay (almost equivalent situations), then the variable may overshoot its original value. An example of negative feedback in agriculture occurs when high agricultural yields reduce prices, resulting in reduced investment and lower yields. Yield and price become variables in a negative feedback loop.

In political life, public demands for some policy change may result in that policy being changed, thus stopping the protest; alternatively, efforts may be made to eliminate the public demand for change by issuance of propaganda or implementation of repressive actions. Both are negative feedbacks that reduce the level of protest, but their long-term consequences may differ dramatically, as this feedback is not an isolated system. In ecosystems, the links in the food chain (predator/prey, parasite/host) give negative feedback loops with length of 2. The production of growth-inhibiting waste also represents a negative feedback loop that links the waste product and the producer.

Positive feedback is a process in which an initial change gives rise to a chain of events that keeps the variable moving in the same direction as its original displacement, thereby enhancing the effects of the original change. For instance, poverty can force people to sell their land, further increasing their poverty. Melting snow results in a darker earth surface; this surface absorbs more solar energy and melts more snow. Disease can undermine nutrition and lead to more disease. In human ecosystems, productivity demands that land be treated with conservation measures, such as balanced use of land for row crops, pasture, and forest, implementation of crop rotation, and use of cover crops. Poor farmers, however, may feel great pressure to bring in cash income from every piece of land. Thus, forests are cleared, savannas are plowed, fallow periods are shortened, rotation is abandoned, and water-hungry crops are planted. The result is a decline in productivity, and further poverty. Vicious circles are positive feedbacks with undesirable results.

The same variables may be linked through different pathways by both positive and negative feedbacks. For instance, if a baby is uncomfortable and cries, actions are usually undertaken by caregivers to remove the source of discomfort and eliminate the crying. Sometimes the crying may evoke anger and abuse by the caregiver, which only increases the baby's crying. Thus, the feedbacks interact in a system in complex ways to determine its behavior. In general, stability requires an average excess of negative over positive feedback for subsystems of every size; in addition, short-pathway negative feedbacks typically predominate over longer negative feedback loops. The detailed mathematics has some sunrises—circumstances in which positive feedbacks stabilize and negative feedbacks destabilize—and we must differentiate disjunct from conjunct cycles in the network's structure.

Resilience and resistance also depend on these feedback structures. A delay in a pathway usually weakens its effect; hence, a stabilizing loop with a delay may produce oscillations and instability. When any event impinges on a system of variables, it percolates throughout the network, being buffered along some pathways and amplified on others. In the end, some variables—not necessarily close to the source of perturbation—will absorb much of the impact, while others—even if they receive the impact directly—may undergo minimal change. The former are "sinks" in the system, or absorbing variables, while the latter are reflecting variables (the impact mostly bounces off them). The metabolic rate in mammals, for example, is a sink in that it varies widely with outside temperature and serves to keep internal temperature relatively constant. The supplies of oxygen and sugar to the brain are also protected variables.

The availability of food for the poor acts as a sink. It absorbs the impact of fluctuations in agricultural production when consumption is mediated through a market, whereas the food intake of the rich is protected by these individuals' ability to pay more. Agricultural decisions are often made on the expectation that prices will remain stable, but the delay between the decision to plant and the start of production for trees or large livestock often generates oscillations in production and prices, to the detriment of development plans.

The qualitative mathematics of systems dynamics can educate the intuition to the extent that some system properties become obvious and we know where to look for unusual phenomena. Systems models, however, take these variables as given. In reality, under some conditions a variable may subdivide into multiple variables, new links may be forged among existing variables, variables may enter or leave the system or merge together, or the dynamic significance of the same variable may change as the context changes.

In simple neoclassical models of the market, "effective demand" is determined from outside the system, expressing the arbitrary preferences or biological needs of the "consumer" and the money available to buy the fruits of production. Production is undertaken to satisfy demand; when it is satisfied, production declines in a simple negative feedback loop. Once the producers act to create or increase demand, the system is changed in a qualitative system. In this case, production itself encourages demand through advertising, longer-term impacts on people's sense of need, decisions by manufacturers about what to offer, or regulations about product standards. Because production both creates and meets demand, the original variables are linked by both positive and negative paths. Although the product—for example, a potato—remains a potato in its physical existence and a price is still a price, the dynamics have been fundamentally altered so that models of consumer sovereignty become inapplicable. Thus, a systems model may represent only a moment in the history of the relevant variables during which they remain relevant. We must always consider changes in the structure of the system as well as its day-to-day, "normal" operations.

Evolution of the System.   The second major question about a system—why it is the way it is instead of very different—deals with the evolution (that is, the history and development) of disequilibrating processes that differentiate systems. It examines the transformation of one system into another. A system reaches a certain state through change, not because it reflects how things have to be and have always been.

Without this second question, systems theory would treat its array of variables and processes as givens—in some sense, as natural and unquestioned. All systems are transient, however, and must be questioned. All systems have external boundaries that must be taken into account and sometimes challenged. In ecology, these questions deal with evolution, coexistence, extinction, migration, and biogeography.

THE INDICATORS OF QUALITATIVE DYNAMICS

If we had a single linear equation in one unknown,

$$dX/dt = A - rX$$

then stability, resilience, and resistance all depend on the same parameter r. If $-r$ is negative, then a stable equilibrium exists at $X^* = A/r$. The time to return halfway back to equilibrium after a perturbation is $\ln2/r$. If A is not constant then

$$x(t) = x(0)e^{-rt} + \int A(t-s)e^{-rs}ds \qquad\qquad 11.11$$

so that the impact of events at time s ago, $A(t - s)$, diminishes with elapsed time at an exponential rate with exponent r.

This relationship can be used in assessment in several ways:

■ If we know r, we can estimate the time to recovery after a proposed perturbation or an uncontrolled event that has just occurred.

- If we know of some past perturbing event, such as the destruction of trees by lumbering or by a hurricane, we can estimate the recovery time r.

- If we know that different variables change at different rates, then a comparison of these variables in the present can date and therefore help to identify past events.

The sensitivity of an equilibrium value to change in the parameter A is

$$\partial X^*/\partial A = 1/r \qquad\qquad 11.12$$

Therefore, increasing r increases both resistance and resilience.

Once we have more variables, however, these three properties become distinct and it is possible to reduce stability while increasing resistance. The parameter r is replaced by n different eigenvalues $\mu\hat{E}$ that must be calculated. These values may be complex numbers. Local stability requires that all of them have negative real parts. The return time depends on the reciprocals of the real parts of the μ, and resistance is related not only to their product but also to the eigenvalues of various subsystems.

LOCAL STABILITY

It is not obvious from the observed pattern of changes in variables whether the system has a stable equilibrium. The equilibrium may be stable, yet external perturbations repeatedly displace variables and the observed changes may mark the trajectories of return toward equilibrium. A system may exist at a stable equilibrium that is itself moving as parameters change slowly enough for the variables to keep up with the equilibrium. The equilibrium may also be unstable, such that with nonlinear dynamics we may observe autonomous periodic or nonperiodic oscillations. Furthermore, some aspects of a system may be at equilibrium while others are not. For instance, a population may be growing and yet the age distribution may remain at equilibrium. Stasis and change, therefore, cannot be regarded as mutually exclusive; equilibrium can be seen as a form of coordinated motion; and the observation of change does not by itself preclude an equilibrium analysis.

If an equilibrium exists, a question arises: Is it stable? Stability has no special virtue. Sometimes the desired state may involve oscillation, and equilibrium would mean a deadening stasis. Sometimes the desired state is growth, and equilibrium means stagnation or even death. Nevertheless, the determination of local stability provides a useful descriptor of the system, a landmark in the space of variables.

The formal theory of local equilibrium is well known. The testing of stability proceeds as follows:

1. Determine the equilibrium values.

2. Find the first partial derivatives of each rate of change with respect to each of the variables, $\partial(dx_i/dt)\partial x_j$, evaluated at equilibrium.

3. Form a matrix in which the element in the jth column of the ith row,

$$a_{ij} = \partial\left(dx_i/dt\right)\partial x_j \qquad 11.13$$

gives the Jacobean matrix of the system of equations,

$$J = \begin{matrix} a_{11} & a_{12} & a_{13} \ldots\ldots\ldots & a_{1n} \\ a_{21} & a_{22} & a_{23} \ldots\ldots\ldots & a_{2n} \\ a_{31} & a_{32} & a_{33} \ldots\ldots\ldots & \\ a_{n1} \ldots\ldots\ldots\ldots\ldots\ldots & & a_{nn} \end{matrix} \qquad 11.14$$

4. Subtract μ from each of the diagonal elements and set the determinant equal to zero as a polynomial equation in the eigenvalues μ.

$$P(\mu) = \mu^n - \cdot D_1\mu^{n-1} + \cdot D_2\mu^{n-2} - \cdot D_3\mu^{n-3} + \ldots (-1)^k D_k\mu^{n-k} + \ldots (-1)^n D_n \qquad 11.15$$

Each coefficient of μ^{n-k} is $(-1)^k$ times the sum of the determinants of all the subsystems of size k.

In a system of differential equations, an equilibrium is stable if all of the eigenvalues (characteristic roots) of the characteristic equation have negative real parts. In a system of difference equations, where time is discrete, the eigenvalues must all lie within the unit circle in the complex plane. A number of theorems about the eigenvalues can be used to interpret the interaction patterns among variables.

Box 11.2 gives some of the tests for local stability and how they may be applied.

Box 11.2 Gershgorin Disks (Gantmacher 1959)

In the complex plane, let each diagonal element of the matrix, a_{ii}, be the center of a circle and let the radius be

$$r_i = \sum \mathrm{abs}\left(a_{ij}\right) \qquad 11.16$$

summed either over i or j (either the row or the column), whichever is smaller. All of the eigenvalues lie in the region bounded by all of the circles. Furthermore, any k circles that are disjoint from the rest contain k eigenvalues. Suppose that all the a_{ii} are negative and that none of the circles crosses into the positive half-plane. Then the real parts of the eigenvalues are negative and the equilibrium is locally

Box 11.2   *(Continued)*

stable. Suppose also that none of the circles overlaps. Because complex roots occur in conjugate pairs, all of the roots are real negative numbers, and we can associate each root with one of the original variables. The root μ_i lies in $a_{ii} \pm r_i$. Thus, the eigenvalue is equal to what it would be in the isolated system at that equilibrium, somewhat modified by the interactions with other variables.

Now let the a_{ii} values approach one another. Once a pair of circles overlaps, two roots exist in their joint area and it is possible to have complex roots, which implies oscillation. Alternatively, let the a_{ii} values remain fixed but allow their interactions to increase, increasing the radii r_i and r_j. Once again, when the circles overlap, complex roots become possible. Therefore, if two variables have very different autonomous rates of change compared with the strength of their interaction, then they can be regarded as almost separate systems. The more similar their autonomous rates, or the stronger their interactions, the easier it is for their disks to overlap and produce complex eigenvalues that can no longer be identified with the separate variables. Therefore, systems of variables that have very different autonomous rates and weak interactions can be treated to a first approximation as independent systems.

If the radii become large enough, they eventually cross into the positive half-plane. In that case, roots may have positive real parts and instability may arise.

In examining any particular system, we can look for autonomous subsystems that would facilitate our analysis and identify processes that might bring the a_{ii} values closer together or increase the strengths of their interactions so that the disks overlap. On the other hand, processes that weaken the interactions or that drastically alter the self-damping of some variables can fragment a system into almost autonomous subsystems. Factors that weaken self-damping include saturation of an enzyme or of the feeding capacity of a species, slowing of input to a system (inflow, immigration), and the removal of variables at some constant rate dictated by external conditions. For instance, if a fishing industry removes a fixed number of fish for its market or to meet some production quota, self-damping is reduced by an amount that is proportional to the fraction of the fish population removed. Factors that strengthen the interactions among variables in an ecosystem include loss of environmental heterogeneity, loss of the preferred food or microhabitat for one member of a pair of loosely competing consumers, and introduced or increased vulnerability to a disease that affects more than one species so that interspecific contagion links the species.

Symmetrized Matrices

We can find bounds on the eigenvalues of a matrix by comparing it with other matrices that are symmetrical. First, symmetrize the matrix with the transformation

Box 11.2 *(Continued)*

$$a_{ij}{}^* = \left(a_{ij} + a_{ji}\right)\big/2 \qquad\qquad 11.17$$

This new matrix is symmetric and, therefore, all of its roots are real. The real parts of the roots of the original matrix are bounded by the roots of the symmetric matrix. If we now apply Gershgorin's disks to this matrix, we see that, if a_{ij} and a_{ji} have opposite signs, then the $a_{ij}{}^*$ values tend to be small. Thus, processes that increase the asymmetry of interactions bring the real parts of the roots closer to a_{ii}. Symmetrical interactions, such as competition, depend on the turnover rate of a resource. If it slows down, then the same degree of feeding overlap results in stronger mutual inhibition. In compartment models, where organisms move from one status to another (susceptible/infected/infective/immune/susceptible, or wet land becomes dry and dry becomes wet, or substances move between physical compartments), one-way flows give tighter bounds on the eigenvalues than symmetrical flow back and forth.

SIGNED DIGRAPHS

The structure of an ecosystem may be represented by a graph in which the vertices are variables and the directed edges represent the first-order interactions, designated by a_{ij}. In most cases, we specify only the signs of these edges. Figure 11.1 shows some of the familiar ecological relationships. The characteristic equation of a system of differential or difference equations can then be expressed in terms of the graph, as shown in Box 11.3.

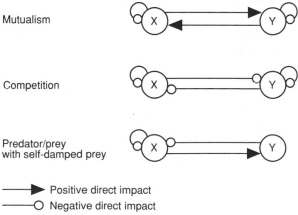

FIGURE 11.1. *Familiar ecological relations.*

Box 11.3 The Method of Signed Digraph

Definitions

The link from variable j to variable i, a_{ij} is represented by the edge X_j to X_i.

A cycle of length k is a sequence of links from X_j back to X_i along which each vertex has only one input and one output. Its value is the product of the a_{ij} values around the cycle. Thus, its sign is the product of the signs of the links. A cycle whose product is negative is a negative feedback loop.

Two cycles are considered disjunct if they have no elements in common. Otherwise they are conjunct.

$L(m, k)$ is the product of m disjunct cycles with a total of k elements.

The *gain* or *feedback* of a system at level k ($\leq n$) is

$$F_k = \sum (-1)^{m+1} L(m,k)$$
11.18

where summation takes place over all combinations of loops for each m and over all $m \leq k$. By definition, $F_0 = -1$. The term $(-1)^{m+1}$ guarantees that if a term $L(m, k)$ consists of all m negative cycles then the gain is negative.

The characteristic equation is

$$P(\mu) = \sum \mu^k F_{n-k}, \quad k = 0, 1, 2, \ldots n$$
11.19

The theory of equations relates the roots of an equation and its coefficients. For all roots to have negative real parts, two sets of conditions must be satisfied:

1. $F_k < 0$ for all k. That is, the negative feedback at every level must exceed the positive feedback.

2. The Routh-Hurwitz inequalities. The first of these is

$$F_1 F_2 + F_3 > 0$$
11.20

Suppose that in a system of three variables, parameter changes cause a transition from stable to unstable. Suppose one real root μ_1 and a pair of complex roots $\mu_2 \pm i\phi$ exist. If the instability arises from a decrease in F_2, then ϕ also decreases. That is, the oscillations become slower. On the other hand, if the instability comes from an increase in the negative feedback F_3, then the frequency of oscillation increases. Thus, an examination of changes in the frequency of oscillation can indicate where the change occurred.

Subsequent inequalities include higher-order feedbacks. Qualitatively they require that the negative feedbacks at higher levels are not too strong compared with the negative feedbacks at lower levels. We can also examine the coefficients from the opposite end of the equation. Divide the entire equation by $\mu^n F_n$ to obtain an equation for i/μ. Its roots will have positive or negative real parts and complex roots only if the original equation has them. Thus we find the criterion for stability

Box 11.3   (Continued)

$$F_{n-2}F_{n-1} - F_{n-3}F_n > 0 \qquad\qquad 11.21$$

That is, a very long negative cycle that includes all elements of the system can destabilize it.

The graphical representation of a system allows us to interpret the Routh Hurwitz criteria in terms of the cycles in the graph. Thus the first inequality becomes

$$-2\sum a_{ii}a_{jj}a_{kk} - \sum a_{ii}^2 a_{jj} - \sum a_{ii}a_{ij}a_{ji} + \sum a_{ij}a_{jk}a_{ki} > 0 \qquad 11.22$$

The first term multiplies unconnected self-damping cycles. The second term multiplies one self-damping term by the square of another. This operation introduces the variance of the self-dampings as a factor promoting stability. The third term sums conjunct pairs, one of length 1 and the other of length 2. It therefore depends on the structure of the network. Each of these terms will be positive if all cycles are negative and thus contribute to stability. The fourth term sums cycles of length 3. If these values are all negative, they destabilize the system. For this reason, a positive feedback loop offsetting an excessive feedback can have a stabilizing effect. Also, if a particular a_{ii} is positive, $a_{ii}a_{ij}a_{ji}$ can be a stabilizing factor if $a_{ij}a_{ji}$ is also positive. Thus, the commonly made claim that negative feedback stabilizes and positive destabilizes is not always true.

The cycles of lengths greater than 1 depend on the structure of the network—in particular, its connectivity. For instance, if every variable is connected to every other variable with probability c, then there will be $cn(n-1)(n-2)/3$ cycles of length 3 and $n(n-1)(n-2)/6$ triplets of self-damping terms. The other terms are of order $n(n-1)$ so that as n increases they become relatively less important. The greater the value of c, and the stronger the negative feedback of the cycles of length 3 compared with the self-damping, the more the system destabilizes as n increases. On the other hand, suppose that the variables can be arranged in some "space" that depends on their biological affinity and that linkages occur mostly among neighbors. In that case, the number of cycles of length 3 is proportional to only n. For a large enough n, the triplets of self-damping terms will swamp these cycles and the system will be stable. Thus, the rules of connection among variables for a particular ecosystem can serve as a guide to how the system responds to fragmentation or increase. In simulation explorations of connectivity, Gardner and Ashby (1970) found that for randomly connected networks connectivities greater than approximately 3 result in unstable equilibria.

The role of the heterogeneity of the variables can be seen from looking at the feedback at level 2, where negative feedback always stabilizes.

$$F_2 = -\sum a_{ii}a_{jj} + \sum a_{ij}a_{ji} \qquad\qquad 11.23$$

Box 11.3   *(Continued)*

The first term is

$$-\left(n(n-1)/2\right)\left(E(a_{ii})^2 - \text{var}(a_{ii})\right)$$

while the second is

$$\left[n(n-1)/2\right]\left\{E(a_{ij})^2 + \text{Cov}(a_{ij},\, a_{ji})\right\} + \text{var}(a_{ij})$$

so that average self-damping increases stability, heterogeneity of self-damping reduces stability, the variance of the links between variables reduces stability, and the symmetry of interactions (positive covariance) also reduces stability.

In compartment models, the contributions of the flow between compartments makes $\Sigma a_{ii} = \Sigma a_{ij}$. If the total flow through each variable is equal, then stability is maximized because it eliminates the variance term. Similarly, if mostly asymmetric flow occurs and if great unevenness characterizes the couplings of variables, then less negative feedback arises.

RESISTANCE

If a permanent or slow change occurs in the environment, it can affect the equilibrium values of all variables. The resistance is the inverse of the sensitivity of the equilibrium value to a unit change in some parameter. This sensitivity is given by

$$dX_i^*/dC = \sum \partial(dX_j/dt)/\partial C\, P_k\left(ijF_{n-k}\,\text{Comp}\{P_k ij\}/F_n\right) \qquad 11.24$$

The first term is the direct impact of a change in parameter C on the rate of change of variable j. The second term, $P_k(ij)$, is the product of the a_{ij} elements along a path from variable j to variable i with k variables. A path is a sequence of links that enters and leaves a variable at most once (that is, a cycle with the link x_i back to x_j removed). If the environmental change impacts directly on variable i, then $P_1 ii = 1$. The sign of a path is the sign of the product of the signs of the links. The third term is the gain or feedback of the complement of the path—that is, of the subsystem that remains when all variables along the path are removed. If the path has k variables, then the complement has $n - k$ variables. Finally, F_n is the feedback of the entire system. Summation occurs along all paths from variable j to variable i. We usually assume that the system is locally stable and therefore that F_n is negative.

The sensitivity of a variable to an environmental change depends on the pathways from the point of entry of the change to that variable and on the complement of each path. If the complement is zero, then the path has no effect on the equilibrium. A zero complement serves as a sink, protecting the target variable from all influences arising outside the complement. If the complement has negative feedback, then the change has the same sign as the path; thus, the outcome corresponds to common-sense expectation. The complement may be envisioned as a reflecting barrier; the stronger its self-damping, the more an impact is reflected back toward its source rather than absorbed. If the complement has positive feedback, then the variable changes in the opposite direction from the sign of the path. In this situation, the investigators may doubt these anomalous results and attribute them to error. If a variable changes in the opposite direction from what makes sense, we should search for a solution involving positive feedback.

Finally, the sensitivity depends on the denominator F_n, which is the same for all variables in the system for a given set of parameters, but can change in nonlinear systems if the environment (parameter set) changes. The stronger the negative feedback F_n, the greater the resistance to external change.

If F_n is always negative regardless of parameter values—for instance, if all cycles in the graph are negative—then the equilibrium values of all variables are continuous functions of the parameters. If some positive feedbacks exist in the network, however, and if these reflections become stronger relative to the negative feedbacks when some parameters change, then F_n can near zero; in this case, the response to environmental change increases. [If F_n becomes zero, it will signal a discontinuous change in response to a small shift of parameter. This case would constitute a type I catastrophe in the sense of René Thom's catastrophe theory (Thom 1970). The equilibrium would become unstable.] Clearly, the positive feedbacks in the system should be identified and changes in them monitored.

The effects of parameter change on all variables generate a correlation pattern among variables that can be diagnostic of the source of change.

The integration of the autoecology of species with their population dynamics provides additional indicators as to the sources of variation and can detect changes in the driving forces. For example, Levins and Adler (1995) examined the relations among population density, fecundity, body size, winter survival, and food supply in insular rodents. Figure 11.2 shows a set of alternative models and the responses of these variables to changes entering the system by way of different variables. Observations of real communities could then both help researchers select the most appropriate model and identify changes in the driving force.

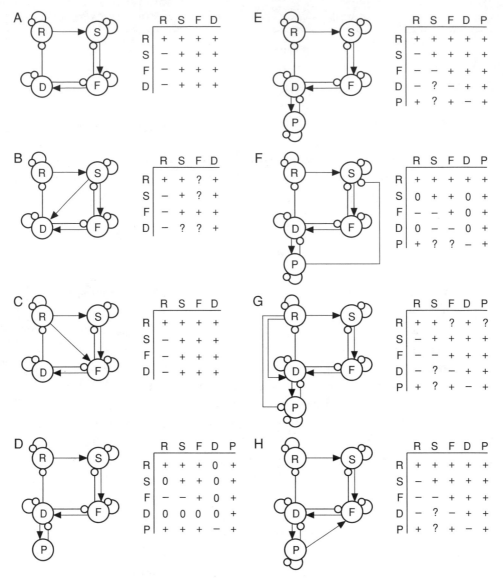

———▶ Positive direct impact
———○ Negative direct impact

FIGURE 11.2. *Models for insular rodent population components. A. Core model: food resource (R) increases body size (S), which increases fecundity (F) and hence density (D) of rodent population, while density reduces the food resource. B. Body size increases density through increased winter survival. C. Food resource increases fecundity directly. D. A predator reduces density and is increased by prey density. E. Same as D, but predator is also self-regulated. F. Predators reduce survivorship of rodents, reducing their age and therefore their body size. G. Food resource reduces predation because with more grass the rodents spend less time foraging in exposed situations. H. Predation keeps the population young and hence more fecund. The tables indicate the responses of the variables shown in the columns to an increase in the row variable. + = positive response (increase); − = negative response (decrease); 0 = no response; ? = response unknown. (Levins and Adler 1993.)*

OSCILLATIONS

The oscillations in autonomous linear systems are associated with the eigenvalues of the characteristic equation. The greater the imaginary part of an eigenvalue, the higher the frequency of the oscillation. Conversely, the smaller the (negative) real part, the more rapidly the oscillation declines and disappears. In a general way, asymmetry of interaction makes oscillation possible. Several indicators of complex eigenvalues can be used (Box 11.4).

Box 11.4   Conditions for Oscillation

Skew-Symmetrized Matrices
Gershgorin's disks can be applied to the skew-symmetrized matrix. Replacing a_{ij} by

$$a_{ij}{}^{\star} = \left(a_{ij} - a_{ji}\right)\big/2 \qquad\qquad 11.25$$

gives a symmetric matrix with zeroes along the principal diagonal. All of the Gershgorin disks are concentric around the origin. The maximum radius bounds the maximum frequency of autonomous oscillation.

Demonstration of a Negative Variance for the Eigenvalues
This technique uses only the pairwise interactions. The variance of the eigenvalues is (Levins 1968)

$$\mathrm{Var}\!\left(\mu\right) = \mathrm{var}\!\left(a_{ii}\right) + \left(n-1\right)\mathrm{Cov}\!\left(a_{ij}, a_{ji}\right) \qquad\qquad 11.26$$

If the pairwise loops are mostly negative (such as in predator/prey or production/price situations), then the variance of μ will be negative in a large system, indicating complex roots with strong imaginary parts. The larger the imaginary part, the greater the frequency of oscillation. If the diagonal elements (self-damping terms) are relatively uniform, then they contribute little to the variance, which will be dominated by the pairwise interactions. Oscillation can be suppressed if the self-dampings differ sufficiently (weakly overlapping Gershgorin disks).

The Harmonic Mean Rule
The harmonic mean of a set of numbers is given by

$$H = n\big/\!\left(\sum 1\big/x_i\right) \qquad\qquad 11.27$$

If all x values are positive real numbers and not identical, then the arithmetic mean (average) of the numbers is greater than the harmonic mean and the discrepancy

Box 11.4   *(Continued)*

increases with the variance of the numbers. If we already know that the numbers have the same sign (usually all negative for a stable system), then if the harmonic mean exceeds the arithmetic mean in absolute value it must be attributed to complex roots. Suppose that we have complex roots $r_j \pm i\phi$. Then the harmonic mean is

$$H = n \Big/ \left(\sum 2r_j \Big/ \left(r_j^2 + \phi^2 \right) \right) \qquad 11.28$$

or

$$H = n \Big/ \left(\sum 2 \Big/ \left(r_j + \phi^2 / r_j \right) \right) \qquad 11.29$$

so that the denominator is reduced by the ϕ^2/r and the harmonic mean correspondingly increased.

In the characteristic equation

$$\sum \mu^{n-k} F_k = 0, \quad k = 0, 1, 2, \ldots n \qquad 11.30$$

the coefficients are related to the roots: $F_k = (-1)^k$ times the sum of the product of the roots taken k at a time. If we divide the equation by μ^n, we get an equation in the reciprocal

$$\sum \left(1/\mu \right)^k F_{n-k} / F_n = 0 \qquad 11.31$$

As $n!/k!(n-k)!$ products of roots taken n at a time exist, the arithmetic mean of k-tuples is

$$A = F_k \Big/ n! \Big/ \left(k!(n-k)! \right) \qquad 11.32$$

and the harmonic mean is

$$H = \left(n! \big/ k!(n-k)! \right) F_n / F_{n-k} \qquad 11.33$$

Therefore, if for any k

$$\left(n! \big/ k!(n-k)! \right)^2 F_n > F_k F_{n-k} \qquad 11.34$$

then there are complex roots and oscillations. Thus, if for any level k the average feedback of subsystems of size k is weak compared with that of the system as a whole, then oscillations will occur.

Finally, as a system becomes unstable, the frequency of oscillation changes. If the instability arises from an increase in the long-loop negative feedbacks then the frequency increases; if it arises from the decrease in short-loop negative feedbacks, the frequency of oscillation diminishes.

CORRELATION PATTERNS

Variables in a complex system are always changing under the impact of external inputs and their own dynamics. The fact of change does not by itself reveal anything about the source of change or its likely outcome. Nevertheless, the pattern of correlations among the variables serves as an indicator of the sources of variation and changes in the system's structure. Figure 11.3 (Lane and Levins 1977) illustrates how this method can be employed to identify the source of perturbation of an ecosystem and to decide among alternative models.

TIME AVERAGING

The method outlined above facilitates the interpretation of differences between sites or over time if the changes occur slowly enough to be understood as moving equilibria. For more rapid changes, the techniques of time averaging (Box 11.5) must be used (Puccia and Levins 1986). The two approaches usually give similar results.

In a simple food chain, all upward links are positive and all downward links are negative. Therefore, if the system is being driven from below, all variables will show positive correlations. If the system is being driven by changes in hours of light, a factor that directly affects the algal level, a negative correlation will exist between algae and minerals but positive correlations will be found among other pairs of variables. If toxic inputs later poison the crustaceans, a positive correlation will exist (upward from crustaceans to fish) but negative downward correlations will characterize the relationships between crustaceans and algae and between algae and minerals.

Similarly, in predator/prey systems, the correlation between predator and prey depends on which variable represents the immediate point of impact of external influences. A review of the insect literature by Stiling (1987) showed that approximately half of all cases involved no correlation between predator and prey abundance; the other half of the cases were divided more or less equally between positive and negative correlations. A change in the correlation pattern would be indicative of a change of the driving force on a system and therefore suggestive of major changes.

The population densities of species are not the only variables of importance in ecosystems. When the numbers change, other properties of the variables become modified as well. Suppose, for instance, that a predator/prey/resource system is driven from the resource end. Positive correlations will, therefore, exist among the variables. When the prey population is abundant, resources will abound. Thus the animals will be healthy, large, and highly fecund. On the other hand, the predators are also more common in

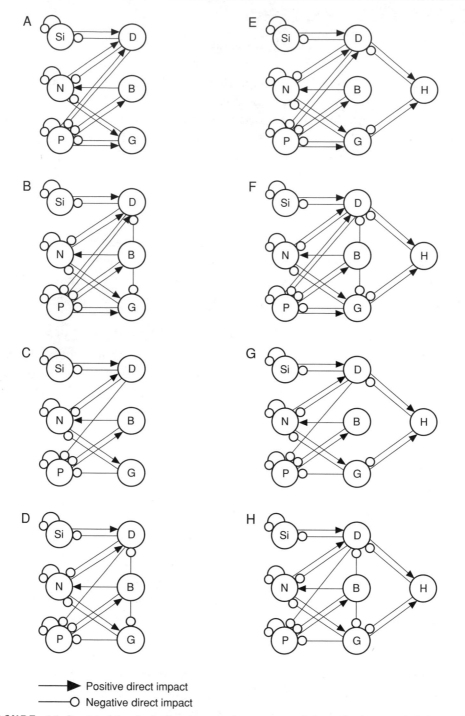

Positive direct impact
Negative direct impact

FIGURE 11.3. *Models of relations among three groups of phytoplankton and their nutrients. Si = silicon; N = nitrogen; P = phosphorus; D = diatoms; B = cyanobacteria; G = green algae; H = herbivore. (Lane and Levins 1977.)*

Box 11.5   Formal Properties of Time Averaging

The statistical definitions of means, variances, and covariances can be applied even in deterministic situations. Thus the time average of a variable is given by

$$E_t(x) = \left(\int_0^t x(s)\, ds \right) \Big/ t \qquad\qquad 11.35$$

$$\mathrm{Var}_t(x) = \left(\int_0^t \left[x(s) - E_s(x) \right]^2 ds \right) \Big/ t \qquad\qquad 11.36$$

and the covariance of x and y is

$$\mathrm{Cov}_t(x, y) = \left(\int_0^t \left[x(s) - E_s(x) \right]\left[y(s) - E_s(y) \right] ds \right) \Big/ t \qquad\qquad 11.37$$

When this equation is applied to the derivative of a bounded variable,

$$E_t(dx/dt) = \left[x(t) - x(0) \right] \big/ t \qquad\qquad 11.38$$

Since x is bounded,

$$E_t(dx/dt) < \left[x_{max} - x_{min} \right] \big/ t \qquad\qquad 11.39$$

so that as t increases without limit we have the long-term result

$$E(dx/dt) = 0 \qquad\qquad 11.40$$

This equation applies to any bounded function of x as well. For example,

$$E(xdx/dt) = E(dx^2/dt) \big/ 2 \qquad\qquad 11.41$$

which is also zero.

This method can now be applied to systems of interacting variables. In a predator/prey system where the prey but not the predator is self-damped (see Fig. 11.2),

$$dx/dt = x(k - x - ay) \qquad\qquad 11.42$$

$$dy/dt = y(ax - b)$$

The long-term averages are

$$E(dx/dt) = E(x)E(k - x - ay) + \mathrm{Cov}(x, k) - \mathrm{var}(x) - a\mathrm{Cov}(x, y) \qquad\qquad 11.43$$

and

$$E(dy/dt) = E(y)E(ax - b) + a\mathrm{Cov}(x, y) - \mathrm{Cov}(y, b) \qquad\qquad 11.44$$

Box 11.5 (*Continued*)

If we divide the equations by x and y, respectively, and then average the results, we get

$$E(d\log x/dt) = E(k - x - ay) \tag{11.45}$$

$$E(d\log y/dt) = E(ax - b) \tag{11.46}$$

Substituting these results, we have

$$\text{Cov}(x,k) = \text{var}(x) + a\text{Cov}(x,y) \tag{11.47}$$

$$\text{Cov}(y,b) = a\text{Cov}(x,y) \tag{11.48}$$

Suppose that b is constant and that the environmental variation enters the system by way of a variable k. Then $\text{Cov}(y, b) = 0$ and therefore

$$\text{Cov}(y,x) = 0 \tag{11.49}$$

Then

$$\text{Cov}(x,k) = \text{var}(x) \tag{11.50}$$

from which it follows that $\text{Cov}(x, k)$ is positive and that

$$\text{Var}(k) > \text{var}(x) \tag{11.51}$$

If k is constant and b varies, $\text{Cov}(x, k) = 0$ and

$$a\text{Cov}(x,y) = -\text{var}(x) < 0 \tag{11.52}$$

$$\text{Cov}(y,b) < 0$$

Finally, if both k and b are constant, $\text{Cov}(x, y) = 0$ and therefore $\text{var}(x) = 0$. Thus x is at a stable equilibrium.

---

this case, which means that mortality is increased. We would expect to observe a population with high birth and death rates and, therefore, a young population that is healthy but at risk of predation. When the resources are rare, the population of prey is low and poorly fed, with a low birth rate and a low death rate. This population will have a greater proportion of older individuals. Thus the oscillation will swing between abundant, young, short-lived animals and rare, ill-fed, older animals. If the system is driven from above (that is, from factors acting directly on the predators), negative correlations will exist between predator and prey and between prey and its resource. In this case, the small prey populations will be young and associated with both abundant food and high mortality; in contrast, the abundant prey populations will be long-lived but with short food supplies and low fecundity.

The techniques of signed digraphs and time averaging are useful for describing the structure of the system, diagnosing where environmental forces enter the system, detecting changes in these forces when a new factor enters the system, finding bounds on the environmental variation, and predicting the likely outcome of interventions at particular places. An intervention may consist of a permanent input to the system, the addition or removal of a variable, the establishment of a new link in the network, or the cutting of an existing link. Removal of a variable does not mean its physical removal, however. If careful monitoring allows us to hold a variable constant, then its physical nature has not been altered but it nevertheless ceases to be a dynamic variable in the system network.

The tools described earlier in this chapter provide a framework for understanding the structure and dynamics of ecosystems even when information remains incomplete. It allows us to identify the sources of particular aspects of behavior in the structure of the process, indicates where precise measurement is needed, and suggests the likely outcomes of interventions. In addition, and perhaps more important, the use of these methods—even without the mathematical arguments—educates our intuition. After we gain some familiarity with diverse examples, we can see at a glance the crucial properties of a system that determine its dynamics and the strategic points for experiment, measurement, and intervention.

REFERENCES

Gantmacher FR. The theory of matrices. New York: Chelsea Publishing Company, 1959.

Gardner MR, Ashby WR. Connectance of large dynamic (cybernetic) systems: critical values for stability. Nature (London) 1970;228:784.

Kreiger N. Epidemiology and the web of causation: has anyone seen the spider? Sco Sci Med 1994;39:887–903.

Lane P, Collins T, Collins M. Food web models of marine plankton community network: an experimental mesocosm approach. J Exper Mar Biol Ecol 1095; 98:41–70.

Lane PA, Levins R. The dynamics of aquatic systems. 2. The effect of nutrient enrichment on model plankton communities. Limnol Oceanogr 1977;23:454–471.

Levins R. Evolution in changing environments. Princeton: Princeton University Press, 1968.

Levins R, Adler GH. Differential diagnosis of island rodent populations. Coenoses 1993;8:131–139.

Puccia C, Levins R. Qualitative modelling of complex systems. Cambridge, MA: Harvard University Press, 1985.

Stiling PD. The frequency of density dependence in insect host–parasitoid systems. Ecology 1987;68:844–856.

Thom R. Topological models in biology. In: Waddington CH, ed. Towards a theoretical biology. Vol. 3. Drafts. Edinburgh, Scotland: Edinburgh University Press, 1970:116.

CHAPTER 12

Validation of Indicators

Walter G. Whitford

ASSESSING the health of any ecosystem requires the measurement of a number of indicators. For an assessment of ecosystem health to have real value, it must not only address status, but also evaluate the risk of that status changing to a less healthy state. Although numerous indicators of ecosystem health have been proposed (Rapport et al 1985; National Research Council 1993), few have been tested to date. Such testing must incorporate evaluation of the indicator's sensitivity so as to provide unambiguous and reliable values for the assessment of ecosystem status. Indicator sensitivity must also be known so as to enable an evaluation of the risk of change in the health status of the ecosystem.

Analogous to indicators of human health, indicators of ecosystem health must change in predictable, repeatable ways when ecosystems shift to a less healthy condition. Sensitivity testing of these indicators must be the first step in evaluating analyzing indicators intended for assessment or monitoring purposes. Sensitivity testing is a necessary calibration step that must be carried out before potential indicators become incorporated into an assessment or monitoring system. The sensitivity of human health indicators (for example, blood tests, X rays, body temperature, heart sounds, pulse, and blood pressure) has been determined by recording many thousands of measurements of healthy and sick patients in order to identify the average values and range of values for healthy humans. Obviously, this approach cannot be used in ecosystem health—we cannot afford to lose ecosystems that are unhealthy and do not have the luxury of beginning to make measurements now in order to demonstrate how sensitive an indicator will be in a century or two. Ecosystem health assessments require sensitive indicators immediately.

Indicators of ecosystem health are classified into one of two categories: ecosystem process measurements or ecosystem property measurements. Measurement of ecosystems processes, such as primary production (rates of energy fixation), energy flow (rates of transfer of energy between trophic levels), and rates of nutrient cycling, requires expensive, complex, and time-consuming analyses.

As a consequence, indicators of the health of ecosystem processes frequently encompass properties that are clearly linked to the ecosystem process. These properties can be measured by point-in-time measurements. Such measurements can generally be made without expensive instrumentation and in a short period of time.

BENCHMARK SITES

Calibration and sensitivity testing of ecosystem health indicators can be performed by substituting space for time. The space (sites where measurements are made) is selected on the basis of available historical records. Places with documented histories of degradation or change in state can be compared with other places with documented histories of little or no degradation or change in state. In many cases, gradients of change may exist—from large change to imperceptible; such sites are very valuable for testing the sensitivity of indicators (deSoyza et al 1997). We obtain values for indicator sensitivity by measuring the responses of indicator variables across a series of sites with documented levels of exposure to environmental stress or with documentation of a variety of ecosystem changes over time that, in total, have resulted in ecosystem degradation.

For indicators of ecosystem health, the timing of measurements may be as critical as the actual measurement. For example, measurements of vegetation parameters that are related to primary productivity are best made at the end or peak of the growing season. Measurements made too early or too late could give erroneous results.

FIELD MEASUREMENTS

When assessing or monitoring the health of ecosystems, the field measurements must be rapid, repeatable, and not subject to observer bias. Such measurements should provide data about the health of both abiotic and biotic parts of the ecosystem. In addition, field measurements that can be used to calculate more than one indicator metric are especially desirable. For example, when assessing or monitoring rangeland health, the species composition and cover (percentage of surface covered by a species) can be used to calculate indicator metrics that relate to ecosystem properties such as productivity, biodiversity, resistance to wind and water erosion, and economic potential. In an aquatic ecosystem, density and species composition of benthos, dissolved oxygen, chlorophyll content, and species composition of the fish community are properties that could be used to calculate several indicator metrics that are related to ecosystem function (Fore and Karr 1996; Small et al 1996).

COMPUTING INDICATOR METRICS

Raw measurements from the field or simple average values may not be the best indicator metrics for assessing or monitoring ecosystem health, however. A more effective tactic may be to examine the statistical properties of certain measurements to determine whether the statistical properties are better metrics than are the mean values. In some cases, variance changes may prove more informative or more sensitive than changes in the mean.

In one example of how the statistical properties of a measurement are used in ecosystem health assessment, the size of unvegetated patches has been determined in desert rangeland ecosystems (deSoyza et al 1997). Their size, which was measured by length of patch intercepted by a line (a percentage of the total length of line), proved to be related to a number of ecosystem processes, such as wind and water erosion, productivity, biodiversity of the soil flora and fauna, water infiltration, and soil nutrient distribution. When this measurement was made on a series of benchmark sites that provided a gradient from completely degraded (unhealthy) to completely functioning (healthy), however, mean bare patch size did not vary with the gradient. The frequency distribution of bare patch size offered a somewhat better metric; when tested against the benchmark gradient, however, it did not provide an interpretable pattern. When the frequency data were log-transformed and used to calculate a skewness value, a pattern emerged with respect to the gradient. The divergence from a normal distribution was indicative of system degradation and could be quantified by the skewness statistic. An early manifestation of disturbance in these desert rangeland ecosystems is a small increase in size of bare patch but a larger increase in the number of large bare patches, which is expressed as skewness to the right. As the ecosystem degrades, larger bare patches become abundant and small bare patches appear only rarely, which skews the frequency to the left (Figure 12.1). An even better correlation between bare patch size and degradation gradient was obtained by multiplying the skewness statistic by the mean bare patch size to obtain a weighted skewness (Figure 12.2).

Indicator metrics need not only involve single measurements. Frequently, combined measurements or ratios emerge as the most sensitive indicators when tested against an unhealthy–healthy benchmark gradient. The selection

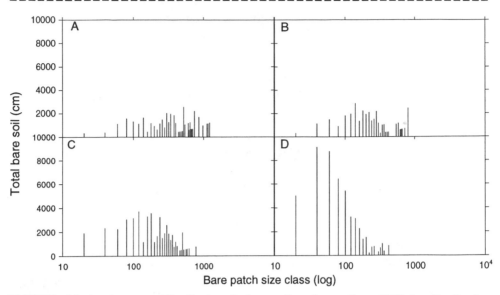

FIGURE 12.1. *Frequency distributions for bare soil patches at Camp Well. A = distribution at CWO, 50 m; B = distribution at CW1, 200 m; C = distribution at CW2, 400 m; D = distribution at CW3, 1050 m. The x-axis is log scale.*

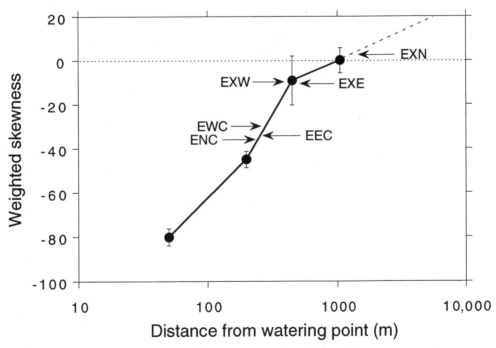

FIGURE 12.2. *Weighted skewness for log-transformed data (skewness statistics × mean bare soil patch size) plotted as the mean for the three disturbance gradients. Also shown (arrows) are the relative positions of bare patch weighted skewness for the ungrazed enclosures and the adjacent grazed pastures. Error bars indicate standard error. ENC = north grazed pasture; EEC = east grazed pasture; EWC = west grazed pasture; EXN = north enclosure; EXE = east enclosure; EXW = west enclosure.*

of metrics and their interpretation with respect to ecosystem function remains dependent upon the type of ecosystem under study. The most important step in the calibration or sensitivity testing of indicators is the selection of as wide a gradient of unhealthy to healthy benchmark ecosystems as possible. The selection of appropriate benchmark ecosystems may depend in part upon the kinds of stressors to which the ecosystem is exposed, but the availability of documentation of the system's degradation history is the most important criterion for selecting benchmark systems.

REFERENCES

deSoyza AG, Whitford WB, Herrick JE. Sensitivity testing of indicators of ecosystem health. Ecosystem Health 1997;3:44–53.

Fore LS, Karr JR. Assessing invertebrate responses to human activities: evaluating alternative approaches. J N Am Benthol Soc 1996;15:212–231.

National Research Council. Rangeland health. Washington, DC: National Academy Press, 1994.

Rapport DJ, Regier HA, Hutchinson TC. Ecosystem behavior under stress. Am Naturalist 1985;125:617–640.

Small AM, Adey WH, Lutz SM, Reese EG, Roberts DL. A macrophyte-based rapid biosurvey of stream water quality: restoration of the watershed node. Restoration Ecol 1996;4:124–145.

CHAPTER 13

Paleoecology: A Diagnostic Approach to Assessing Ecosystem Health

John P. Smol

A FREQUENT complaint of ecosystem managers is that direct measurements of long-term environmental trends are not available, making it difficult to manage ecosystems effectively. Important questions include: Is the system deteriorating? If so, how rapidly and to what extent? What were preimpact conditions like—that is, how can we set realistic targets for mitigation if we don't know what the state of the system before anthropogenic impact? What critical loads of pollutants and stressors can the system handle before it begins to show signs of deterioration? Without answers to these questions, it is difficult to assess ecosystem health, to determine whether the system is actually under stress, and to prescribe effective treatment (see Smol 1990, 1992, 1995 for reviews).

To effectively answer these questions, managers need long-term data. Environmental assessments, however, are usually performed after the fact (that is, after a problem is recognized), so long-term environmental monitoring and preimpact data are not directly available. Fortunately, paleoenvironmental techniques can be used to infer these missing data sets.

A large amount of important paleoenvironmental information is stored in a variety of natural records. These archives include ice cores (see, for example, Boutron et al 1991), peat bogs (see, for example, Gorham and Janssens 1992), museum specimens (see, for example, Monitoring and Assessment Research Centre 1985), and sediments (Smol and Glew 1992). This section focuses on the use of the latter.

PALEOLIMNOLOGY

Paleolimnology is the study of lake histories using the information stored in lake sediment profiles (Smol and Glew 1992). Under ideal conditions, sediments slowly but steadily accumulate at the bottom of lakes without disruptions. Certainly, in some cases, problems may arise (for example, excessive bioturbation), but they can usually be assessed. Over time, a depth/time profile is deposited. Incorporated in these sediments is a surprisingly large library of information on the conditions present in the lake (from autochthonous indicators) as well as environmental conditions that existed outside the lake (from allochthonous indicators). Paleoenvironmental data include the morphological and biogeochemical fossils of organisms (for example, diatom valves, invertebrate parts, fossil algal pigments, and pollen grains), chemical precipitates, soil particles, and so forth (see, for example, review articles compiled by Gray 1986 and Warner 1989). The paleolimnologist's job is to study this information and interpret it in a reliable way that is meaningful to environmental managers (Smol 1992). Considerable progress has been made in the field of paleolimnology over the last decade, and it is now widely recognized as a robust environmental management tool.

Paleolimnological Approach

The overall paleolimnological approach is summarized in Figure 13.1. Once the lake to be studied is chosen, a sediment core is removed, usually from near the center of the lake. In general, this part of the lake basin integrates indicators from across the lake, thereby archiving a more holistic record of past environmental change. Paleolimnological approaches have been subjected to a large amount of "quality assurance and quality control" considerations; if undertaken carefully and correctly, this method is both robust and reproducible.

A large number of different types of coring apparatus are available to retrieve sediment cores (several are illustrated in Smol and Glew 1992). The choice of equipment is largely dependent upon the type of lake under study and the temporal resolution required. In many lake systems, close to annual (and sometimes subannual) resolution is possible. The resolution also depends on the type of sectioning techniques and equipment used. Close-interval sectioning equipment and techniques (see, for example, Glew 1988) can provide lake managers with a high degree of resolution.

Once the core is retrieved and sectioned, the depth/time profile must be established. This determination requires "dating" a sufficient number of sediment layers to attain a reliable chronology. For most paleolimnological studies dealing with recent anthropogenic impacts, ^{210}Pb dating is most often used (Oldfield and Appleby 1984), as the half-life (22.26 years) of this natu-

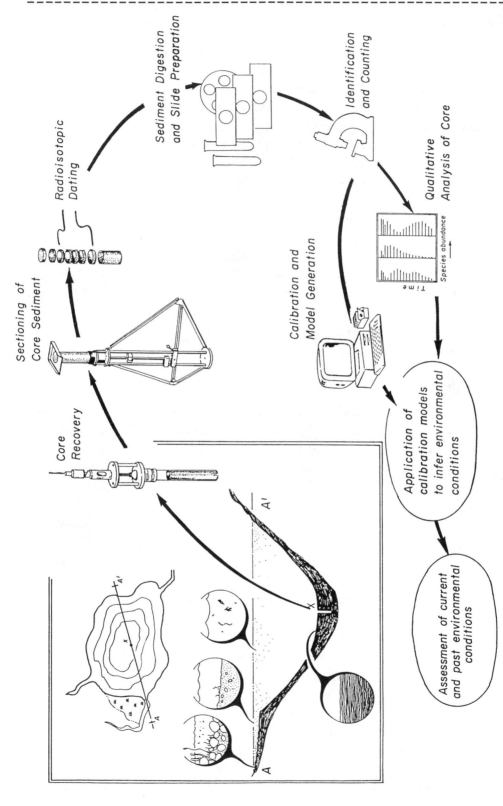

FIGURE 13.1. *Schematic diagram showing the major steps used in a typical paleolimnological analysis of a lake system (from Dixit et al 1992a).*

rally occurring isotope enables one to date, with reasonable certainty, approximately the last century or so of sediment accumulation.

Interpreting Information in Sediment Cores

The next step is to identify the paleoecological information preserved in the dated sediment cores. As noted earlier, lake sediments contain many different types of proxy data. For example, past changes in terrestrial vegetation can be inferred from the analyses of fossil pollen grains (the field of palynology). Magnetic measurements and other techniques can be used to estimate past erosion rates (see, for example, Dearing et al 1987). Chemical constituents, such as some metals and other contaminants (for example, PCBs and DDT), can be analyzed from the sedimentary profiles (Autenrieth et al 1991).

The fossils of organisms that once lived in the lake can be used to reconstruct past limnological characteristics (such as lakewater pH, eutrophication variables, salinity, and so forth). Virtually every organism that lived in the lake leaves some sort of morphological or chemical fossil in the sedimentary record. Foremost amongst these fossils are diatom algae, whose siliceous cell walls are species-specific and which are usually both abundant and well preserved. Because many diatom taxa have specific ecological optima and tolerances to important environmental variables (such as pH), past limnological conditions can be inferred from the percentages of taxa that lived at different times of the lake's history (Dixit et al 1992a, 1992b).

Quantitative Inference

Until recently, paleolimnologists simply used qualitative assessments of past environmental change. For example, one might know that the taxa dominating at a certain period in the lake's history tend to be found more commonly in nutrient-poor waters, acid waters, or cold waters. These qualitative assessments are rapidly being replaced by quantitative inferences of past environmental change, using transfer functions derived from surface sediment calibration sets and the application of a variety of multivariate statistical techniques. These approaches are statistically robust and ecologically sound (reviewed in Charles and Smol 1994; Birks 1995).

Briefly, a "calibration" or "training set" is constructed by choosing a suite of lakes that have been well studied and that span the gradient to be reconstructed. For example, while transfer functions were being developed to infer past lake water acidity levels in a suite of Adirondack (New York) lakes, a calibration set of 71 lakes was chosen that ranged in present-day pH from 4.4 to 7.8 (Dixit et al 1993). From each of these lakes, the surface sediments (for example, the top 1 cm of sediment accumulation, representing the past few years of sediment deposition) were removed using a gravity corer (Glew 1991). The indicators preserved in these sediments—in this case, diatom valves (Dixit et al 1993) and chrysophyte scales (Cumming et al 1992)—were analyzed (identified to the species level, counted, and expressed as relative frequencies) from the surface sediments of the calibration lakes. This operation provided one of the matrices (the 71 lakes and the percentages of the taxa found in the recent sediments of lakes) that will be required for the calibration (the species matrix). The second matrix is the present-day environmental data collected for the 71 calibration lakes (in this example, 21 limnological variables, such as lakewater pH, nutrient levels, and depth, were recorded).

The paleolimnologist then uses a variety of statistical techniques (see, for example, Charles and Smol 1994; Birks 1995) to combine these matrices, and construct robust inference equations that can infer lakewater characteristics (with known errors) from the diatoms or chrysophyte assemblages recorded in the sediments. Such approaches have been used in a variety of management issues (see, for example, Anderson and Battarbee 1994; Smol 1992, 1995). A case study showing how these approaches were used to study the effects of acidic precipitation on Adirondack lakes is presented in chapter 19.

REFERENCES

Anderson NJ, Battarbee RW. Aquatic community persistence and variability: a paleolimnological perspective. In: Giller PS, Hildrew AG, Raffelli D, eds. Aquatic ecology: scale, pattern and process. Oxford, UK: Blackwell Scientific Press, 1994:233–259.

Autenrieth R, Bonner J, Schreiber L. Aquatic sediments. Res J Water Poll Control Fed 1991;63:709–725.

Birks HJB. Quantitative palaeoenvironmental reconstructions. In: Maddy D, Brew JS, eds. Statistical modelling of Quaternary science data. Technical guide 5. Cambridge: Quaternary Research Association, 1995:161–254.

Boutron CF, Görlach U, Candelone JP, Bolshov MA, Delmas RJ. Decrease in anthropogenic lead, cadmium and zinc in Greenland snows since the late 1960s. Nature 1991;353:153–156.

Charles DF, Smol JP. Long-term chemical changes in lakes: quantitative inferences using biotic remains in the sediment record. In: Baker L, ed. Environmental chemistry of lakes and reservoirs. Washington, DC: American Chemical Society, 1994:3–31.

Cumming BF, Smol JP, Birks HJB. Scaled chrysophytes (Chrysophyceae and Synurophyceae) from Adirondack (N.Y., USA) drainage lakes and their relationship to measured environmental variables, with special reference to lakewater pH and labile monomeric aluminum. J Phycol 1992;28:162–178.

Dearing JA, Håkansson H, Liedberg-Jönsson B, Persson A, Skansjö S, Widholm D, El-Dahousy F. Lake sediments used to quantify the erosional response to land use changes in southern Sweden. Oikos 1987;50:60–78.

Dixit SS, Cumming BF, Kingston JC, Smol JP, Birks HJB, Uutala AJ, Charles DF, Camburn K. Diatom assemblages from Adirondack lakes (N.Y., USA) and the development of inference models for retrospective environmental assessment. J Paleolim 1993;8:27–47.

Dixit SS, Cumming BF, Smol JP, Kingston JC. Monitoring environmental changes in lakes using algal microfossils. In: McKenzie DH, Hyatt DE, MacDonald VJ, eds. Ecological indicators. Vol. 2. Amsterdam: Elsevier Applied Sciences, 1992a: 1135–1155.

Dixit SS, Smol JP, Kingston JC, Charles DF. Diatoms: powerful indicators of environmental change. Environ Sci Tech 1992b;26:22–33.

Glew JR. Miniature gravity corer for recovering short sediment cores. J Paleolim 1991;5:285–287.

———. A portable extruding device for close interval sectioning of unconsolidated core samples. J Paleolim 1988;1:235–239.

Gorham E, Janssens J. The paleorecord of geochemistry and hydrology in northern peatlands and its relation to global change. Suo 1992;43:117–126.

Gray J, ed. Paleolimnology. Amsterdam: Elsevier, 1988.

Monitoring and Assessment Research Centre. Historical monitoring. MARC report no. 31. London: University of London, 1985.

Oldfield F, Appleby PG. Empirical testing of 210Pb-dating models for lake sediments. In: Haworth EY, Lund JWG, eds. Lake sediments and environmental history. Minneapolis: University of Minnesota Press, 1984:93–124.

Smol JP. Paleolimnological approaches to the evaluation and monitoring of ecosystem health: providing a history for environmental damage and recovery. In: Rapport DJ, Gaudet CL, Calow P, eds. Evaluating and monitoring the health of large scale ecosystems. NATO ASI Series. Vol. 128. Berlin: Springer-Verlag, 1995: 301–318.

———. Paleolimnology: An important tool for effective ecosystem management. J Aquat Ecos Health 1992;1:49–58.

———. Paleolimnology—recent advances and future challenges. Mem Ist Ital Idrobiol 1990;47:253–276.

Smol JP, Glew JR. Paleolimnology. In: Nierenberg WA, ed. Encyclopedia of Earth system science. Vol. 3. San Diego: Academic Press, 1992:551–564.

Warner BG, ed. Methods in Quaternary ecology. Geoscience Canada, reprint series 5. St. John, Newfoundland: Geological Association of Canada, 1990.

CHAPTER 14

Ecological Risk Assessment: A Predictive Approach to Assessing Ecosystem Health

John Cairns, Jr.

Once harm has been done, even a fool understands it.
Homer, The Iliad, *Book XVII, l.32*

Nothing that is done, either on a personal basis or as a society, has zero risk. On a personal level, risks come from making judgments about how likely a consequence is, how unpleasant that consequence will be, and how easily the risk can be avoided. Certain risks may be judged acceptable; either they are unlikely to happen, easy to fix, or difficult to avoid. Using a car to move from one place to another is a substantial personal risk that is accepted each day because the alternatives are substantially less convenient and not proportionately less risky. Other personal risks are judged as excessive, unrewarding, or both. Relatively few people skydive, and those who undertake this activity enjoy the thrill enough to justify the higher risk. As a society, an analogous process occurs when we evaluate risks that affect communities, regions, or landscapes instead of individuals.

Webster's Third New International Dictionary defines "risk" as "the possibility of loss, injury, disadvantage, or destruction." By characterizing the risks inherent in alternative actions, societies can make better choices about what to do and when to take those steps. Formal risk assessment is an activity common to many disciplines, such as the engineering of buildings, bridges, or elevators, investment counseling, and medical treatment, among many others. Ecological risk assessment consists of the estimation of the probability of injury to ecological components resulting from human actions or natural disasters (see, for example, USEPA 1992; Norton et al 1992; Suter

1993; Calabrese and Baldwin 1993). This type of assessment evaluates the probability of damage to human health from similar stressors (NRC 1983, 1994). The most common ecological risk assessments characterize the probability of unacceptable damage resulting from the environmental release of chemicals or from habitat alterations.

Public debate about many environmental issues has become polarized, with seemingly contradictory information provided by various parties. Easterbrook (1995) claims that the environmental movement has been such a spectacular success that its mission has been accomplished and the planet no longer needs such activities. At the same time, *The World Scientists' Warning to Humanity* (Union of Concerned Citizens 1992), which was signed by more than 1600 senior scientists, warns that "A great change in our stewardship of the Earth and the life on it is required." This polarization of opinion points out a genuine need for a process that can provide rational and consistent information as the basis for comparing alternative actions. The process of risk assessment exists to meet this need.

While an ideal ecological risk assessment would be an objective, scientific pursuit, independent of ideology or social goals, in reality it is impossible to estimate the probability of an undesirable event without first reaching some consensus about what is "undesirable." Risk assessment cannot occur outside the context of societal goals, and values are built into the process with the adoption of each operational definition. A certain vigilance is required to prevent the intrusion of *non-consensual* values into the process. At its best, the science that contributes to risk assessments makes probabilistic statements about the nature of the world, but does not tell society what it ought to do. Nevertheless, strong moral and ethical components are clearly needed in deciding what steps should be taken (VanDeVeer and Pierce 1994). Do wild systems have intrinsic value? What should be preserved for future generations? How much should preservation of the common good intrude upon individual freedoms or property rights, and vice versa? How can environmental costs and benefits be distributed fairly? These questions are intimately entwined with risk management and must form part of the public policy debate. In addition, the number of issues in this category will decrease as scientific data continue to demonstrate the clear human dependence on robust, healthy natural systems.

STAGES IN AN ECOLOGICAL RISK ASSESSMENT

Any risk assessment must provide three kinds of information. Risk itself has two components—the relative likelihood of a consequence and its relative unpleasantness. The process of assessing risks adds an additional kind of information—the relative quality of the information used to characterize the risk. The uncertainty that results from predicting future consequences from imperfect information is an essential part of the assessment because risk managers must compare the relative risks, benefits, and uncertainties for various alternative actions (Figure 14.1). In some cases, the best choice between alternatives may be obvious. When persuasive evidence indicates that something very unpleasant will likely happen following some human action and alternatives without comparable risks are available, then the alternatives are clearly more attractive. In contrast, when a consequence of low magnitude with no beneficial alternatives has a low probability, then society is likely to judge the risk as being acceptable. In fact, a basic assumption underlying risk assessment as it is presently practiced is that a "loading capacity" exists for ecosystems, below which no observable deleterious effects will occur. Cairns (1977) has termed this ecosystem loading capacity *assimilative capacity*. Decisions are most difficult where weak evidence exists for either low probability/high consequence risk or high probability/low consequence risk.

An early consensus among representatives of industry, regulatory agencies, and academia on the steps needed to assess the risks of existing environmental releases of chemicals to aquatic ecosystems can be found in Cairns et al (1978). The following four steps were recommended:

1. Screening or range-finding tests

2. Predictive tests (that is, those intended to produce predictive models)

3. Confirmatory or validation tests

4. Monitoring to ensure that previously established quality control conditions have been met

At each successive stage, information about the expected environmental concentration is compared with the best estimates of a concentration having no harmful effects (Figure 14.2). To reduce uncertainty about environmental risk, multiple lines of evidence are gathered in a systematic and orderly fashion. When independent approaches result in the same conclusions, then confidence in that conclusion is high.

In addition, the uncertainty contributed by relying on a particular predictive model is characterized in step 3. Model error is measured by comparing the predictions of biological response with the observed response in the system of interest. Some models will give predictions that correspond more closely to observed biological responses than others. Obviously, these models will provide the most useful information for risk managers. Comparisons of error make it easy to choose between models. If the system of interest is very large or if the biological response occurs over a very long time span, however, it may be impossible to measure model error in the system. In the absence of empirical determinations of the size of predictive errors, some indication of

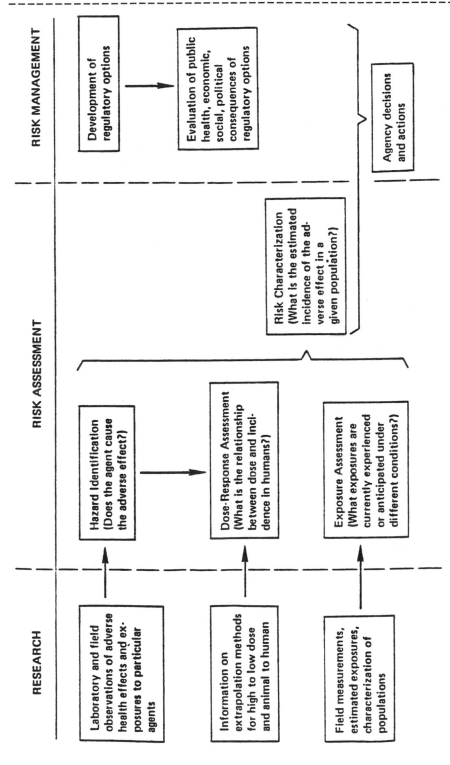

FIGURE 14.1. *Elements of risk assessment and risk management.*

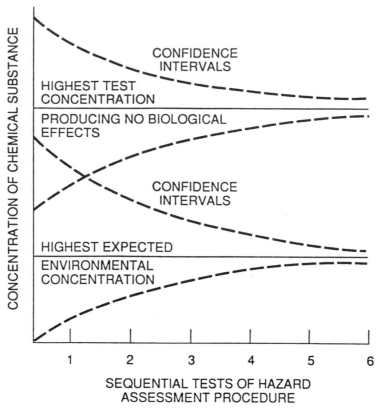

FIGURE 14.2. *Tiered testing and assessment in the hazard assessment paradigm.*

the relative credibility of alternative models can be obtained by comparing the predictive errors of their testable components (Aber et al 1993) or by peer review (Suter 1993).

Although some have suggested that validation is not necessary (Chapman 1995), this view appears to reflect continuing confusion about the purpose of validation (Cairns 1986). The outcome of a validation study is not a "yes" or "no" determination of the validity of a single test technique. Instead, it provides the information necessary to compare alternative models and to guide the choice between models or the construction of new models. Even models with large predictive errors can offer benefits if they are the best tools available. Obviously, validation remains essential when it provides a continual evaluation of whether the assessment tools are meeting the needs of risk managers and whether the best tools available are being chosen. As knowledge of ecosystem dynamics improves, then more inclusive, more informative models can be developed. And they should. A similar view of the importance of validation has been recently emphasized in the National Research Council's comprehensive review of risk assessment methods (NRC 1994).

General schemes for risk assessment have been described (Table 14.1). Remarkably similar schemes have been described for ecological effects by the U.S. Environmental Protection Agency (USEPA 1992) and for human health effects by the National Research Council (NRC 1983, 1994), especially considering the difference between the underlying tasks of predicting effects on one species and predicting effects on many millions of species. As would be expected, ecological risk assessments possess more possible end-points and show more possible variations in motivating values (Norton et al 1992).

A recent review of health effects risk assessments, based on more than a decade of use, identifies several concerns about risk assessment as it is currently practiced (NRC 1994). These concerns seem equally relevant for the evaluation of risk to ecological components:

- The lack of scientific data quantitatively relating chemical exposure to risks

- Some divergence of opinions within the scientific community about the merits of the underlying scientific evidence

- A lack of conformity among methods in studies used for risk characterizations

- The uncertainty of results produced by theoretical modeling used in the absence of measurements

- The use of conservative assumptions in the absence of applicable empirical information

While the report contains many recommendations for improving the process, the overall conclusions nevertheless confirm the utility of risk assessment. Even uncertain estimates of risk are useful.

TABLE 14.1   The components of a risk assessment (after NRC 1983, 1994, USEPA 1992)

Component	Purpose
Hazard identification or problem formulation	Asking the question "Is there a potential problem?"; operationally defining key concepts and describing alternative scenarios (that is, determining whether an agent can cause an adverse effect)
Effects assessment	Characterizing the relationship between the intensity of exposure to an agent and the magnitude of ecological effects
Exposure assessment	Measuring or estimating the intensity, frequency, and duration of exposures to an agent
Risk characterization	Combining effects and exposure information to estimate the incidence of adverse effects given the various exposure conditions

UNCERTAINTY IN RISK ASSESSMENTS

Both the most basic and the most detailed scientific information about a chemical are useful in making decisions, and each must be applied with a healthy respect for its limitations, which are expressed as uncertainties. This caution must not be construed as an excuse not to consider the ecological consequences of human actions until perfect data become available. Ecological risk assessments are needed at every stage of planning. The alternative to using imperfect information is not waiting for better information but rather proceeding with planning without consideration of ecological factors. Experience has shown that even imprecise predictions can still aid in the decision-making process. For example, weather reports and leading economic indicators both prove erroneous much of the time, but still have some value. Uncertainty is not unique to the ecological information in any given decision about risk. The economic and sociological data entering into the same decision also have uncertainties that are important to the decision.

Given practical limits on resources devoted to gathering and evaluating ecological risk, many assessments will be based on less than ideal databases. The best databases include laboratory toxicity tests and field studies along with chemistry to integrate these two sets of data. They also evaluate multiple end-points at several levels of biological hierarchy. The best databases will satisfy the environmental equivalent of Koch's postulates, convincingly demonstrating a cause-and-effect relationship between a contaminant and harm (Suter 1993). In actuality, however, many ecological risk assessments—especially those carried out to prioritize risks—are conducted on whatever data are available. No new data are collected.

At any point, using the best available model and conservative assumptions represents a reasonable approach to evaluating risk, as long as the uncertainties inherent in these estimates are communicated (NRC 1994). Reducing these uncertainties by addressing the many shortcomings in underlying data will inevitably increase the influence of ecological considerations in decisions. The problem lies in determining who will pay to decrease the uncertainties. It has been suggested that industries have no incentive to fund this activity, and regulatory agencies are inadequately equipped for this task. While risk assessment can be a useful tool, it will be irretrievably biased if equal effort is not expended to characterize risks and benefits.

UNCERTAINTY AND SCALE

Uncertainties that relate to problems affecting larger areas and longer time-frames cannot be addressed as easily as those related to local environmental problems. In many cases, one cannot make relevant observations in the system of interest or characterize the predictive error of models for large-scale environmental problems. As mentioned previously, some credibility for large-scale predictive models can be established by measuring the error of testable parts of the model or by undertaking peer review, but it is simply impossible to test empirically the accuracy of models for problems on a global scale. Dobzhansky (1945) stated:

> We like to believe that if we secure adequate data bearing on a scientific problem, then anybody with normal intelligence who takes the trouble to become acquainted with these data will necessarily arrive at the same conclusion regarding the problem in question. We like to speak of conclusions demonstrated, settled, proved and established. It appears, however, that no evidence is powerful enough to force acceptance of a conclusion that is emotionally distasteful.

When uncertainty is high, the distinction between ethical and scientific conclusions becomes clouded. A decision that seems empirically based to some scientists, who judge the uncertainty of the scientific data to be acceptable, may seem to be based on ethics to other scientists, who judge the uncertainty as being excessive. Relative tolerances of scientific uncertainty and risk are proper subjects of societal debate. The relationship between them must be noted, however—decisions made with an intolerance of uncertainty imply a higher tolerance for any risks associated with delaying action (Cairns 1994).

The controversy over burying high-level radioactive waste at Yucca Mountain, some 90 miles northwest of Las Vegas, clearly illustrates the difficulties of assessing risk over long periods of time (Batt 1995; Manning 1995; Schrader-Frechette 1993). No scientific basis exists for evaluating many of the crucial factors, such as the likelihood of human intrusion into the repository thousands of years after its establishment. Also, the question remains of whether the risk should be established for individuals—especially when one is uncertain about what an individual's general condition will be thousands of years later—or for populations.

ECOSYSTEM HEALTH AND HUMAN SELF-INTEREST

Uncertainty of risk assessments represents only one factor limiting the influence of ecological factors in decision making. An even larger factor may be a failure to communicate successfully the intimate links between the protection of ecosystem services and capital and human self-interest to risk managers, the general public, and their political representatives.

Ecosystem services are those functions of ecosystems that contribute to the quality of human life, yet occur with no or minimal human intervention. They include the capture of solar energy to provide food, energy, and building materials; the decomposition of waste; the regeneration of breathable air; the storage, purification, and redistribution of potable water; and many others. If ecosystems no longer performed these services, human engineering would have to replace them. When this task was attempted in the Biosphere 2, the cost was estimated at $9 million per person per year (Avise 1994). Clearly, robustly functioning ecosystems are extraordinarily valuable.

Better estimates of ecological values must be included in all decisions (see, for example, Costanza 1989). Commodity valuation of ecological components has proved to be biased—futures are discounted, natural resources are not depreciated, and economic costs of waste products are not assessed. Given the divergence of opinion among economists about valuation of those few natural resources currently given commodity value (summarized by Sagoff 1995), however, fundamental problems with attempts to integrate ecological and economic values into the same assessment schemes become readily apparent.

Some mainstream economists believe that nature sets no limits on economic growth (Solow 1973, 1974). Essentially, they believe that the Earth's carrying capacity for humans cannot be measured empirically because it is a function of the state of knowledge and technology rather than natural resources. Reserves of natural resources are envisioned as a function of technology; thus, the more advanced the technology, the more reserves become known and recoverable (Lee 1989; Gianturco 1994). In addition, the use of natural resources becomes more efficient in this scheme (Solow 1992). In contrast, Daly (1995) believes that current technological systems could not be recreated today because all of the easily accessible resources used in their creation, such as surface oil and metals, have been exhausted. While these arguments may be oversimplified here, it is worth noting that these kinds of issues strongly color the valuations in risk management.

Piasecki (1994) notes that the actual cost of environmentally driven expenditures in the United States now hovers around $4 billion annually—a substantial sum of money in this budget-conscious era, particularly in organizations that are "downsizing." In corporations, many environmental managers have traditionally assumed that the value of their strategic environmental programs was obvious and significant. What is obvious to environmental practitioners has not been obvious to the business side of the organization, however. Shelton (1994) describes "hitting the Green Wall" as the point at which the overall corporate organization refuses to move forward with its strategic environmental management program. As a result, the environmental initiative stops dead in its tracks. Negative or deferred decisions may follow from a lack of management support for strategic environmental management concepts and programs, environmental programs that lack focus and, finally, the inability to demonstrate attractive returns on further investments in the environmental programs. As examples of companies that have hit the Green Wall, Shelton (1994) lists Apple Computer, Warner-Lambert, ABB, and McDonald's. Shelton urges that environmental programs pay greater attention to factors such as the net present value of avoided fines, lost market share as a percentage of industrial fines, the decreased backlog of remediation time, leveraging of staff from engineering and operations to minimize the head count in the environmental group, and the addition of environmental issues that are important to business customers into the product development process. If risk assessment is to make serious inroads into the corporate world, clearly these and other issues must be given serious attention.

As mentioned earlier, ecological information's influence on decision making is limited by its ability to reassign the burden of proof in risk assessment. Both uncertainty and risk have associated costs, but different people often bear the burden of these expenses. One approach is to require that the individual or organization proposing a new course of action be mandated to provide evidence of the probable harm to human health and the environment. For example, the Toxic Substances Control Act (TSCA) states that "evidence shall be provided for all new chemicals or old chemicals used for a new purpose regarding the effects on human health and the environment." This statement was intended to include environmental concerns in planning by considering environmental benefits and risks. The present U.S. Congress seems to be shifting the burden of proof, however. Instead of holding responsible those proposing new courses of action for demonstrating a lack of harm, those few individuals with legal standing may be required to show the presence of harm.

This shift in the burden of proof is expected to free industry and individuals from undue government regulation. It will almost certainly bring short-term financial benefits for the industries that are relieved of this obligation. Certainly, not all regulatory enforcement has been carried out with good judgment. On the other hand, the rights of future generations to enjoy the same amenities enjoyed by present generations (that is, the "sustainable use of the planet" concept) and the right of society as a whole not to be deprived of ecosystem services upon which its well-being ultimately depends should be balanced against the benefits of freedom from regulatory constraints.

The problem of cumulative effects also arises if ecological risk assessments are conducted in a piecemeal fashion. The adverse ecological effects of filling in a single acre of wetland on the Mississippi River would almost certainly be impossible to detect. The filling in of thousands of acres of such wetlands on the Mississippi and its tributaries would have cumulative effects that would be relatively easy to measure, however. The same cumulation occurs when risk assessment proceeds chemical by chemical, with these chemicals being eventually discharged into the same system.

CONCLUSIONS

Ecological risk assessment is a useful tool. In its ideal state, it gathers, evaluates, and links ecological information relevant to societal choices. It enables environmental considerations to enter into planning at earlier stages and facilitates sound choices about human actions. The final form of this type of assessment does not yet exist, however. Obviously, we cannot retreat from the call for validation, as this concept means continual reevaluation of whether the assessment tools are meeting the needs of risk managers. Better, more informative, more efficient tools can be developed, and they should. Some ways in which the methods of ecological risk assessments will be improved include the following:

- Making the link between human interest and ecological systems more obvious to risk managers, the general public, and their political representatives

- Focusing on preserving health, rather than preventing damage

- Finding more equitable ways to compare ecological and economic components contributing to risk management decisions

- Treating the uncertainties from ecological and economic components equitably

The fact that ecological risk assessment methods can be improved does not justify waiting for better methods. Ecological consequences of actions must still be considered at every stage of planning. The uncertainty level for estimates of ecological risk is often reasonable, and uncertainty is clearly not unique to the ecological components. Economic and sociological information, for example, is also associated with substantial uncertainties. Nevertheless, the usefulness of risk assessment as a tool for making rational decisions will be completely compromised if the amount of effort expended to characterize benefits is disproportionate to the amount of effort expended to characterize risks. The crucial question remains: Who will pay to fund the research that will reduce uncertainties in ecological risk assessments? The usefulness of risk assessments in guiding societal decisions depends on the resolution of this question.

ACKNOWLEDGMENTS

I am indebted to Lisa Maddox for transcribing the dictated copy of the first draft and for making subsequent corrections, to Darla Donald for editorial assistance in preparing the manuscript for the publisher, and to B. R. Niederlehner for providing comments on early drafts.

REFERENCES

Aber JD, Driscoll C, Federer CA, Lathrop R, Lovett G, Melillo JM, Steudler P, Vogelmann J. A strategy for the regional analysis of the effects of physical and chemical climate change on biogeochemical cycles in northeastern (U.S.) forests. Ecol Model 1993;67:37–47.

Avise J C. The real message from Biosphere 2. Conserv Biol 1994;8:327–329.

Batt TD. Panel: risk standards, not radiation leak limits, should be used at Yucca. Las Vegas Rev J, Washington Bureau, Aug 2, 1995.

Cairns J Jr. Eco-societal restoration: re-examining human societies' relationship with natural systems. Abel Wolman Distinguished Lecture. Washington, DC: National Academy of Sciences, Dec 5, 1994.

———. What is meant by validation of predictions based on laboratory toxicity tests? Hydrobiologia 1986;137:271–278.

———. Aquatic ecosystem assimilative capacity. Fisheries 1977;2:5–7.

Cairns J Jr, Dickson KL, Maki A, eds. Estimating the hazard of chemical substances to aquatic life. STP 657. Philadelphia: American Society for Testing and Materials, 1978.

Calabrese EJ, Baldwin LA. Performing ecological risk assessments. Boca Raton, FL: Lewis Publishers, 1993.

Chapman PM. Do sediment toxicity tests require field validation? (letter) Environ Toxicol Chem 1995;14:1451–1453.

Costanza R. What is ecological economics? Ecol Econ 1989;1:1–7.

Daly HE. Reply to Mark Sagoff's "Carrying capacity and ecological economics." BioScience 1995;45:621–624.

Dobzhansky TA. Evolution, creation, and science. By Frank Lewis Marsh (a review). Am Nat 1945;79:73–75.

Easterbrook G. A moment on the Earth: the coming age of environmental optimism. New York: Viking Press, 1995.

Gianturco M. Seeing into the earth. Forbes 1994;153:120.

Lee TH. Advanced fossil fuel systems and beyond. In: Ausubel J, Sladovich H, eds. Technologic and environment. Washington, DC: National Academy Press, 1989: 114–136.

Manning M. Scientists admit Yucca risks top 10,000 years. Las Vegas Sun Aug 2, 1995.

National Research Council (NRC). Science and judgment in risk assessment. Report of the Committee on Risk Assessment of Hazardous Air Pollutants. Washington, DC: National Academy Press, 1994.

———. Risk assessment in the federal government. Report of the Committee on Institutional Means for Assessment of Risks to Public Health. Washington, DC: National Academy Press, 1983.

Norton S, Rodier DJ, Gentile JH, van der Schalie WH, Wood WP, Slimak MW. A framework for ecological risk assessment at the EPA. Environ Toxicol Chem 1992;11:1663–1672.

Piasecki B. Editorial. Corp Environ Strat 1994;2:2–3.

Sagoff M. Carrying capacity and ecological economics. BioScience 1995;45:610–620.

Schrader-Frechette KS. Burying uncertainty: risk and the case against geological disposal of nuclear waste. Berkeley, CA: University of California Press, 1993.

Shelton RD. Hitting the green wall: why corporate programs get stalled. Corp Environ Strat 1994;2:5–11.

Solow RM. An almost practical step toward sustainability. Washington, DC: Resources for the Future, 1992.

———. The economics of resources or the resources of economics. Am Econ Rev 1974;64:1–14.

———. Is the end of the world at hand? In: Weintraub A, Schwartz E, Aronson J, eds. The Economic Growth Controversy. New York: International Arts and Sciences Press, 1973:38–61.

Suter GW II. Ecological risk assessment. Chelsea, MI: Lewis Publishers, 1993.

Union of Concerned Citizens. World scientists' warning to humanity. Cambridge, MA. 1992.

United States Environmental Protection Agency (USEPA). A framework for ecological risk assessments. EPA 630/R-92-001. Washington, DC: Risk Assessment Forum, 1992.

VanDeVeer D, Pierce C. The environmental ethics and policy book. Belmont, CA: Wadsworth Publishing, 1994.

PART III

Ecosystem Health and Sustainability

This section develops the relationship between ecosystem health and sustainable use of the Earth's resources. It examines emerging concepts from the field of "ecological economics" that describe how we assess the overall health of both ecological and economic systems, the relationship of these concepts to sustainability, and the cultural context within which they offer relevance and usefulness. "Sustainability" refers to the longevity of a system; "health" indicates the overall quality of its performance, one important aspect and integrator of which is its longevity.

System health represents a desired management end-point, but it requires adaptive, ongoing definition and assessment. Models, assessments, and indices are most useful within the context of a broadly defined model of social decision making that includes three components: (1) stakeholder dialog and consensus building; (2) evolving, adaptive, qualitative, and quantitative assessment; and (3) integrated modeling. A concept of ecosystem health and sustainability that is pluralistic, comprehensive, multiscale, dynamic, adaptive, and hierarchical is required. In chapter 15, we propose that a key criterion for a healthy ecosystem is that it be sustainable—that is, that it have the ability to maintain its structure (organization) and function (vigor) over time in the face of external stress (resilience). In chapter 16, we examine methods to quantify these three ecosystem attributes, and illustrate how they can be incorporated into a quantitative assessment of ecosystem (and economic system) health.

An improved social decision-making system that can adequately mediate between conflicting interests to avoid social traps and reach long-term sustainability goals is also necessary. A two-tiered decision-making framework is proposed in chapter 17. In this framework, social consensus about long-term social goals is achieved first (tier 1), and then these goals are used to mediate conflicts on the individual, short-term scale. A three-step process for using models to build this consensus is also proposed.

CHAPTER 15

What Is Sustainability?

Robert Costanza
Michael Mageau
Bryan Norton
Bernard C. Patten

ALL complex systems are, by definition, composed of a number of interacting parts. In general, these components vary in their type, structure, and function within the whole system. Thus, a system's behavior cannot be summarized by simply adding up the behavior of its individual parts. Consider the differences between a simple physical system (for example, an ideal gas) and a complex biological system (for example, an organism). The temperature of the gas represents a simple aggregation of the kinetic energy of all individual molecules in the gas. The temperature, pressure, and volume of the gas are related by simple relationships with little or no uncertainty. An organism, however, consists of complex cells and organ systems. Its state cannot be surmised by merely adding together the states of the individual components, as these components are themselves complex and perform different, noncommensurable functions within the overall system. Indicators that might be useful for understanding heart function—pumping rate and blood pressure, for instance—are meaningless for assessing skin or teeth.

To understand and manage complex systems, however, we need some way of assessing the system's overall performance (its relative "health"). The U.S. Environmental Protection Agency (EPA) has recently begun to shift the stated goals of its monitoring and enforcement activities from protecting only "human health" to protecting overall "ecological health." Indeed, EPA's Science Advisory Board (SAB 1990) recently stated:

The EPA should attach as much importance to reducing ecological risk as it does to reducing human health risk. These very close linkages between human health and ecological health should be reflected in national environmental policy. When EPA compares the risks posed by different environmental problems in order to set priorities for Agency action, the risks posed to ecological systems must be an important part of the equation.

Although this statement gives the concept of ecological health importance as a primary EPA goal, it begs the question of what ecosystem health *is*, while tacitly defining it as analogous to human health. The dictionary definitions of health are as follows: (1) the condition of being sound in mind, body, and spirit; and (2) flourishing condition or well-being. Both are rather vague, however. To meet the mandate for effectively managing the environment, we must construct a more rigorous and operational definition of health that is applicable to all complex systems at all levels of scale, including organisms, ecosystems, and economic systems. Past explicit or implicit definitions of ecosystem health have described health as:

- Homeostasis

- The absence of disease

- Diversity or complexity

- Stability or resilience

- Vigor or scope for growth

- Balance between system components

All of these concepts represent pieces of the puzzle, but none is comprehensive enough to serve our purposes here. We will elaborate on the concept of ecosystem health as *a comprehensive, multiscale, dynamic, hierarchical measure of system resilience, organization, and vigor.* These concepts are embodied in the term "sustainability," which implies that the system can maintain its structure (organization) and function (vigor) over time in the face of external stress (resilience). A healthy system must also be defined in light of both its context (the larger system of which it is a part) and its components (the smaller systems that form it).

In its simplest terms, then, "health" is a measure of the overall performance of a complex system that is constructed from the behavior of its parts. Such measures of system health imply a *weighted* summation or a more complex operation over the component parts, where the weighting factors incorporate an assessment of the relative importance of each component to the functioning of the system as a whole. This assessment utilizes "values," which can range from subjective and qualitative to objective and quantitative

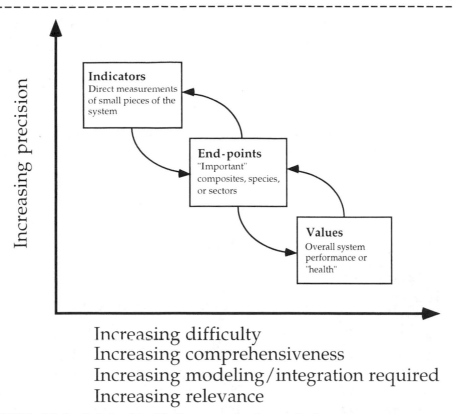

FIGURE 15.1. *Relationship of indicators, end-points, and values.*

as we gain more knowledge about the system under study. In the practice of human medicine, these weighting factors or values are contained in the body of knowledge and experience embodied in the medical practitioner.

Figure 15.1 shows the progression from directly measured "indicators" of a component's status, through "end-points" that are composites of these indicators, to health with the help of "values." Measures of health are inherently more difficult and more comprehensive, require more modeling and synthesis, and involve less precision, but are more relevant than the end-points and indicators from which they are built. It remains to determine which general approaches to developing these measures of health for ecosystems are most effective.

DEFINING AND PREDICTING SUSTAINABILITY

Much recent discussion has focused on how one "defines" sustainability, sustainable development, and related concepts (see, for example, Pezzey 1989; World Commission on Environment and Development 1987; Costanza 1991;

Pearce and Atkinson 1993). Critics argue that the concept is useless because it cannot be "adequately defined." Much of this discussion is misdirected, for two reasons:

- It casts the problem as definitional, when in fact it relates more to predicting what will last and achieving consensus on what we want to last.

- It fails to account for the range of interrelated time and space scales over which the concept must apply.

The basic idea of sustainability is quite straightforward: *a sustainable system is one that survives or persists.* Three complicating questions arise, however:

- What system, subsystems, or characteristics of systems persist?

- How long do they last?

- When do we assess whether the system, subsystem, or characteristic has persisted?

This chapter attempts to address these questions by acknowledging that sustainability can be assessed only after the fact, that one must consider systems and subsystems as hierarchically interconnected over a range of time and space scales, and that each of these systems and subsystems has a necessarily finite lifespan.

When Should Sustainability Be Assessed?

Biologically, sustainability means avoiding extinction, and living to survive and reproduce. Economically, it means avoiding major disruptions and collapses, hedging against instabilities and discontinuities. At its most basic level, sustainability always concerns temporality and, in particular, longevity.

Like "fitness" in evolutionary biology, however, determinations of sustainability can be made only after the fact. An organism alive right now is fit to the extent that its progeny survive and contribute to the gene pool of future generations. The assessment of fitness today must wait until tomorrow. Likewise, the assessment of sustainability must wait.

What may pass as definitions of sustainability are, therefore, often merely predictions of actions taken today that one hopes will lead to sustainability. For example, keeping harvest rates of a resource system below rates of natural renewal should, one could argue, lead to a sustainable extraction system— that is a prediction, however, and not a definition. In fact, this argument forms the foundation of MSY (maximum sustainable yield), which served as the basis for management of exploited wildlife and fisheries populations for many years (Roedel 1975). As learned in these fields, a system can be known to be sustainable only after enough time has passed to observe whether the prediction held true. Usually so much uncertainty surrounds the estimation of natural rates of renewal, and observation and regulation of harvest rates, that such a simple prediction, as Ludwig et al (1993) correctly observe, is always highly suspect, especially if it is erroneously put forth as a definition.

Similarly, the sustainability of any economic system can be observed only after the fact. Many elements of sustainability definitions are really predictions of system characteristics that one hopes lead to sustainability. Like all predictions, they are uncertain and should rightly be the subject of much elaboration, discussion, and disagreement.

What System Should Be Sustained?

Another question deals with the selection of the system, subsystem, or characteristics of these systems to be sustained. Is it a particular ecological system? A particular species or the total of all species (biodiversity)? The current economic system? A particular culture? A particular business or industry? In this situation, definitions of sustainability usually become transformed into a list of preferred characteristics, most often pertaining to the global socioeconomic system in the context of its ecological life-support system. For example, most definitions of sustainable development (WCED 1987; Pezzey 1989; Costanza 1991) contain the following elements:

■ A sustainable *scale* of the economy relative to its ecological life-support system

■ An equitable *distribution* of resources and opportunities between present and future generations

■ An efficient *allocation* of resources that adequately accounts for natural capital

It is important that we achieve consensus on these characteristics as desirable social goals. This process will be aided by separating the consensus-building process from the definition of sustainability and the other related questions described here. For example, in addition to their desirability as social goals, the three general characteristics listed above are predictors of the necessary characteristics that will allow the system to be sustained. They are thought to be both necessary conditions (predictors) for sustainability and desirable social goals. Choosing particular systems, subsystems, or specific characteristics as the objects to sustain (presumably forever) hides the hierarchical interactions between systems and subsystems over a range of scales in space and time. This development brings us to the final question, which relates to spatial and temporal extent.

How Long Should the System Be Sustained?

A third problem arises when we say that a system has achieved sustainability; we then have to specify the time span involved. Some would argue that sustainability means "maintenance forever." Nothing lasts forever, however—not even the universe as a whole. Sustainability thus cannot mean an infinite lifespan, or nothing would be sustainable. Instead, we argue that it means a lifespan that is consistent with the system's time and space scale. Figure 15.2 illustrates this relationship by plotting a hypothetical curve of system life expectancy on the y-axis versus time and space scale on the x-axis. We would expect a cell in an organism to have a relatively short lifespan, the organism to have a longer lifespan, a population of organisms to have an even longer lifespan, an economic system to have an even longer lifespan, and the planet as a whole to have a longer lifespan. No system (even the universe), however, is expected to have an infinite lifespan. A sustainable system in this context is thus one that attains its expected lifespan within the nested hierarchy of systems in which it is embedded. This nested hierarchy of systems and subsystems over a range of time and space scales represents the "metasystem."

Individual humans are sustainable in the metasystem if they achieve their "normal" maximum lifespan. At the population level, average life expectancy is often used as an indicator of health and well-being of the population. The population itself is expected to have a much longer lifespan than any individual, and would not be considered to be sustainable if it crashed prematurely,

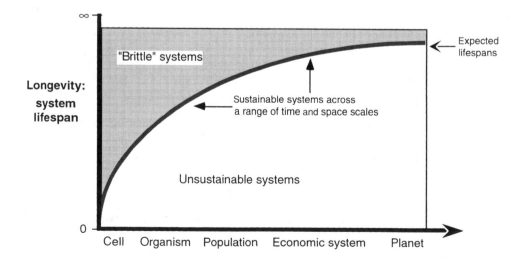

FIGURE 15.2. *Health and sustainability as scale (time and space)-dependent concepts (after Costanza and Patten 1995).*

even if all the individuals in the population were living out their full "sustainable" lifespans. Consider what would happen to sustainability if all individual humans actually lived forever. As we argue below, immortality of any subsystem is not sustainable because it cuts off evolutionary adaptation.

Because ecosystems experience succession as a result of changing climatic conditions and internal developmental changes, they also have a limited (albeit very long) lifespan. The key is differentiating between changes attributable to normal lifespan limits and changes that truncate the system's lifespan.

Under this definition, anything that reduces a system's natural longevity also reduces its sustainability. Thus, in humans factors like cancer, AIDS, accidents, and a host of other causes decrease sustainability. Human interventions in ecosystems frequently have deleterious consequences:

■ Cultural eutrophication of water bodies decreases the longevity of oligotrophic states, degrades water quality, and accelerates the arrival of dystrophic senescence.

■ Commercial lumbering ahead of sustainable schedules necessitates later sacrifice of nonextractive social values when "saw timber" forestry must be converted to a "pulp and chip" industry, and remnant old-growth forests must be harvested to "preserve jobs."

■ High-technology agriculture can be sustained only through exogenous subsidies of energy, fertilizers, pesticides, and gene pools.

These and other uses of natural capital interrupt normal function and both truncate natural longevity and hasten senescent decline. The exact consequences realized depend on the nature and state of the impacted system and the kind and severity of disturbance.

More formally, this aspect of sustainability can be described in terms of the system and its component parts' longevity (Patten and Costanza, in press):

■ A system is sustainable if and only if it persists in nominal behavioral states as long as or longer than its expected natural longevity or existence time.

■ Neither component- nor system-level sustainability, as assessed by the longevity criterion, confers sustainability on the other level.

One can now ask, Why should small-scale systems have shorter lifespans than larger-scale systems? Why don't cells or individual organisms last forever? We suggest that this situation is an outcome of the nested hierarchical interrelationship of systems across scales (the metasystem) that is necessary for evolutionary adaptation. Evolution cannot occur unless the component parts have limited longevity, allowing new alternatives to be selected. This longevity must increase hierarchically with scale (see Fig. 15.2). Larger systems can attain longer lifespans because their component parts have shorter lifespans, which allows the system to adapt to changing conditions. Without "death" at the lower scale, no evolutionary change can occur at the higher scale. Sustaining life requires death.

Systems with an improper balance of longevity across scales can become either "brittle" when their parts last too long and they cannot adapt fast enough (Holling 1992) or "unsustainable" when their parts do not last long enough and the higher-level system's longevity is cut unnecessarily short.

CONCLUSIONS

In this chapter, we have separated the problem of defining sustainability from three other, more basic questions, and have provided some tentative answers to those questions to motivate further discussion.

First, one must ask the question of what system, subsystem, or characteristics of systems persist. To answer this question, a nested hierarchy of systems over a range of time and space scales must be considered (the metasystem). Within the socioeconomic subsystem, a social consensus on desired characteristics that are consistent with the relationship of these subsystems with other subsystems in the hierarchy (notably ecosystems) must be developed. These characteristics also function as predictors of the kind of system that will actually be sustainable.

Second, we asked how long the system must endure to be considered "sustainable." All systems are of limited longevity, so sustainability cannot mean "maintenance forever." To maintain a sustainably evolving metasystem, we hypothesize that a particular relationship between the longevity of component subsystems and their time and space scales (see Fig. 15.2) may be necessary.

Third, one must determine the point in time at which the system or subsystem can be said to have persisted. This type of assessment can be carried out only after the fact, so the emphasis shifts to methods that enable us to better predict what configurations will persist, and to policies and instruments that deal with the remaining uncertainty. Given the huge uncertainties involved at the scale of the socioeconomic system, it is particularly important to select policies that are precautionary (Costanza and Perrings 1990; Costanza and Cornwell 1992)—that is, that do not take unnecessary risks with sustainability and do not count on unrealistic technological fixes for their success.

REFERENCES

Costanza R, ed. Ecological economics: the science and management of sustainability. New York: Columbia University Press, 1991.

Costanza R, Cornwell L. The 4P approach to dealing with scientific uncertainty. Environment 1992;34:12–20, 42.

Costanza R, Patten BC. Defining and predicting sustainability. Ecological Economics 1995;15:193–196.

Costanza R, Perrings C. A flexible assurance bonding system for improved environmental management. Ecol Econ 1990;2:57–76.

Holling CS. Cross-scale morphology, geometry and dynamics of ecosystems. Ecol Monographs 1992;62:447–502.

Ludwig D, Hilborn R, Walters C. Uncertainty, resource exploitation, and conservation: lessons from history. Science 1993;260:17, 36.

Patten BC, Costanza R. A rigorous definition of sustainability. 1996 (In press)

Pearce DW, Atkinson GD. Capital theory and the measurement of sustainable development: an indicator of "weak" sustainability. Ecological Economics 1993; 8:103–108.

Pezzey J. Economic analysis of sustainable growth and sustainable development. Environmental Department Working Paper No. 15. Washington, DC: The World Bank, 1989.

Roedel PM. Optimum sustainable yield as a concept in fisheries management. Special Publication No. 9. Washington, DC: American Fisheries Society, 1975.

Science Advisory Board (SAB). Reducing risk: setting priorities and strategies for environmental protection. SAB-EC-90-021. Washington, DC: EPA, 1990.

World Commission on Environment and Development. Our common future. Oxford, England: Oxford University, 1987.

CHAPTER 16

Predictors of Ecosystem Health

Robert Costanza
Michael Mageau
Bryan Norton
Bernard C. Patten

How can we create a practical concept of system health that is applicable with equal facility to complex systems at all scales and that incorporates the ideas of sustainability outlined in chapter 15? It is clear that we seek a set of indicators that can serve as both measures of the overall performance of the system at a point in time and predictors of its ultimate sustainability. Such a set of measures should fulfill four criteria.

First, the indicators should offer some combined measure of system resilience, life expectancy, balance, organization (diversity), and vigor (metabolism). Second, the concept should provide a comprehensive description of the metasystem along with its component systems. Looking at only one part of the system implicitly assigns a weight of zero to the remaining parts. Third, the indicators will require the use of weighting factors to compare and aggregate different components in the system. They should weight components based on their links to the functional dependence of the system's sustainability, and the weights should vary as the system changes to account for "balance." Fourth, the definition should be hierarchical to account for the interdependence of various time and space scales.

Costanza and colleagues (1992) developed the following definition of ecosystem health that attempts to combine these criteria:

> An ecological system is healthy and free from "distress syndrome" if it is stable and sustainable, that is, if it is active and maintains its organization and autonomy over time and is resilient to stress.

240

Ecosystem health is thus closely linked to the idea of sustainability, which we defined in chapter 15 as a comprehensive, multiscale, dynamic measure of system resilience, organization, and vigor. This definition is applicable to all complex systems, from cells to ecosystems to economic systems (hence it is comprehensive and multiscale); it also allows for the fact that systems may be growing and developing as a result of both natural and cultural influences. According to this definition, a diseased or unhealthy system is one that is not sustainable and will not achieve its maximum lifespan. The time and space frame are obviously important considerations in this definition. "Distress syndrome" (Rapport et al 1985, 1992) refers to the irreversible processes of system breakdown that lead to the termination of the system before its normal lifespan. To be healthy and sustainable, a system must maintain its metabolic activity level as well as its internal structure and organization (a diversity of processes effectively linked to one another); it must also prove resilient to outside stresses over a time and space frame relevant to that system.

What do these criteria mean in practice? Table 16.1 describes the three main components of this proposed concept of system health— resilience, organization, and vigor—along with related concepts and measurements in

TABLE 16.1   Indices of vigor, organization, and resilience in various fields

Component of Health	Related Concepts	Related Measures	Field of Origin	Probable Method of Solution
Vigor	Function Productivity Throughput	GPP, NPP, GEP GNP Metabolism	Ecology Economics Biology	Measurement
Organization	Structure Biodiversity	Diversity index Average mutual information (Ulanowicz 1986) Predictability (Turner et al 1989)	Ecology	Network analysis
Resilience		Scope for growth (Bayne 1987) Population recovery time (Pimm 1984) Disturbance absorption capacity (Holling 1987)	Ecology	Simulation modeling
Combinations		Ascendancy (Ulanowicz 1986) Index of biotic integrity (Karr 1981)	Ecology	

various fields. The assessment should combine these three basic aspects of system performance and sustainability. To operationalize these concepts (especially organization and resilience) will require a heavy dose of systems analysis, synthesis, and modeling, combined with broad-based input from the full range of stakeholders involved in the management of ecosystems.

Later in this chapter, we elaborate on a system-level assessment of ecosystem health that is reasonably easy to measure and incorporates values in a general manner allowing for the possibility of reaching a consensus. More specifically, we identify three components of ecosystem health (vigor, organization, and resilience), describe the quantification of these components, illustrate how they can be incorporated into a quantitative assessment of ecosystem health that satisfies the criteria given above, examine some initial testing of the assessment, discuss the unique opportunities for future testing, and comment on the potential of this method of assessment in ecological systems.

The *vigor* of a system is simply a measure of its activity, metabolism, or primary productivity. Examples include net primary productivity (NPP) and organism metabolism in ecological systems, and gross national product (GNP) in economic systems. It has been hypothesized that a system's ability to recover from or use stress relates to its overall metabolism, energy flow (Odum 1971), or "scope for growth" (Bayne et al 1987); the latter factor is the difference between the energy required for system maintenance and the energy available to the system for all purposes. Each of these measures targets the system's capability to respond to generalized stress.

The *organization* of a system refers to the number and diversity of interactions between its components. Measures of organization are affected by both the diversity of species and the number of pathways of material exchange between each component. For example, a highly organized system is characterized by a high diversity of specialized components and their corresponding specialized exchange pathways. Organization decreases as the diversity of species and the specialization of exchange pathways diminish. For any given level of species diversity, organization can vary with the pattern of exchange pathways between those species. A system containing species that feed on only one or two specific prey items and are themselves prey for only one or two other species, for example, will have higher values of organization than a system containing the same number of generalist feeders with multiple pathways of exchange between them. Organization, therefore, extends traditional measures of diversity by considering the patterns of exchange between system components.

The *resilience* of a system refers to its ability to maintain its structure and pattern of behavior in the presence of stress (Holling 1986). In the context of this chapter, it may refer to the system's ability to maintain its vigor and organization in the presence of stress. A healthy system possesses adequate resilience to survive various small-scale perturbations. The concept of system resilience incorporates two main factors. The first and most commonly used aspect refers to the length of time it takes a system to recover from stress. A second aspect refers to the magnitude of stress from which the system can recover, or the system's specific thresholds for absorbing various stresses. A related point involves the alternative system states that arise when thresholds are crossed; these states may vary from total system collapse to a stable state that may actually have beneficial effects. The limits of ecosystem stability or resilience are currently the source of much debate. Holling (1986) argues that they range from the assumption of complete global stability, implicit in many of humanity's past efforts to manage ecosystems, to the idea that ecosystems are extremely fragile.

The three components of system health can be illustrated in a three-dimensional plot (Figure 16.1). The two-dimensional planes formed when each of the components is zero are labeled. The first plane describes systems characterized by various combinations of organization and resilience, but no vigor. Systems with little or no vigor (such as ice, rocks, and minerals) are "crystallized." The second plane describes systems characterized by various combinations of resilience and vigor, but no organization. Systems with little or no organization, such as nutrient-enriched lakes, streams, and ponds, or early successional ecosystems dominated exclusively by *r*-selected species, are

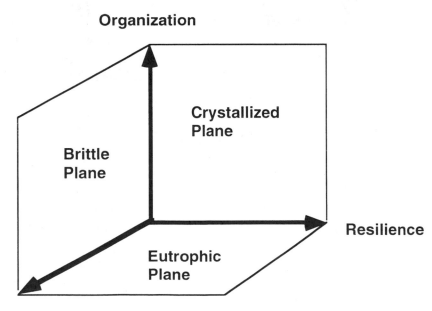

FIGURE 16.1. *A three-dimensional plot of system vigor, organization, and resilience. Each of the planes formed when one component is zero is labeled.*

considered "eutrophic." The third plane indicates systems that are characterized by various combinations of vigor and organization, but no resilience. Natural variation in external environments preserves resilience, thereby preventing systems from reaching the extreme of this plane. Nevertheless, certain highly managed systems, such as agriculture, aquaculture, and plantations, approach this plane and are "brittle."

Crystallized, eutrophic, and brittle systems are not healthy. Instead, a healthy system is characterized by some balance between vigor, organization, and resilience. We propose that a "healthy" system is one that can maintain an efficient diversity of components and exchange pathways (high organization) while maintaining some resilience as insurance against stress and substantial vigor to quickly recover or utilize stress in a positive manner.

MEASURING VIGOR

Vigor is the most straightforward of the three components to assess. In most systems, it can be measured directly and relatively easily by existing methods. These empirical measures quantify the magnitude of input (material or energy) available to an ecosystem (NPP) or the overall activity (measured in dollars per unit time) of an economic system.

Two potential quantitative estimates of vigor may be calculated using network analysis. Ulanowicz (1986) described the calculation of total systems throughput (TST) and net input (NI), which may be used as proxies for quantifying a system's vigor. Given a systems exchange network, Ulanowicz (1986) calculates TST by simply summing the magnitude of material transferred along the various individual exchange pathways in any unit of time (TST $= \cdot\cdot T_{ij}$); the NI to the system can be easily separated out of the TST. As investigators in several fields have long recognized, however, vigor alone is not an adequate measure of health.

MEASURING ORGANIZATION

It is more difficult to quantify organization than vigor because the former task requires measuring both the diversity and magnitude of system components and the exchange pathways between them. Diversity indices and multispecies indices fail to incorporate exchange pathways connecting system components.

One potential approach to the problem of measuring organization is network analysis, which involves the quantitative analysis of interconnections between components of a system (species) and their connections with the larger encapsulating system (their abiotic environments). Practical quantitative analysis of interconnections in complex systems began with the economist Wassily Leontief (1941), using what is known as input-output (I/O) analysis. Recently, these concepts have been applied to the study of interconnections in ecosystems (Hannon 1973, 1976, 1979, 1985a, 1985b, 1985c; Costanza and Neill 1984). Related ideas, found under the heading of compartmental analysis, were also developed (Barber et al 1979; Finn 1976; Funderlic and Heath 1971). Walter Isard (1972) was the first to take advantage of the similar methodology by attempting a combined ecological/economic system I/O analysis, and several others have since proposed ecological/economic mass-balance models (Daly 1968; Cumberland 1987). Ulanowicz (1986) has used information theory to develop a specialized suite of system-level, quantitative "network analysis" indices that may be used to calculate a system's organization. These indices can be used to develop more comprehensive, quantitative indicators of system organization that transcend the traditional diversity indices used in ecology by estimating not only the number of different species in a system, but also—and more importantly—their organization.

Because Ulanowicz (1986) describes the quantification of system-level information indices in detail, we will provide only a brief summary here. First, we must estimate a matrix of material and energy exchanges between system components. Each cell in the matrix carries the label T_{ij}, designating a specific transfer from a particular component in row I to a particular component in column j. Estimating matrices of this type for ecosystems is a difficult task, but our ability to realize this goal is growing with the completion of field experiments directed at estimating trophic transfers (such as various tracer experiments and feeding patterns) and the emergence of improved simulation modeling. We plan to use the simulation models we have calibrated using data from MEERC mesocosm experiments to estimate these matrices of material and energy exchange. The system-level information indices can then be quantified using the conditional probabilities calculated from these matrices. One can gain valuable insight into ecosystem structure and function by studying the changes in these indices that accompany ecosystem perturbations.

Ulanowicz (1996) identifies mutualism or autocatalysis between system components, connected by cyclic flow, as a nonmechanistic, ecosystem phenomenon that provides evolution and ecological succession with a sense of direction. This natural process dictates the behavior of the system-level information indices. In autocatalysis, an increase in the activity of any component increases the activity of all other members in the cycle and ultimately itself, resulting in configurations that enhance growth via positive feedback. These autocatalytic configurations also exert selection pressure on their members. If a more efficient species enters the cycle, its influence on the cycle will be positively reinforced; if the species is less efficient, negative reinforcement will serve to decrease its role. In addition, as the activity of the autocatalytic cycle grows, the cycle adsorbs resources from its surroundings. Therefore, as ecosystems undergo the process of succession in the absence of stress, auto-catalysis increases the amount of material being transported throughout the system and the efficiency by which its members transfer material and energy. Finally, different members may come and go, but the fundamental structure of the autocatalytic cycle remains stable, keeping the loop independent of its constituents.

We believe that average mutual information—a series of system-level, network analysis indices developed by Ulanowicz (1986)—may be used as a measure of system organization. Ulanowicz (1986) argues that autocatalysis streamlines the topology of interconnections, favoring those transfers that more effectively engage in autocatalysis at the expense of those that do not, and thereby resulting in networks that tend to become dominated by a few intense flows. For example, as specialists replace generalists through ecological succession, each species or system component exchanges material along fewer pathways.

Ulanowicz (1996) has described how these effects can be quantified using a modified average mutual information equation. The statement $p(ai, bj)$ refers to the probability that a unit of medium leaves component I and enters component j (T_{ij}). Because T is the aggregate of all such system transfers, we can estimate $p(ai, bj)$ by T_{ij}/T. Similarly, $p(b_j)$—the probability that a quantum enters element j—will be estimated by $\cdot T_j/T$. Finally, the conditional probability $p(b_j I a_i)$ that a quantum enters j after leaving I is approximated by $T_{ij}/(\cdot T_i)$. Substituting these estimators into the equation for average mutual information yields an equation that quantifies the degree to which autocatalysis has organized or streamlined the system's flow structure:

$$I = \cdot T_{ij}/T * \log\left(T_{ij} * T / T_j * T_i\right)$$

Ulanowicz (1986) also develops two related concepts. First, he scales the average mutual information equation by the total system throughput (because autocatalysis tends to increase T) to yield a network property called system ascendancy (A).

$$A = T * I = \cdot T_{ij} * \log\left(T_{ij} * T / T_j * T_i\right)$$

In addition, average mutual information could be scaled by net input to yield a modified ascendancy value (A^*). Ulanowicz (1980) hypothesized that, in the absence of major perturbations, autonomous systems tend to evolve in a direction of increasing network ascendancy—first via an increase in total systems throughput, and then via increasing average mutual information as competition for limiting resources begins to streamline the network of system exchanges. The autocatalytic process tends to increase overall system throughput, efficiency, and organization, all of which results in increased system ascendancy values. Odum (1969) reached similar conclusions, arguing that more developed systems usually contain a larger number of elements that exchange more material and energy among themselves over less equivocal routes. In addition, Odum (1969) found more developed systems tended to internalize or recycle waste products more efficiently, thereby decreasing their losses to the external world and their dependence on imported resources.

In addition, Ulanowicz (1986) identifies a third information indices "system uncertainty" (H). This upper bound on the total uncertainty arises when we have no information regarding material exchange. Uncertainty also reflects the total complexity of the system,

$$H = \cdot\left(T_{ij}/T\right)^{*}\log\left(T_{ij}/T\right)$$

or the total number of potential pathways of material exchange between system components. As natural systems develop and autocatalysis streamlines the exchange network, I increases and approaches H as information replaces uncertainty. Ulanowicz scales H by total system throughput (T) to yield "development capacity" (C). Alternatively, H could be scaled by net input to yield a modified capacity (C^{*}). Therefore, with ecosystem development in the absence of perturbation, as I approaches H, A approaches C, or A^{*} approaches C^{*}.

MEASURING RESILIENCE

Measuring the resilience of a system is very difficult because it implies the ability to predict the dynamics of that system under stress. Predicting these ecosystem impacts requires computer simulation models that synthesize the best available understanding of the way these complex systems function dynamically (Costanza et al 1990).

Figure 16.2 illustrates the two components of resilience that can be estimated using simulation models. The recovery time (R_T) can be estimated simply by measuring the time it takes for a system to recover from a wide variety of stresses to some previous steady state. The maximum magnitude of stress (MS) from which a system can recover can be measured by progressively increasing simulated stress until the system reverts to some new steady state, and then documenting the magnitude of the stress that caused the shift. We propose that an overall measure of resilience can be obtained from the ratio of MS/R_T (see Fig. 16.2).

When calculating this measure of system resilience, the choice of indicators to be tracked over time becomes very important. The ordinate axis in Figure 16.2 indicates the candidates for this function. Although the popula-

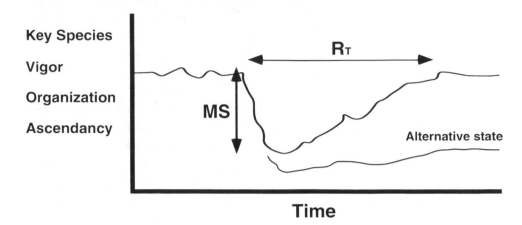

$$\text{Resilience} = MS / R_T$$

FIGURE 16.2. *The two components of resilience, and how they are integrated into a single quantitative measure. Candidates for tracking system performance through time are listed on the vertical axis. The lower line indicates the alternative state of a system that was unable to completely recover from stress.*

tion of a single species would be easiest to track, it would provide the least information about the entire system's response to stress. As discussed earlier, increasingly complicated measures such as those suggested for vigor, organization, and their combination (ascendancy) reveal more about the system's response, but at the expense of ease of measure and reliability. We plan to use MEERC experiments and parallel modeling to test these various indices and determine which is best for the purpose of measuring resilience.

Once we obtain a reliable measure of resilience, we can test the ability of two indicators to serve as proxies or quantitative measures of resilience. The first potential proxy is the ratio of photosynthesis per unit respiration (P/R). This ratio measures the excess vigor or energy available to the system. We have suggested that, as this ratio increases, a system gains more energy in reserve, which will allow it to recover from stress more easily and quickly. To test this hypothesis, we plan to study the correlation between the empirically measured value of resilience and the P/R ratio.

The second potential proxy—systems overhead (L), a third information index developed by Ulanowicz (1986)—is calculated as $C - A$ or $C^* - A^*$. Overhead, which quantifies the number of redundant or alternative pathways of material exchange, may be envisioned as a system's ability to absorb stress without dramatic loss of function. Ulanowicz (1986) suggests that higher values of system overhead tend to be associated with systems in earlier stages of development—before autocatalysis has eliminated alternative, redundant, less efficient pathways of material and energy transfer (low I value). Overhead values can also be high in systems containing large seed populations that have the potential to maintain system vigor and efficiency under different conditions (high H value). As with the P/R ratio, the correlation between systems overhead and our empirical measure of resilience must still be determined. If systems overhead can serve as a useful proxy for resilience, network ascendancy and overhead could potentially provide the ecologist with a method to quantify the change in ecosystem status resulting from anthropogenic and natural stress.

In conclusion, experiments and parallel modeling can be used to generate exchange networks representing each of the mesocosm spatial scales. We can then use these exchange networks to calculate Ulanowicz's (1986) system-level information indices (T, I, A, C, and L), and test their ability to serve as proxies for measuring Costanza's (1992) three main components (vigor, organization, and resilience) of system health. If the initial tests prove unsuccessful, we can test several other potential proxies for these three components in a similar manner. If the correlations are strong, however, these indices can be integrated into a potential assessment of ecosystem health. This measure can then be tested for its ability to track the "health" of each experimental ecosystem before, during, and after the nutrient and chemical additions.

REFERENCES

Barber M, Patten B, Finn J. Review and evaluation of I-O flow analysis for ecological applications. In: Matis J, Patten B, White G, eds. Compartmental analysis of ecosystem models. Statistical ecology, vol. 10. Bertonsville, MD: International Cooperative Publishing House, 1979.

Bayne BL, et al. The effects of stress and pollution on marine animals. New York: Praeger, 1987.

Costanza R. Toward an operational definition of health. In: Costanza R, Norton B, Haskell B, eds. Ecosystem health: new goals for environmental management. Washington, DC: Island Press, 1992.

Costanza R, Neill C. Energy intensities, interdependence, and value in ecological systems: a linear programming approach. J Theor Biol 1984;106:41–57.

Costanza R, Sklar FH, White ML. Modeling coastal landscape dynamics. Bioscience 1990;40:91–107.

Cumberland JH. Need economic development be hazardous to the health of the Chesapeake Bay? Marine Res Econ 1987;4:81–93.

Daly H. On economics as a life science. J Polit Econ 1968;76:392–406.

Finn J. The cycling index. J Theor Biol 1976;56:363–373.

Funderlic R, Heath M. Linear compartmental analysis of ecosystems. ORNL-IBP-71-4. Oak Ridge, TN: Oak Ridge National Laboratory, 1971.

Hannon B. Ecosystem flow analysis. Can Bull Fish Aquat Sci 1985a;213:97–118.

Hannon B. Conditioning the ecosystem. Math Biol 1985b;75:23–42.

———. Linear dynamic ecosystems. J Theor Biol 1985c;116:89–98.

———. Total energy costs in ecosystems. J Theor Biol 1979;80:271–293.

———. Marginal product pricing in the ecosystem. J Theor Biol 1976;56:256–267.

———. The structure of ecosystems. J Theor Biol 1973;41:535–546.

Holling CS. Simplifying the complex: The paradigms of ecological function and structure. Eur J Oper Res 1987;30:139–146.

———. The resilience of terrestrial ecosystems: local surprise and global change. In: Clark WC, Munn RE, eds. Sustainable development of the biosphere. Cambridge, UK: Cambridge University Press, 1986.

Isard W. Ecologic–economic analysis for regional development. New York: Free Press, 1972.

Karr JR. Assessment of biotic integrity using fish communities. Fisheries 1981;6: 21–27.

Leontief W. The structure of the American economy, 1919–1939. New York: Oxford University Press, 1941.

Odum EP. The strategy of ecosystem development. Science 1969;164:262–270.

Odum HT. Environment, power and society. New York: Wiley, 1971.

Pimm SL. The complexity and stability of ecosystems. Nature 1984;307:321–326.

Rapport DJ. Evaluating ecosystem health. J Aquat Ecosys Health 1992;1:15–24.

Rapport DJ, Regier HA, Hutchinson TC. Ecosystem behavior under stress. Am Naturalist 1985;125:617–640.

Turner MG, Costanza R, Sklar FH. Methods to compare spatial patterns for landscape modeling and analysis. Ecological Modeling 1989;48:1–18.

Ulanowicz RE. The propensities of evolving systems. In: Khalil EL, Boulding KE, eds. Social and natural complexity. London: Routledge, 1996.

———. Growth and development: ecosystems phenomenology. New York: Springer-Verlag, 1986.

———. A hypothesis on the development of natural communities. J Theor Biol 1980;85:223–245.

CHAPTER 17

Social Decision Making

Robert Costanza
Michael Mageau
Bryan Norton
Bernard C. Patten

To operationalize this concept of ecosystem health, a "two-tiered" decision structure must be implemented. Such a structure eliminates inconsistencies between short- and long-term goals as well as between local and global goals, a phenomenon described in the social psychology literature as a "social trap" (Platt 1973; Cross and Guyer 1980). First, general, democratic consensus must be reached on the broad, long-term goals of society. At this level, "individual sovereignty" is taken into account in that the rights and goals of all individuals in society are considered, albeit in the context of a shared dialog aimed at achieving broad consensus. Ecosystem health and sustainability are two such broad social goals.

Once these broad goals are democratically determined, they can be used to limit and direct activities at lower levels. For example, once general consensus on the goal of sustainability has been reached, with agreement by all major stakeholders in society, then society is justified in taking action to change local behaviors that are inconsistent with this goal. For example, society may attempt to change either people's preferences for driving automobiles or the price of doing so (or both) so as to modify behavior to become more consistent with the longer-term sustainability goals. In this way, we utilize foresight to modify short-term cultural evolutionary forces with a view toward achieving our shared long-term goals. If economics and other social sciences are to address problems of sustainability adequately, evolutionary models must be developed that incorporate preference formation and reformation as endogenous parts of the analysis; we must also develop mechanisms to modify short-term cultural evolutionary forces in the direction of long-term health and sustainability goals.

USING MODELS TO BUILD CONSENSUS

One tool that can be used to implement this two-tiered decision structure is computer modeling. We do not mean computer modeling in its current manifestation of "expert analysis," however, but rather used in a new role—as a tool for building a broad consensus not only across academic disciplines, but also between science and policy. More broadly, this process of stakeholder involvement is a key issue in achieving sustainability. Ethicist John Rawls (1971, 1987) has argued persuasively that policies that represent an "overlapping consensus" of the interest groups involved in a problem will be fair, thereby making them both effective and resilient. Thus, solutions to the problems of sustainability will be robust (resilient) and effective only if they appear fair and equitable (that is, are fairly distributed) to all of the interest groups involved, including future generations and other species. How can we identify and promote such solutions?

Integrated modeling of large systems—from individual companies to industries to entire economies—on spatial scales ranging from watersheds, to continent-scale systems, and ultimately to the global scale, requires input from a very broad range of people. The modeling process must be seen as involving not only technical aspects, but also sociological aspects; this approach will help build consensus about the way in which the system works and about which management options are most effective. This consensus must extend across the gulf separating the relevant academic disciplines as well as across the even broader gulf separating the science and policy communities from the public. Appropriately designed and appropriately used integrated ecological–economic modeling exercises can help to bridge these gaps.

The process of modeling can (and must) also serve this consensus-building function. It can help to build mutual understanding, solicit input from a broad range of stakeholder groups, and maintain a substantive dialog between members of these groups. In the process of adaptive management, integrated modeling and consensus building are essential components (Gunderson et al 1995).

A THREE-STEP MODELING PROCESS

We advocate the use of a three-step modeling process. The first stage involves the development of a high-generality, low-resolution *scoping and consensus building* model that includes broad representation of all stakeholder groups affected by the problem. Easy-to-use, icon-based simulation software makes this process feasible. This first-stage scoping model can be used to answer preliminary questions about the dynamics of the system, especially its main areas of sensitivity and uncertainty, thereby guiding the research agenda in the second stage.

The second-stage *research* models consist of more detailed and realistic attempts to replicate the dynamics of the particular system. This stage involves collecting large amounts of historical data for calibration and testing; a detailed analysis of the areas of uncertainty in the model is also carried out. Resolution is medium to high, depending on the results of the scoping model.

The third stage aims to develop models that can be used to answer particular *management* questions. These models must be based on the previous two stages, and can simply represent the exercising of the research models to produce future scenarios. Alternatively, they can be a further elaboration of the research models that allow them to be applied to management questions. In general, these models will offer medium to high resolution.

Each of these stages in the overall modeling process produces useful end-results, but the process is most useful and effective if followed in the order described. All too often, we jump to the research or management stage of the process without first building adequate consensus about the nature of the problem and without seeking the input of the appropriate stakeholder groups. In the following sections, we discuss these three stages in more detail before describing some case studies of how they have been used.

Scoping and Consensus-Building Models

The potential to use modeling as a way to build consensus has greatly expanded in recent years, thanks to the advent of new, easier-to-use computers and modeling software. When working with graphic, icon-based modeling software packages such as STELLA II, it is possible for a group of relative modeling novices to construct relatively complex models, with a few people competent in modeling acting as facilitators.

Using STELLA II and projecting the computer screen onto the wall, the process of model construction can be made transparent to a group of diverse stakeholders. Participants can then both follow the model construction process and contribute their knowledge to the process. After the basic model structure is developed, the program requires more detailed decisions about the functional connections between variables. This process is also transparent to the group, through the use of well-designed dialog boxes and the potential for both graphic and algebraic input. Once preliminary versions of the model have been constructed, it can be run to develop understanding of its dynamics and sensitivity, to compare its behavior with data for the system, and to decide where to put additional effort into improving the model. This phase can be described as an initial "scoping" step that facilitates broad-based input and consensus.

Research Models

After the initial consensus-building model development stage, which focuses on generalism, it may be appropriate and desirable to move to a more realistic or precise modeling stage. This phase, which could incorporate the input of more traditional "experts," is more concerned with analyzing the details of the historical development of a particular system with an eye toward developing specific scenarios or policy options in the next stage. It is still critical to maintain stakeholder involvement and interaction in this stage, with regular workshops and meetings being held to discuss model progress and results.

While integrated models aimed at realism and precision are large, complex, and loaded with uncertainties of various kinds (Costanza et al 1990; Groffman and Likens 1994; Bockstael et al 1995), our ability to understand, communicate, and deal with these uncertainties is improving rapidly. While increasing the resolution and complexity of models increases how much we can say about a system, however, it also limits how accurately we can say it. Model predictability tends to decline with increasing resolution because of compounding uncertainties (Costanza and Maxwell 1993). The desired models should optimize their effectiveness (Costanza and Sklar 1985) by choosing an intermediate resolution where the product of predictability and resolution (effectiveness) is maximized.

Management Models

If it is to be effective, the modeling process must be placed within the larger framework of adaptive management (Holling 1978). We need to view the implementation of policy prescriptions in a more adaptive way that acknowledges the uncertainty embedded in the models and allows participation by the various stakeholder groups. Adaptive management views regional development policy and management as experiments, where interventions at several scales are made to achieve understanding and to identify and test policy options (Holling 1978; Walters 1986; Lee 1993; Gunderson et al 1995). Thus, models and the policies based on them are taken not as the ultimate answers, but rather as guiding an adaptive experimentation process with the regional system. Greater emphasis is placed on monitoring and feedback to check and improve the model, rather than on using the model to obfuscate and defend a policy that does not correspond to reality. Continuing stakeholder involvement is essential in adaptive management.

This third stage of management models focuses on producing scenarios and management options in this context of adaptive feedback and monitoring, based on the earlier scoping and research models.

TOWARD GLOBAL ECOSYSTEM HEALTH AND SUSTAINABILITY: THE IMPORTANCE OF ENVISIONING

A broad, overlapping consensus is forming around the goals of health and sustainability, including its ecological, social, and economic aspects. Movement toward this goal is being impeded not so much by lack of knowledge or even lack of "political will," but rather by a lack of a coherent, relatively detailed, shared vision of a sustainable society. Developing this shared vision is an essential prerequisite for generating any movement toward it. Although the default vision of continued, unlimited growth in material consumption is inherently unsustainable, we cannot abandon this vision until a credible and desirable alternative becomes available. The process of collaboratively developing this shared vision can also help to mediate many short-term conflicts that will otherwise remain unresolved.

An impressive amount of success has already been realized by using envisioning and "future searches" in organizations and communities around the world (Weisbord 1992; Weisbord and Janoff 1995). This experience has shown that it is quite possible for disparate (even adversarial) groups to collaborate on envisioning a desirable future, given the right forum. The process has worked well in hundreds of cases, ranging from the level of individual firms and communities up to the size of large cities. The challenge is to scale it up to entire states, nations, and the world.

Meadows (1996) discusses why the processes of envisioning and goal setting are so important (at all levels of problem solving), why envisioning and goal setting remain underdeveloped in our society, and how we can begin to train people in the skill of envisioning and construct shared visions of a sustainable society. She tells the personal story of her own discovery of that skill and her attempts to use the shared envisioning process in problem solving. From this experience, several general principles emerged.

First, to envision effectively, it is necessary to focus on what one really wants—not what one will settle for. For example, the lists below show the kinds of things people really want, compared with the kinds of things they often find merely acceptable.

Really Want:	*Settle for*:
Self-esteem	Fancy car
Serenity	Drugs
Health	Medicine
Human happiness	GNP
Permanent prosperity	Unsustainable growth

Second, a vision should be judged by the clarity of its values, not the clarity of its implementation path. Holding to the vision while remaining flexible about the path is often the only way to find the best course of action. Third, responsible vision must acknowledge, but not become crushed by, the physical constraints of the real world. Fourth, visions must be shared because only shared visions can be responsible. Fifth, vision must be both flexible and continually evolving.

Probably the most challenging task facing humanity today is the creation of a shared vision of a sustainable and desirable society—one that can provide permanent prosperity within the biophysical constraints of the real world in a way that is fair and equitable to all of humanity, to other species, and to future generations. This vision does not now exist, although the seeds have been planted. We all have our own private visions of the world we really want; now we need to overcome our fears and skepticism and begin to share and build on these visions until we achieve the world we want.

Throughout Part 3 of this book, we have sketched out the general characteristics of this world—it is ecologically healthy and sustainable, fair, efficient, and secure. Our next task is to fill in the details so as to make it tangible enough to motivate people across the spectrum to work toward achieving it. The time to start is now.

Nagpal and Foltz (1995) have begun this task by commissioning a range of individual visions of a sustainable world from around the world. They laid out the following challenge for each of their "envisionaries":

> Individuals were asked not to try to predict what lies ahead, but rather to imagine a positive future for their respective region, defined in any way they chose—village, group of villages, nation, group of nations, or continent. We asked only that people remain within the bounds of plausibility, and set no other restrictive guidelines.

The results were quite revealing. While these independent visions were difficult to generalize, they shared at least one important point. The "default," Western vision of continued material growth was not what people envisioned as part of their "positive future." Instead, they indicated a desire for a future with "enough" material consumption, but with the focus shifted toward maintaining high-quality communities and environments, education, culturally rewarding full employment, and peace.

Much more work is necessary to implement living democracy and, within that system, to create a truly shared vision of a healthy and sustainable future. This ongoing work needs to engage all members of society in a substantive dialog about the future they desire and the policies and instruments necessary to bring it to fruition.

REFERENCES

Bockstael N, Costanza R, Strand I, Boynton W, Bell K, Wainger L. Ecological economic modeling and valuation of ecosystems. Ecol Econ 1995;14:143–159.

Costanza R, Maxwell T. Resolution and predictability: an approach to the scaling problem. Landscape Ecol 1993;9:47–57.

Costanza R, Sklar FH. Articulation, accuracy, and effectiveness of mathematical models: a review of freshwater wetland applications. Ecol Modeling 1985; 27:45–68.

Costanza R, Sklar FH, White ML. Modeling coastal landscape dynamics. Bioscience 1990;40:91–107.

Cross JG, Guyer MJ. Social traps. Ann Arbor, MI: University of Michigan, 1980.

Groffman PM, Likens GE. Integrated regional models: interactions between humans and their environment. New York: Chapman and Hall, 1994.

Gunderson LH, Holling CS, Light S. Barriers and bridges to the renewal of ecosystems and institutions. New York: Columbia University, 1995.

Holling CS, ed. Adaptive environmental assessment and management. London: Wiley, 1978.

Lee K. Compass and the gyroscope. Washington, DC: Island, 1993.

Meadows DH. Envisioning a sustainable world. In: Costanza R, Segura O, Martinez-Alier J, eds. Getting down to Earth: practical applications of ecological economics. Washington, DC: Island Press, 1996.

Nagpal T, Foltz C. Choosing our future: visions of a sustainable world. Washington, DC: World Resources Institute, 1995.

Platt J. Social traps. American Psychologist 1973;28:642–651.

Rawls J. The idea of an overlapping consensus. Oxford J Legal Stud 1987;7:1–25.

———. A theory of justice. Oxford, UK: Oxford University Press, 1971.

Walters CJ. Adaptive management of renewable resources. New York: McGraw Hill, 1986.

Weisbord M, ed. Discovering common ground. San Francisco: Berrett-Koehler, 1992.

Weisbord M, Janoff S. Future search: an action guide to finding common ground in organizations and communities. San Francisco: Berrett-Koehler, 1995.

PART IV

Case Studies

Ecosystem health is a major organizing paradigm for protecting and sustaining the quality of the environment and our own well-being. Though concepts and theories related to an ecosystem health approach are rapidly emerging, practical examples of the utility and results of this approach are critically important in understanding its application to real-life issues. In this final section, we illustrate the approach using actual examples representing a range of ecosystems, issues, and methodologies.

It is important to note that the case studies vary broadly in approach. On the surface, this may seem to point to a lack of rigor in scientific method, but in fact, it illustrates a key point. Ecosystem health is not meant to be prescriptive in terms of method; the approach is as much a way of thinking as a precise way of doing things. It opens the door to a variety of innovative methods and techniques aimed at the integrated, holistic assessment and management of the environment and our role in it. An ecosystem health approach seeks to address not just the physical and biological dimensions of a system, but all of the interrelated factors contributing to ecosystem health (or lack thereof). The following cases demonstrate a cross-section of approaches that have put these principles into practice.

CHAPTER 18

The Chesapeake Bay and Its Watershed: A Model for Sustainable Ecosystem Management?

Robert Costanza

Jack Greer

T HE Chesapeake Bay, the largest estuary in North America, has been the subject of more scientific study and political wrangling than any other body of coastal water in the world. Although what happens in the Bay is largely a function of activities in the drainage basin, the focus of most studies has traditionally been rather narrow. We are only now beginning to develop a comprehensive picture of the Chesapeake Bay and its connections to its watershed, not only in ecological terms, but also in demographic, cultural, political, and economic terms.

In this chapter, we first develop a historical and spatial perspective of human activities in the Chesapeake Bay watershed. Fundamental in gaining this perspective is the conceptualization of the watershed *system* as the combination of the drainage basin and the Bay itself. We have assembled and mapped past and present human activities in the Chesapeake watershed to gain this perspective.

The Chesapeake watershed is an ecological system whose beauty and productivity have led to high human population growth rates. This growth has both directly and indirectly caused its infirmity, including declining fisheries, receding wetlands, vanishing seagrasses, and a devastated oyster industry. These trends have also produced a decline in the quality of human life in the area. Traffic congestion, disappearing natural and agricultural areas, swelling landfills, and overtaxed water treatment facilities are but a few of the effects. This chapter addresses the larger-scale terrestrial trends within the entire watershed that have produced these effects, the barriers to effective environmental management capable of reversing these effects, and the bridges necessary to overcome these barriers.

In many ways, efforts to manage the Chesapeake can be viewed as a "best case scenario" for ecosystem management. The situation has been characterized by relatively early and widespread recognition of the Bay's decline, significant scientific analysis about the causes and solutions, and a broad consensus for remedial action. These aspects of the Bay's management may serve as models for other areas.

Nevertheless, significant barriers to implementation of the restoration strategy have appeared. These barriers include the absence of realistic, comprehensive, long-term goals for managing complex ecological and economic systems, the lack of systems for translating these goals into local incentives that can adequately change and guide human behavior, and the built-in, but misplaced arrogance about our level of understanding of the system.

The bridges that could overcome these barriers include developing and applying:

■ More realistic models of the functioning of ecosystems

■ More realistic acknowledgment of the severe limits of our knowledge

■ Incentive systems that can translate this understanding (and especially the limits and uncertainty of this understanding) into local decision making

A SUMMARY OF THE PROBLEM

Because of the special characteristics of the Chesapeake Bay and its watershed, it is simultaneously productive, unpredictable, resilient, sensitive to stress, and hard to understand and manage with traditional methods.

It is productive because it is a broad, shallow, large estuary where freshwater and nutrients running off a large watershed interact with seawater in very complex and unpredictable patterns. Its shallowness means nutrients are recycled easily and constantly, maintaining its high productivity. Because the Chesapeake Bay watershed is so large and already had plenty of nutrients in its pre-European settlement state, it attracted many settlers; it was also sensitive to the further addition of nutrients that such settlement brought, however. Because it has remained a completely open access resource for most of the last 300 years, the Bay's tremendous productivity has been tragically overused and abused. Its complexity and unpredictability have kept the ecosystem relatively resilient. Indeed, if we can develop ways to intelligently manage its use and reduce the stresses we put on it, it can recover and remain resilient and productive for a long time to come.

We can accomplish this goal only by first understanding the history and current status of the Bay and its watershed as an integrated system, involving the interaction of humans and a unique ecological life-support system. We must also better understand how human institutions affect the situation, how they typically fail when dealing with resources like the Chesapeake Bay, and how to fix them or replace them with better alternatives.

THE CHESAPEAKE BAY AND ITS WATERSHED

The early European settlers who arrived on the shores of the Chesapeake Bay came looking for a better life. They settled in Jamestown and Williamsburg in Virginia, and in St. Mary's and Londontowne in Maryland. For all of these immigrants, the Chesapeake Bay represented a new kind of "promised land."

Settlers came to begin a new enterprise, but ultimately founded a new country. Even the native Americans who greeted the settlers had come from somewhere else, crossing a land bridge—where Alaska once touched Asia—many centuries before. This first group of immigrants, however, had preserved a lifestyle that had not taken a great toll on the natural environment or the countryside. They took oysters, but did not destroy the oyster beds. They used logs for canoes, but did not clear-cut the forests. The settlers who came from Europe, Africa, and Asia would have a much greater impact on the forests, streams, and rivers as they created their new country.

That new nation had its beginnings in land grants from the kings of Europe, who were the great powers of their day—for example, the kings of England issued land grants to men such as Lord Baltimore. The country would flourish through commerce as well, with the backing of businesses such as the Virginia Company (the equivalent of today's large international corporations, such as Exxon or Toyota). The life the settlers made for themselves generated profits not so much from their European backing, however, as from the riches of the land they had found.

While explorers farther south were exploiting gold from the ancient South American cultures, the colonists of the Chesapeake Bay region found another kind of wealth: teeming schools of fish, hardwood forests, fertile soil, and wild game. The riches of the Chesapeake were the riches of nature. As the centuries passed, the Bay region, with its large farms and plantations, became known as "the Land of Pleasant Living."

Of course, living in this colonial milieu was not pleasant for everyone all the time. Indentured servants labored for years to gain their freedom. Slaves labored for a lifetime, and many never became free. War came to the Chesapeake region more than once: native Americans fought the settlers;

slaves revolted against their masters; and major wars were fought between the settlers and their European forebears, and then finally among the settlers themselves.

As the nation suffered its pangs of growth and transition, the Chesapeake watershed served repeatedly as the scene for bloodshed. The Revolutionary War's final campaign took place in the Bay region with the surrender of the British general Cornwallis at Yorktown on the shores of the York River. The War of 1812 saw the British return to the Bay, with warships guarding its entrance and raiding parties making their way up Bay rivers to attack Washington and Baltimore. The bloodiest battle of this nation's bloodiest war was fought in the Chesapeake watershed along the edges of Antietam Creek.

Throughout this period, the Chesapeake Bay continued to provide for the humans who lived in the region. People took its productivity for granted, like the air or the rain. Nevertheless, changes were taking place in the Chesapeake and the impacts of settlement began to be felt. The Chesapeake changed slowly at first, but it would never be the same.

History of the Bay and Its Watershed

While the humans were writing their own history of hardship and prosperity, and war and peace, the Chesapeake Bay was evolving according to a history of its own. Its history is first and foremost geologic. Without the rhythms of the earth—the shifting of the continents, the spread and retreat of glaciers—the Bay would not have existed. Actually, as J. R. Schubel has pointed out, we should remember that the Chesapeake Bay of today is merely one of many bays that have come and gone over long geological epochs, growing and shrinking with the rise and fall of the sea (Schubel 1981).

In its current form, the Chesapeake Bay is a mosaic of estuaries that constitute the largest single estuary in the United States. It was formed from Atlantic waters eroding and drowning the mouths of the rivers that feed it. The Bay is 193 miles long and 3 to 25 miles wide; with its tidal tributaries, it covers an area of 4400 square miles and has 8100 miles of shoreline (Tippie 1984). The drainage basins that feed it contain 64,000 square miles in six states, and three major cities lie on the banks of its tidal system (Figure 18.1). The Chesapeake watershed is an ecosystem of enormous historical, political, and economic significance. As a result, it has been among the most well-studied and heavily managed large ecosystems in the world. It is both a model and an experiment from which we can learn.

The Chesapeake Bay we see today began nearly 10,000 years ago, when the glaciers that had advanced as far south as present-day New York City finally began to recede as the world, in a planetary seasonal rhythm that we still do not fully understand, began to warm. The changes in temperature as

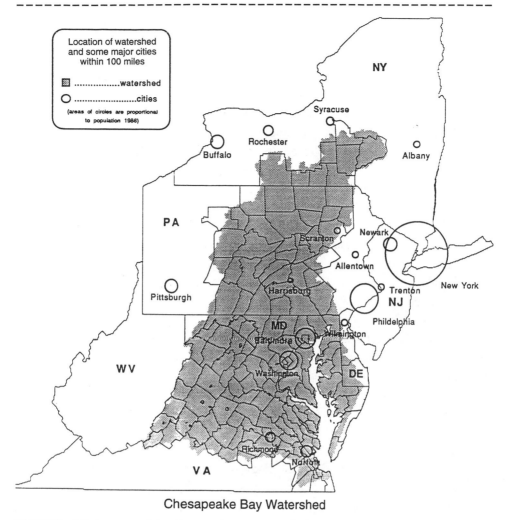

Chesapeake Bay Watershed

FIGURE 18.1. *Map of the Chesapeake Bay watershed.*

the Earth swings through these imperfectly understood, long-term seasons are slight—often only a few degrees. The change, however, is sufficient to affect the entire face of the globe. When the Earth experiences an Ice Age, much of the planet's moisture becomes caught up in ice. The level of the oceans falls for two reasons: less water is available, and water molecules contract when colder.

For 90% of the recent geologic past, the sea has remained in a well-defined basin with a sharp drop-off at its edge. When the glaciers melt, however, the sea rises. The overflowing ocean covers the edge of that basin, flooding the long, flat shoreline with a shallow sheet of water. That shoreline, now covered with seawater, is called the continental shelf.

If a small crew were to sail out of the Chesapeake Bay with a depthfinder on board, they would find that the depths remain fairly shallow—approximately 40 feet—for 50 miles out to sea. They would be sailing over the former shoreline, the flat plain that once bordered the coast. The depthfinder would then drop to more than 100 feet and would soon stop reading altogether, because the boat would have passed over the edge of a great cliff. If the depthfinder's pulsing signal could reach that far, its numbers would rise rapidly to more than 1000 feet, then 2000, then 3000. Nearly 120 miles out of the Bay, off the Virginia capes, the boat would sail beyond the continental shelf, past the slope of the shelf, and over the ancient sea bottom. The depthfinder, if it could read that far, would say 9000 feet.

In the Chesapeake Bay, the patterns are the same even if the depths are much less—the Chesapeake, taken as a whole, is a *very* shallow body of water. Its average depth is less than 10 feet. Many of its tidal flats may be only a few feet deep. For much of its history, the Chesapeake Bay was not a bay at all, but rather the Susquehanna River.

The Susquehanna River, which is one of the longest rivers in North America, begins in New York State and wends its way through Pennsylvania. In the past, it continued straight to the sea, emptying into the ocean basin at a point that now lies far from the coast. Today, the sea meets the river, rising and backfilling into Virginia past Norfolk, past Reedville, into Maryland past the Potomac, past the Patapsco. Tidal waters now reach all the way to Havre de Grace, where they flood a broad area only a foot or two deep called the Susquehanna Flats.

In the 10,000 years since the glaciers receded, a very special place has been created in the Chesapeake Bay. The Bay is an estuary where saltwater from the sea mixes with freshwater from the rivers. This natural mixing bowl, fed with nutrients off the land, provides a fertile feeding ground and nursery ground for fish and shellfish. It made a great breeding ground, for example, for oysters. It seems remarkable that a system that developed more than 10,000 years ago could be changed in a matter of a few centuries, perhaps even a few decades, by the settlers that came to its shores. To understand how this dramatic change could have occurred, consider the different uses humans have made of the Bay over the years and some of the problems that resulted from those applications.

Principal Uses and Problems of the Chesapeake Bay

As a Waterway. The settlers first used the Bay to land their ships. According to historical writer Donald Shomette, the Spanish came even before the British, giving the Chesapeake a new name: the Bahia Santa Maria. Sadly, some of the ships that arrived in the Bay after difficult Atlantic crossings ran aground in the shallows and broke apart.

Those who arrived safely rode the relatively protected waters of the Bay north and west, founding settlements and towns. For them, the Bay became their connection with the Old World—that is, a supply line and a chance for trade and commerce. The Bay has continued as a major waterway to the present day, with ocean-going ships calling at Baltimore and Norfolk, which rank among the United States' busiest ports.

From the beginning, ships on the Bay have probably created some pollution in the form of raw sewage, garbage dumped overboard, and river bottoms disturbed and sediment stirred up by anchors. These impacts were quite small—until modern times. Garbage from a square-rigger is a relatively minor issue; flushing the oily bilges of a 500-foot freighter is far more serious. While the colonial period may have seen a few watermen fishing or hunting for oysters, and perhaps a few hearty souls rowing for recreation, today the Bay is home to hundreds of thousands of recreational boats and thousands of watermen when the catch is good. These modern boats rarely drift silently with the wind. Many of them have powerful engines and throw substantial wakes, which then beat against the shoreline and further erode the banks of rivers and creeks. These modern craft also use petroleum—gasoline, diesel fuel, and oil—which may find their way into the estuary, as do chemicals for cleaning and polishing, painting and varnishing, and antifouling agents to keep marine growth off of hulls.

As a Dumping Ground. All animals—whether humans or striped bass or blue crabs—continually cycle food, taking in nutrients and expelling wastes in the process. When very few people lived on the shores of the Bay, human sewage did not create a major problem. Even when more people moved into the area, settling cities like Baltimore, the biggest problems from sewage were health-related, as sewage can carry bacteria that cause diseases such as cholera. Once modern sewage treatment began—and modern medicine evolved—these water-borne diseases became less of a threat. Only in areas of the world where sewage treatment is not properly managed (as in underdeveloped countries) does human waste still pose a serious health problem.

The Chesapeake Bay, however, faces yet another problem from sewage, one related to environmental health: too many nutrients. Sewage is rich in nutrients, including phosphorus and nitrogen. These nutrients can over-enrich the Bay and make it too productive for its own good. The presence of too many nutrients fuels explosive growth in blue-green algae and other undesirable phytoplankton that block light from other more desirable plants. When these algae finally die and drop to the bottom, they begin to decompose. As bacteria break down the fibers and minerals of the dead algae, they consume large amounts of oxygen, especially in deeper areas such as channels. This lack of oxygen makes it difficult for other animals that also need oxygen (such as oysters, fish, and crabs) to thrive. Sometimes the waters have very little oxygen (hypoxic waters) or virtually no oxygen at all (anoxic waters), and fish and other sea life that are unable to escape may suffocate and die.

As a Fishing Ground. Despite the other ways in which humans have used the Chesapeake Bay, its reputation and its tradition center on its value as a fishing ground. In its heyday, its shallow waters provided, acre for acre, more fish and shellfish than any other body of water in the world.

For many years, for example, Maryland's oyster harvest held steady at roughly 2 million bushels a year—a mere pittance compared with the 15 million bushel harvests of the nineteenth century, but enough to sustain a long-standing fishery and the bayside communities that depend on oysters to round out an annual cycle of working the water. During the 1980s, however, oyster harvests in Maryland dropped to less than 2 million bushels, then to less than 1 million bushels, and finally to less than 500,000 bushels.

Other changes have also affected the Chesapeake's water trades. The blue crab has become king of the commercial fisheries, surpassing oysters as the Bay's most lucrative harvest. Because of relatively low fuel prices during the 1980s and high demand for crab meat, watermen hauling in blue crabs saw a 250% rise in their profits during the 1980s. In many ways, the blue crab, once an undesirable by-catch, has kept the Bay's seafood harvesting industry alive.

Equally as dramatic as the blue crab's rise has been the fall of the striped bass fishery. Generally known as rockfish in the Bay region, striped bass once traveled by the tractor truckload from ports like Rock Hall, headed for Lexington Market in Baltimore or Fulton Market in New York. Harvest figures for 1973 exceeded 4 million pounds in Maryland alone. A precipitous decline in striped bass stocks, coupled with an appreciation of the Chesapeake's importance as the Atlantic Coast's major spawning ground, brought about stiff restrictions, followed by a total ban on fishing for (or even possession of) striped bass in Maryland in 1985.

A recent reopening of Maryland's fishery, based on promising surveys of juvenile stripers, has meant a short season for both commercial and recreational anglers. It will be a long time, if ever, before the Chesapeake sees a return of the great harvests of 20 years ago.

The Bay and Its Watershed as a System

A Stressed Ecosystem. Because the Chesapeake Bay has been so productive in the past, some have called it a food factory or compared it with a powerful engine that runs on nutrients. The Bay is not a factory or an engine, however; it is an ecosystem. Instead of machinery, it is composed of living parts: animals, plants, and microorganisms that depend on one another. If some of these living parts are removed or changed, the entire ecosystem will feel the effects in both direct and indirect ways—ways that are sometimes obvious and sometimes impossible to predict, and ways that depend on the subtle interplay of the system's parts at several spatial and temporal scales that determine whether it is healthy and resilient, or brittle to the breaking point.

This concept is not simply an abstract notion. Consider, for example, one of the Bay's prized shellfish, the oyster. For years the reputation of the Chesapeake Bay oyster has spread across the country. In the Bay region, oysters have meant money—some $20 million per year in Maryland alone during some years. As the oyster populations in the Bay declined, largely due to disease and overfishing, many watermen and natural resource managers realized that this trend would lead to a significant loss of livelihood and a shrinking economic resource for the tidewater region. What many may not have realized, however, were the ecological effects of the oyster's demise.

Like the Bay itself, the oyster bar is an ecosystem. The oyster is a gregarious animal: it prefers to grow in groups. Some scientists believe that young oyster larvae can actually detect certain chemical signals that draw them to other oysters. Because of this attraction, oyster larvae set and grow in clusters, ultimately forming large aggregations called oyster bars (or rocks or reefs).

After years of harvesting by oyster tongs or dredge, an oyster bar may lie low and scattered across the bottom of the Bay. In contrast, during the Colonial period, the oyster bars were said to reach from the Bay bottom to the water's surface. These bars actually formed oyster reefs, and like coral reefs in the tropics, they undoubtedly created rich ecosystems.

Imagine, for a moment, oyster reefs stretching up both sides of the Chesapeake Bay, along the shallow margins. Fish and other marine animals would have gathered around the reefs to feed, and one could probably see these fish because the water would be relatively clear. The water would have been less murky during those early years for two reasons. First, prior to European settlement and the introduction of intensive agriculture, the land surrounding the Bay and its rivers was covered with forests, which protected the soil and prevented run-off. Second, the Bay would be more transparent because the oysters themselves were actually cleaning the water.

Oysters are filter feeders. One oyster pumps approximately 50 gallons of water each day to filter out algae (also called phytoplankton), the tiny floating plants that serve as its primary food source. As oysters feed, they act like filters in a swimming pool, drawing out algae and clearing the water. Roger Newell has estimated that the Bay once had so many oyster reefs that the oysters could pump through a volume of water equal to the entire Chesapeake Bay in less than one week. Because the oyster populations have dwindled to such low levels, today's oyster population would need a year or more to filter the same amount of water.

Disturb one part of the ecosystem, and the entire ecosystem changes. The Bay now contains too much phytoplankton. Not only does the water look murky, but this lack of clarity has kept out light needed by underwater grasses. In many areas, these grasses—which may provide food for diving ducks or shelter for molting crabs—have died as a result of too much algae, which cloud the water and cover the submerged grasses with slime. Overwhelmed, many grasses have disappeared, leaving large stretches of the Bay bottom in a bare state.

In short, the Bay has changed in a number of ways because of the actions of humans. First, humans have harvested oysters and destroyed the oyster reefs. Most of the Bay's oysters were taken during the end of the last century. During that time, watermen hauled up in a single year what it would now take more than ten years to harvest. Second, we have increased the amount of algae in the water by adding nutrients. These nutrients come from sewage treatment plants, septic systems, farm fertilizers, and residential lawns. They produce even larger blooms of algae, which susbsequently die and decompose on the Bay floor, a process that draws life-sustaining oxygen out of the water.

Third, humans have increased the amount of sediments in the water. Although the Bay naturally receives a heavy load of sediment, especially during heavy spring rains and storms, human uses of the land have torn away the protective forests and allowed the dirt to wash away. This increased runoff began many years ago, when farmers cleared vast amounts of land for agriculture. Soil erosion continues today, not only because of agriculture, but also because of housing construction and other land development in the watershed. Every construction site has the potential to release tons of sediment into the Bay and its tributaries.

Fourth, we have added new chemical compounds to the Bay. Heavy metals (such as zinc and mercury) from industrial applications, pesticides from farms and suburban lawns, cleaning solutions from households, and a host of petroleum products and other compounds all wash off the land or down storm drains and into the streams and rivers that feed the Bay. Although we do not fully understand the effects of these substances, significant doses of many of these compounds are toxic to fish and other marine organisms. Researchers continue to try to determine the effect that such toxic compounds have at very low levels in our waterways—levels so low that they may be difficult to measure.

Toxic compounds sometimes interact to create additional problems for Bay organisms. For example, when rain becomes polluted with nitrous oxides (from automobiles) and carbon dioxide (from coal-burning industries and electrical power plants), it becomes acidic. When this acidic rain falls on the Bay, it has two effects. First, it adds nutrients to the Bay. Some researchers think that 25% or more of the nutrients entering the Bay derives from airborne sources. Second, according to some researchers, acid rain causes elements like aluminum to leach out of the soil. If this leaching occurs in a tributary where fish are spawning, it can kill the delicate larvae. Some scientists suggest that this double threat from acid rain and aluminum has hurt the reproduction of striped bass, for example.

Some problems may draw immediate public attention. For instance, the decline of striped bass (a popular sport and commerical fish) has caused an uproar. Other effects of toxic compounds on the Bay's plants and animals—which may not be as visible to the public—may go largely unnoticed, however. As one researcher has said, it may be that the Bay has a giant "headache" caused by toxic compounds, and we just don't realize it.

The Chesapeake Bay's ailing ecosystem also serves as something of an indicator for the health of the entire region. As it lies at the base of an enormous watershed—some 64,000 square miles of mountains, foothills, and coastal plain—the watershed gathers much of what we put on the land or pour into the water throughout the area. Because it supports such a rich and productive ecosystem, the Bay provides ample opportunities for us to witness changes and trends such as significant decreases in the populations of animals and plants.

The Bay has a story to tell about how human life can affect the environment. It is not simply a pool of water where oysters and crabs grow, but rather the most visible part of a vast network of plant and animal life. Human beings rely on this environment as much as any animal; we also bear a special responsibility, as our actions can have a greater affect on the ecosystem than the actions of any other creature.

Population Growth and Change in the Bay Watershed. During the Revolutionary War, when George Washington traveled through Annapolis, approximately 500,000 people lived in Maryland. In the past two centuries, Maryland's population has grown almost 10 times. Now 4.7 million people live in the state, and the Maryland Office of Planning predicts that another 800,000 will settle in Maryland by the year 2020. Maryland's growth illustrates how dramatic a change the Chesapeake watershed has undergone. Today, more than 14 million people live in the watershed; that number is expected to top 17 million by the year 2020.

Population changes in the watershed are shown in Figure 18.2. In the figure, the number of dots is proportional to population in each county and distributed randomly within each county. Between 1940 and 1986, the population of the watershed increased 87%. Roughly 20% of this increase

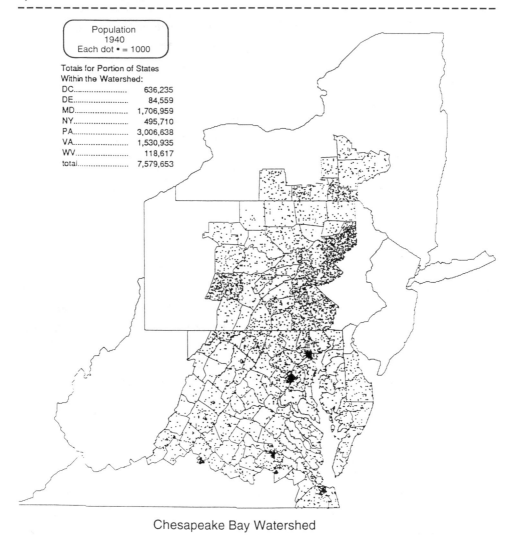

Population
1940
Each dot • = 1000

Totals for Portion of States
Within the Watershed:

DC	636,235
DE	84,559
MD	1,706,959
NY	495,710
PA	3,006,638
VA	1,530,935
WV	118,617
total	7,579,653

Chesapeake Bay Watershed

FIGURE 18.2. *Population in the Chesapeake Bay watershed: (A) 1940; (B) 1952; (C) 1972; and (D)1986.*

represented net migration into the watershed, the majority of it into the areas surrounding Baltimore and Washington in the states of Maryland and Virginia. The population of the watershed grew at an average annual rate of 1.6% between 1952 and 1972, almost the same as the U.S. average of 1.5% for the same period. The growth was concentrated in the Maryland and Virginia portions of the watershed, which averaged 2.6% growth; the remainder of the watershed grew more slowly, at only 0.4%.

The most striking changes occurred in three areas: Richmond, the Norfolk-Virginia Beach area, and the Baltimore-Washington corridor. Growth can, in part, be attributed to increases in industry related directly and

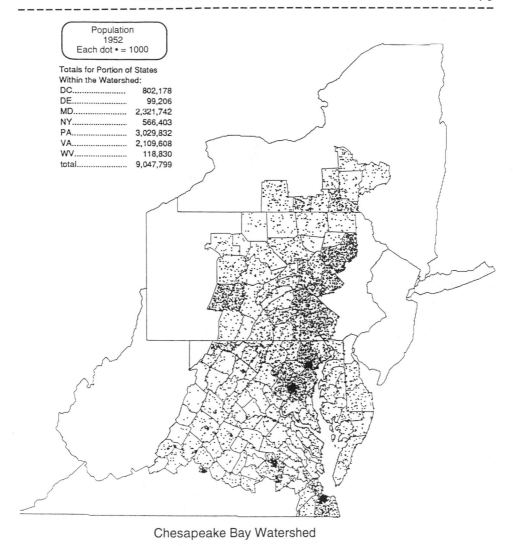

Chesapeake Bay Watershed

FIGURE 18.2. *Continued (B).*

indirectly to the expansion of the U.S. government as well as to the increasing fashionableness of the Chesapeake Bay as a recreation area. The latter forces are amplified by the high immigration rates to the area.

These areas also illustrate the "urban flight–suburban sprawl" phenomenon that has at once undermined the more natural and rural atmosphere for which many inhabitants originally left the city, while simultaneously removing both businesses and middle- and upper-income residents that served as a revenue base for the cities. The resulting deterioration of services and infrastructure worsens as people move farther away from the cities. The increasing travel times required to reach work, accompanied by a degeneration of traffic

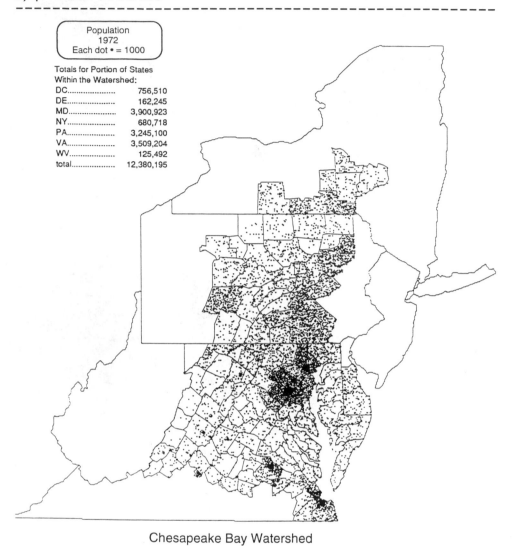

Population
1972
Each dot • = 1000

Totals for Portion of States
Within the Watershed:

DC......................	756,510
DE......................	162,245
MD......................	3,900,923
NY......................	680,718
PA......................	3,245,100
VA......................	3,509,204
WV......................	125,492
total..................	12,380,195

Chesapeake Bay Watershed

FIGURE 18.2. *Continued (C).*

conditions, and soaring property values are some of the resistive forces that quell the further spread of suburban development.

These forces may be approaching an equilibrium, at least for the present. Between 1972 and 1986, the growth of the Maryland watershed population slowed to an annual rate of 0.9%; in Virginia, the corresponding population growth rate declined to 1.8%. Emigration rates from the cities have slowed as well. For example, Washington's emigration rate decreased from an annual average of 1.3% in the 1960s to less than 0.8% in the early 1980s. The spatial extent of the sprawl also appears to be limited to the counties immediately surrounding the cities in question (see Fig. 18.2D). The information

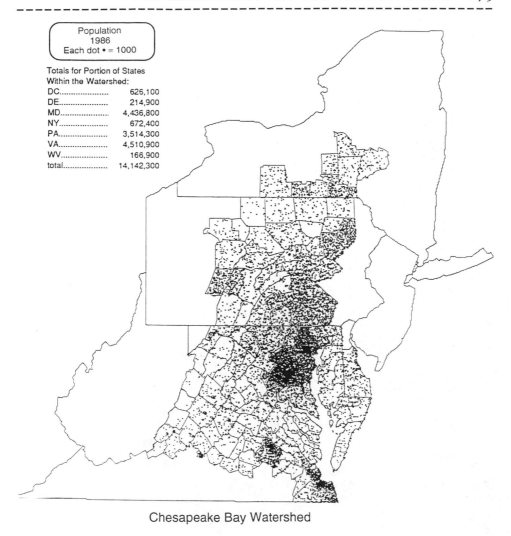

Population 1986 Each dot • = 1000	
Totals for Portion of States Within the Watershed:	
DC......................	626,100
DE......................	214,900
MD......................	4,436,800
NY......................	672,400
PA......................	3,514,300
VA......................	4,510,900
WV......................	166,900
total...................	14,142,300

Chesapeake Bay Watershed

FIGURE 18.2. *Continued (D).*

contained in these maps is only suggestive, however, as demographic trends can result from any number of factors. Variations in birth rates, cultural heritage, local versus long-distance moves, political climate, and zoning laws all muddy the waters. In addition, as sprawl and growth occur simultaneously, secondary economic centers inevitably spring up, initiating their own cycles.

As the human population in the watershed has increased, both land and water have felt its effects. Forests and farmland have been lost to development, and the Chesapeake Bay's underwater meadows of aquatic vegetation have died off, largely because of an influx of too many nutrients entering the estuary (see the earlier discussion). As the underwater grasses have dwindled,

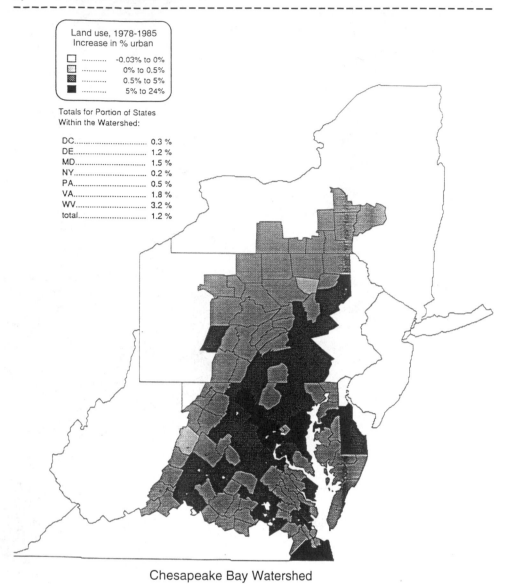

Land use, 1978-1985
Increase in % urban

☐ .... -0.03% to 0%
▨ .... 0% to 0.5%
▦ .... 0.5% to 5%
■ .... 5% to 24%

Totals for Portion of States
Within the Watershed:

DC.............................. 0.3 %
DE.............................. 1.2 %
MD.............................. 1.5 %
NY.............................. 0.2 %
PA.............................. 0.5 %
VA.............................. 1.8 %
WV.............................. 3.2 %
total.......................... 1.2 %

Chesapeake Bay Watershed

FIGURE 18.3. *Percent change in land-use from 1978 to 1985 in the Chesapeake Bay watershed: (A) urban; (B) cropland; (C) pasture; and (D) woodland.*

widgeons and other diving ducks that grazed on the grasses have also disappeared. The population growth has also meant heavy harvesting of the Chesapeake's fisheries, such as oysters, which were already feeling the strain of turbidity and disease.

Patterns of settlement and resettlement broadly affect land use in other ways as well, as shown in maps detailing the percent change in land-use (Figure 18.3). For example, a slight but noticeable increase has recently

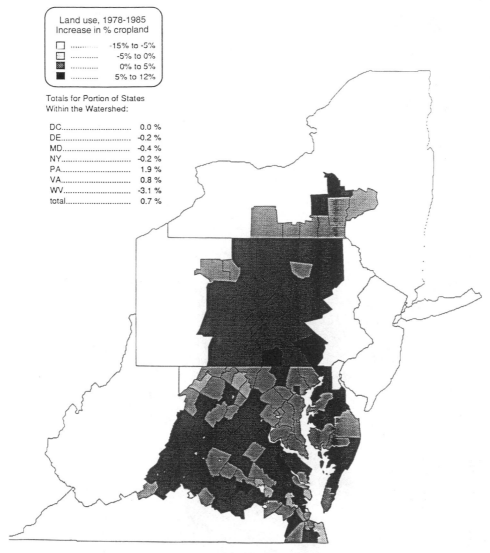

Chesapeake Bay Watershed

FIGURE 18.3. *Continued (B).*

occurred in woodland in the New York portion of the watershed (see Fig. 18.3C). The existence of a concurrent loss of cropland and pasture (see Figs. 18.3A and 18.3B) only indicates that net changes occurred in the transformation from one type of land-use to another; it does not explain why those changes arose. In fact, the marked differences in changes in agricultural and forested acreage from one state to another strongly suggest that these trends relate to differences in state agricultural policies and tax laws. On the other

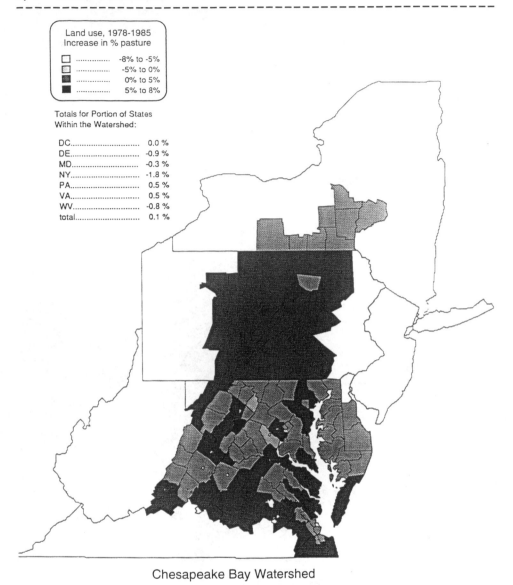

Land use, 1978-1985
Increase in % pasture

☐ .... -8% to -5%
▨ .... -5% to 0%
▩ .... 0% to 5%
■ .... 5% to 8%

Totals for Portion of States
Within the Watershed:

DC............................ 0.0 %
DE............................ -0.9 %
MD............................ -0.3 %
NY............................ -1.8 %
PA............................ 0.5 %
VA............................ 0.5 %
WV............................ -0.8 %
total............................ 0.1 %

Chesapeake Bay Watershed

FIGURE 18.3. *Continued (C).*

hand, the pattern of urbanization appears more straightforward given the population changes discussed earlier (see Fig. 18.3D).

Increases in evidence of human activity accompany the increases in population in both magnitude and distribution. Maps of manufacturers, energy consumption, housing units, water use, solid-waste production, and air pollutants all closely resemble the maps of population. On the other hand, changes in lifestyle have actually accelerated the increases in consumption and waste production. From 1952 to 1986, the following increases occurred in

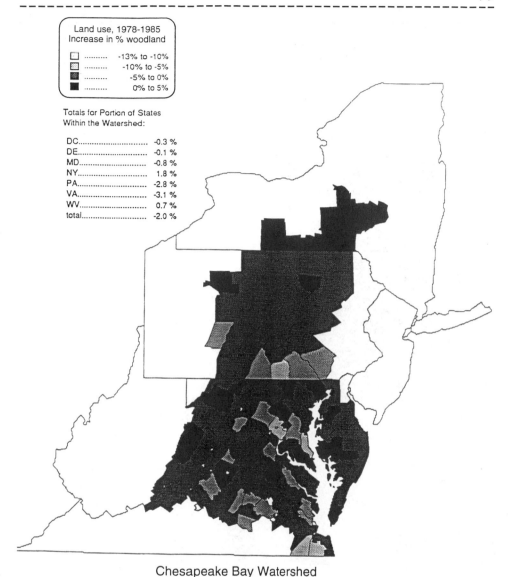

Land use, 1978-1985
Increase in % woodland

☐ .... -13% to -10%
▨ .... -10% to -5%
▦ .... -5% to 0%
■ .... 0% to 5%

Totals for Portion of States
Within the Watershed:

DC............................ -0.3 %
DE............................ -0.1 %
MD............................ -0.8 %
NY............................ 1.8 %
PA............................ -2.8 %
VA............................ -3.1 %
WV............................ 0.7 %
total.......................... -2.0 %

Chesapeake Bay Watershed

FIGURE 18.3. *Continued (D).*

the watershed: per capita energy consumption rose from 567,000 BTU per day to 744,000 BTU per day; NO_2 emissions by vehicles increased from 1.0 lb per week per person to 1.7 lb per week per person; and per capita solid-waste production grew from 2.2 lb per day to 3.7 lb per day. In contrast, public and industrial per capita water use decreased over the same period for the watershed as a whole—from 334 gallons per day to 278 gallons per day. This decline largely reflects lower total use in Pennsylvania during this period; the state's water use dropped from 1.46 billion gallons per day in 1952 to only

1.18 billion gallons per day in 1986. In the Maryland and Virginia portions of the watershed, the per capita use rate held nearly constant at about 237 gallons per day. The former change may reflect changes in heavy industry in Pennsylvania, while the latter lack of change may represent a relatively constant public demand for water in a region that is primarily characterized by residential use and light industry.

Agriculture.   Changes associated with agriculture are tied to population increases in terms of their overall magnitude, and to cultural and historical practices in terms of their distribution and local magnitude. Heavily populated areas

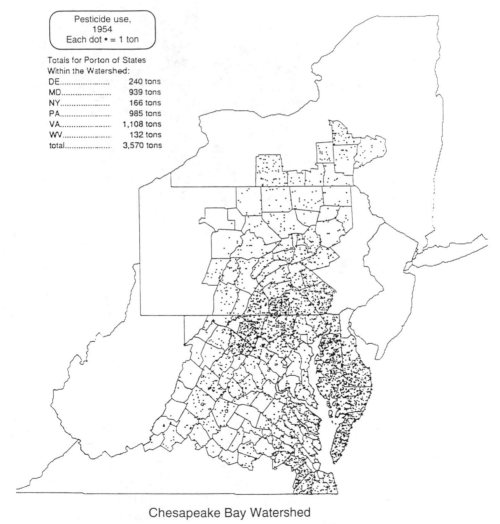

Chesapeake Bay Watershed

FIGURE 18.4. *Pesticide use in the Chesapeake Bay watershed: (A) 1954; (B) 1974; and (C) 1987.*

necessarily exclude agriculture. It is unfortunate that some of the best agricultural land in the United States, which was the basis for the initial local growth, is rapidly being converted to residential developments, industrial parks, and shopping malls as rising property values render farming unprofitable. It is not clear, however, whether agriculture itself is more benign than urban and suburban development for an ecosystem such as the Chesapeake Bay. Both land-uses load the system with wastes and nutrients and consume "natural" areas that could have absorbed some of that input.

While total farm acreage has fallen from 23.2 million acres in 1954 to 14.4 million acres in 1987 in the watershed, cropland has decreased from 10.5 million acres to 9.0 million. Meanwhile, over the same period, the average

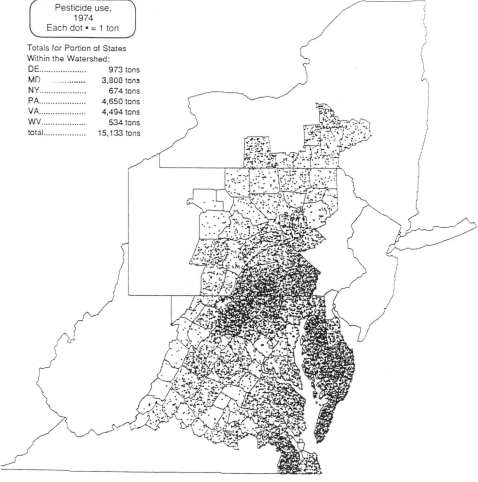

Pesticide use,
1974
Each dot • = 1 ton

Totals for Portion of States
Within the Watershed:
DE....................	973 tons
MD	3,808 tons
NY....................	674 tons
PA....................	4,650 tons
VA....................	4,494 tons
WV....................	534 tons
total..................	15,133 tons

Chesapeake Bay Watershed

FIGURE 18.4. *Continued (B).*

farm size increased from 126 acres to 190 acres. Thus, larger percentages of farms are being devoted to crops (45% and 64% in 1954 and 1987, respectively) while less land lies fallow, pastured, and wooded. Today's farming operations tend to be both larger and more intensive. Irrigation increased from 40,000 acres to 180,000 acres from 1954 to 1987, and fertilization rates have increased from nearly 210 lb per cultivated acre to 250 lb per cultivated acre during the same period. Figure 18.4 illustrates trends in pesticide use, which was almost nonexistent in the early 1950s, peaked in the late 1970s, and decreased after 1974, when 15,000 tons were used, to 13,000 tons in 1986. Of course, this smaller amount also reflects the greater specificity of the pesticides used in today's farming practices.

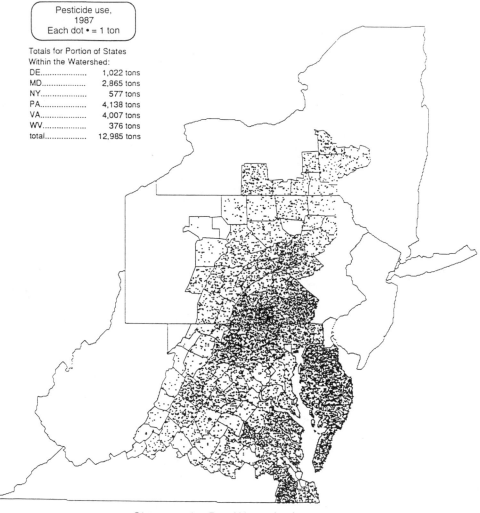

Chesapeake Bay Watershed

FIGURE 18.4. *Continued (C).*

If we examine the application of fertilizer nitrogen (Figure 18.5), a pattern becomes evident. Nitrogen use follows the general distribution of farmland but shows an overall increase in its intensity. Unlike the phosphorus content of fertilizers used in the watershed, which remained relatively constant over the years (at approximately 10%), the nitrogen content steadily rose from approximately 7% in the 1950s to 15% in the mid-1980s.

Today, the nutrient loading rates for the Chesapeake Bay are 78,700 tons nitrogen per year and 3600 tons phosphorus per year (Boynton et al 1990). If one assumes that the loading rate is proportional to the amount of nutrients produced and applied within the watershed, then this use level implies that

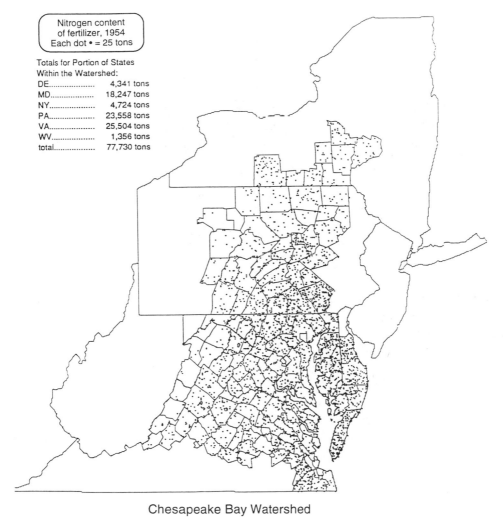

Nitrogen content
of fertilizer, 1954
Each dot • = 25 tons

Totals for Portion of States
Within the Watershed:

DE..................	4,341 tons
MD..................	18,247 tons
NY..................	4,724 tons
PA..................	23,558 tons
VA..................	25,504 tons
WV..................	1,356 tons
total..................	77,730 tons

Chesapeake Bay Watershed

FIGURE 18.5. *Nitrogen content of fertilizer use in the Chesapeake Bay watershed: (A) 1954; (B) 1974; and (C) 1987.*

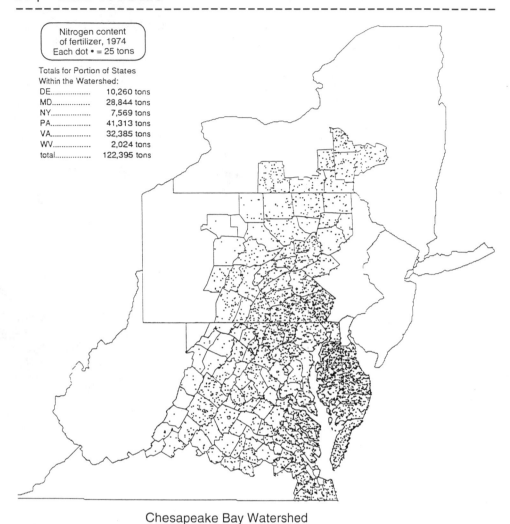

Nitrogen content
of fertilizer, 1974
Each dot • = 25 tons

Totals for Portion of States
Within the Watershed:

DE.................   10,260 tons
MD................   28,844 tons
NY.................   7,569 tons
PA.................   41,313 tons
VA.................   32,385 tons
WV................   2,024 tons
total............... 122,395 tons

Chesapeake Bay Watershed

FIGURE 18.5. *Continued (B).*

phosphorus loading has increased by 4% to 10% and nitrogen loading has
increased between 34% and 125% since 1954. This increase would not only
represent a major change in amount of nutrients, but (an even more crucial
development) it would also signal a huge change in the relative proportion of
nutrients. The increased concentrations of nutrients in specific localities such
as Lancaster County means that this problem may be more exaggerated in
certain parts of the system.

Summary

The Chesapeake Bay has undergone very rapid population growth (Table
18.1) with its associated environmental impacts. We have mapped some
of these changes as they are reflected in the characteristics of the Bay's

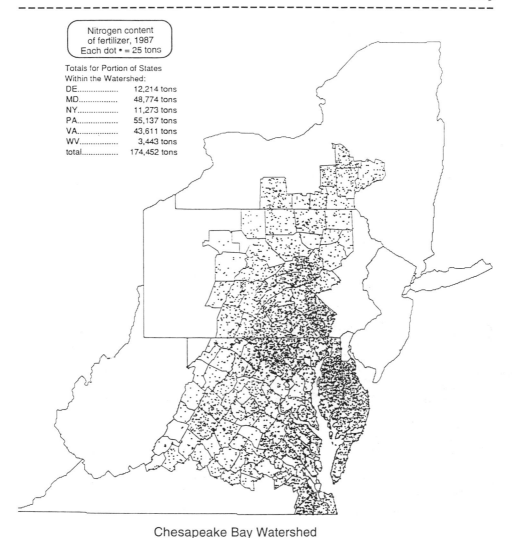

Chesapeake Bay Watershed

FIGURE 18.5. *Continued (C).*

watershed. The impacts of these activities in the watershed on the Bay itself are known to be large, but their specific interconnections are only now becoming the focus of scientific investigation. The Chesapeake Bay has 200,000 people living in its drainage basin for every km^3 of water in the Bay (in contrast, the Baltic Sea supports a population of 4000 people/km^3, and the Mediterranean has 85 people/km^3). Even if all of these people minimized their environmental impacts, their sheer numbers would represent a daunting challenge for a system as sensitive as the Chesapeake. If these numbers continue to increase as they have in the past, the prospects for the United States' largest estuary seem bleak.

TABLE 18.1   The watershed's population at a glance (only those portions of each state within the watershed are included)

State or District	1940	1986	2020 (projected)
District of Columbia	636,235	626,100	626,100
Delaware	84,559	214,900	215,000
Maryland	1,706,959	4,436,800	5,496,600
New York	495,710	672,400	700,000
Pennsylvania	3,006,638	3,514,300	3,854,500
Virginia	1,530,935	4,510,900	6,229,800
West Virginia	118,617	166,900	166,900
Total Watershed	**7,579,653**	**14,142,300**	**17,288,900**

Sources: U.S. Bureau of the Census 1952, 1988; Maryland Office of Planning.

EVOLUTION OF CHESAPEAKE BAY MANAGEMENT

Efforts to comprehensively understand and manage the Bay ecosystem began to gain momentum in the late 1960s, in tandem with the general increase in environmental awareness in the United States. Several major studies of the Bay have been undertaken; indeed, the Chesapeake Bay is arguably the most intensively studied ecosystem in the world. It is also generally acknowledged that the system is a "best case scenario" of how science, government, the media, and private environmental interest groups may be integrated in attempts to achieve rational and effective ecosystem management. As such, it is worthwhile to examine in some detail how this situation arose, how well it has really worked, whether it serves as an appropriate model for application to the management of other ecosystems, how it might continue to evolve and improve, and what major improvements are still lacking.

Although the Bay has always been an ecosystem of concern for its inhabitants, our chronology of Bay "management" will begin in 1965. In that year, the U.S. Army Corps of Engineers conducted a major study of the Bay, and two important pieces of federal legislation were passed—the Federal Water Pollution Control Act and the Federal Rivers and Harbors Act. By this time, both scientists and citizens had begun to notice significant declines in some key indicators. The water had become more turbid, algae blooms were occurring more frequently, and it was becoming harder to catch certain kinds of fish.

1965–1976: Era of Shared Experience and Raised Consciousness

The signs of a declining ecosystem are frequently ignored, but a unique feature of the Chesapeake is the broadly shared experience and enjoyment of the Bay ecosystem by a large cross-section of the population. The Bay was (and is) everyone's "backyard." Government bureaucrats from Washington sailed and fished on the Bay side-by-side with scientists, watermen, local politicians, and others. The result was a shared commitment that is often missing from attempts to protect remote ecosystems that few people actually experience on a daily basis. In 1966, the Chesapeake Bay Foundation was established as a nonprofit conservation organization committed to help "Save the Bay" through public education and advocacy.

Another notable factor was the role of the media in publicizing the Bay's plight and the efforts to save it. In 1976, William Warner published his Pulitzer Prize–winning book about crabbing on the Bay, *Beautiful Swimmers*. Tom Horton, then of the *Baltimore Sun*, was a key environmental journalist in the effort to bring the Bay's problems to the public eye. His job was made easier by the fact that everyone could directly experience the problems he described, including the sheer number of people living in the watershed. By 1970, there were 3.9 million people living in Maryland and 12.4 million in the overall watershed.

Another defining event came when Maryland State Senator Bernie Fowler waded into the waters of the Patuxent River near Broome's Island and discovered that he could no longer see his feet when chest-high in water, as he could as a boy growing up in the area. Even compensating for the fact that his chest and toes were farther apart, the water quality in the Patuxent had obviously deteriorated significantly. This kind of shared experience spurred politicians into action and made the "Save the Bay" program much more than an academic exercise.

1977–1983: Era of Intense Scientific Analysis with Political Backing

Once a consensus began to form that the Bay was indeed in trouble, a phase of detailed applied scientific analysis was initiated. Unlike many purely scientific studies, this effort had political backing from the start and was aimed directly at providing answers to two questions:

■ What was responsible for the Bay's decline?

■ What should be done about it?

In 1975, Senator Mac Mathias toured the Bay and requested that the EPA study the Bay's problems. In 1977, a six-year, $26 million EPA-funded study began. This study was unique in both its size and its express goal of supporting ecosystem management. In 1978, the Chesapeake Bay Legislative Advisory Commission (CBLAC, an intergovernmental coordinating committee) was created to develop a method for managing the Bay's resources. In 1980, the Chesapeake Bay Commission was created as an outgrowth of the CBLAC. The commission was composed of 10 state legislators from Maryland and Virginia who were responsible for developing a cooperative arrangement between the two states to clean up the Bay. In 1983, the EPA study was completed and methods to implement its findings were being studied.

Meanwhile, the population in the region continued to increase. By 1980, 4.2 million people lived in Maryland, and 13.4 million in the watershed.

1983–Present: Era of Implementation and Monitoring

The next nine years saw both implementation of the EPA findings and a unique effort to monitor and model the effects of these implementation plans. In December 1983, the first Chesapeake Bay Agreement was signed by Maryland, Virginia, Pennsylvania, Washington, D.C., and the EPA. The agreement is considered a compact between all of the parties.

In 1984, the Maryland Critical Area Law was enacted and several other clean-up initiatives were launched. The law restricted residential and urban development in the area within 1000 feet of tidal water. In 1985, the state of Maryland released its Restoration Plan, aimed at "restoring and maintaining the Bay's ecological integrity, productivity, and beneficial uses and to protect public health."

In 1987, a new Chesapeake Bay Agreement was signed by Maryland, Virginia, Pennsylvania, Washington, D.C., the EPA, and the Chesapeake Bay Commission. This agreement extended and expanded the 1983 compact and called for achieving more specific goals by certain deadlines. For example, a 40% reduction of nitrogen and phosphorus is to be achieved by the year 2000.

In 1989, the "Year 2020" panel released its findings on the effects of increased population growth within the watershed on the Chesapeake Bay and made recommendations to avert further adverse effects on the Bay's water quality. By 1990, 4.7 million people lived in Maryland and 14.5 million in the watershed.

In 1991, the Chesapeake Bay Growth and Preservation Act was presented before the Maryland legislature. It represents the first bill to address statewide growth management and to focus on human population growth as the Bay's major problem. The bill was not passed, however, but was deferred for further study.

Not until the implementation era was the full extent of the Bay's problems recognized, and in particular the necessity to treat the entire Bay watershed as a system. Nutrient reductions required changes in activities conducted throughout the watershed. While point-sources of nutrients, such as industries and sewage treatment plants, proved relatively easy to control, it soon became evident that non–point-sources, such as residences and agriculture, were responsible for a large part of the problem; these sources of population are much more difficult to manage. The growing population of the watershed itself also came to be recognized as a primary cause of the Bay's problems.

The Bay is now primed for new approaches to management that can go beyond the traditional command and control methods (which were successful with point-sources) to complete the implementation phase on non–point-sources. A later section will discuss these measures in more depth.

Monitoring. The development of a comprehensive monitoring program has proved essential to managing the Chesapeake. All estuaries are in constant flux, and the Chesapeake Bay is no exception. With 150 rivers and streams feeding it and several thousand miles of shoreline, the Bay system continually responds to nature's changing conditions: rains, tides, drought, shoreline erosion, storm run-off. It also responds to human activity—discharges of nutrients from waste treatment plants, fertilizer run-off from farmland, toxic chemicals in industrial outfalls, fish and shellfish harvesting. How do we distinguish between natural and human impacts on the Bay? And why is this step necessary?

The Chesapeake's vital signs—its temperature, dissolved oxygen, pH, and water clarity—serve as indicators of the Bay's environmental health. These and other indicators, such as the presence (or absence) of underwater grasses or the numbers and diversity of fish, measured week by week, month by month, and year by year, give researchers and resource managers insight into the health of the Bay by tracking changes in the system.

Monitoring is charged with keeping track of these changes. To restore the health of the Chesapeake to the fullest extent possible, we need as complete an understanding of its ecology as we can obtain. Bay monitoring programs, funded by state and federal governments, help chart emerging trends, enabling us to see whether the ecosystem is responding to programs designed to reverse the decline of water quality or populations of fish or shellfish.

Government agencies in the Bay region (for example, the U.S. EPA, the Maryland Department of the Environment, and the Maryland Department of Natural Resources) are cooperating by regularly monitoring many different water quality factors, including toxic chemicals, metals, and nutrients. They are also noting the abundance of key fish and shellfish species, the presence of diseases in fish and shellfish, and the health and population of underwater grasses and waterfowl. Especially important are samplings of dissolved oxygen and chlorophyll, as they indicate whether overgrowths of algae are occurring. (Virtually all plants, including algae, contain chlorophyll. Thus, if high levels of chlorophyll emerge, then scientists can assume that high numbers of algae are present.) Interpretation of these data can help scientists and resource managers determine whether management programs to restore living resources to the Bay—such as underwater grasses, striped bass, and oysters —are working, need to be revised, or need to be scrapped in favor of new strategies.

The Bay is enormously complex, and monitoring is too costly for government agencies to take all of the samples necessary to obtain a thorough understanding of the ecosystem. In this regard, citizen monitoring programs in the Bay are highly helpful. For several years, students and volunteers in citizen organizations have been regularly monitoring streams and rivers throughout the watershed. With assistance from organizations such as the Alliance for the Chesapeake Bay and Save Our Streams, volunteers learn how to take water quality measurements and record the data so that they will be reliable and useful for data managers.

Each week volunteers head for an assigned site along the shore of a stream or river. At that location, they take samples and compile a number of measurements: pH (a measure of alkalinity or acidity), dissolved oxygen level, salinity, temperature, and turbidity (a measure of water clarity). These data are then entered into the Chesapeake Bay database, along with data gathered by state and federal monitoring programs. When collected over a period of years, these measurements of water quality will enable resource managers to better analyze and describe the Bay's shifting ecosystem. They will also help us to determine the success of our efforts to curb pollution.

Modeling the Bay and the Watershed. To combine the monitoring data with an understanding of how the complex pieces of the watershed and the Bay interact, we need to synthesize and integrate our knowledge in the form of models. A model is simply an abstract representation of the system that we can manipulate. For example, architects build cardboard models of their projects to aid the design process. The mental picture we conjure up when someone mentions the Chesapeake Bay is also a model. More recently, hydrologists, ecologists, economists, and other scientists have begun to build mathematical models of various parts of the Chesapeake system that can be simulated via computers. They range from hydrodynamic models of the water-flow dynamics in the Bay itself (Dortch et al 1988; HydroQual 1987, 1989) and other estuaries in the Chesapeake system (Hwang 1990), to models of water and nutrient run-off from the watershed (Summers 1990), to integrated models of the ecological and economic components of the system with spatial articulation (DeBellevue and Costanza 1991).

Because ecosystems are being threatened by a host of human activities, protecting and preserving them requires an understanding of the direct and indirect effects of human activities over long periods of time and over large areas. Computer simulations are emerging as important tools that can investigate these and other interactions. Without sophisticated global atmospheric simulations, for example, our understanding of the potential impacts of increasing CO_2 concentrations in the atmosphere due to fossil fuel burning would be much more primitive. Computer simulations can now be used to understand not only human impacts on ecosystems, but also our economic dependence on natural ecosystem services and capital, and the interdependence between ecological and economic components of the system (see, for example, Braat and Steetskamp 1991; Costanza et al 1990).

Several recent developments have made such computer simulation modeling feasible, including the accessibility of extensive spatial and temporal databases and advances in computer power and convenience. Computer simulation models are potentially one of our best tools for illuminating the complex functions of integrated ecological economic systems like the Chesapeake watershed.

Even with the best conceivable modeling capabilities, however, we will always be confronted with large amounts of uncertainty about the response of the environment to human actions (Funtowicz and Ravetz 1991). Learning how to manage the environment effectively in the face of this uncertainty is critical (Costanza and Perrings 1990; Perrings 1991).

To use quantitative computer modeling to understand and manage complex ecological economic systems like the Chesapeake watershed, we need an integrated, multiscale, transdisciplinary, and pluralistic approach. This strategy should acknowledge the large remaining uncertainty inherent in modeling these systems and enable the development of new ways to effectively deal with this uncertainty. One important addition to our suite of models of the Chesapeake watershed is the "4 box" model of systems, described later in this chapter.

BARRIERS AND BRIDGES TO IMPROVED MANAGEMENT

Social Traps

One fundamental reason for system collapse and mismanagement is the inherent mismatch in many systems between the fundamental time and space scales of the ecological system on the one hand and of the human institutions developed to manage it on the other. When it comes to human institutions, one critical feature is the incentive structures produced by these institutions. These incentive structures often lead to behavior that runs directly counter to the long-term health of the whole system, and often even to the stated goals of the institution itself. How does this situation arise and how can we fix it?

This process of short-run and local incentives moving out of sync with long-term and global goals has been well studied in the last decade under

----

several rubrics (see, for example, Hardin 1968; Axelrod 1984). One especially appealing study yielded John Platt's notion of "social traps" (Platt 1973; Cross and Guyer 1980; Teger 1980; Brockner and Rubin 1985; Costanza 1987). In such cases, the decision maker becomes "trapped" by the local conditions into making what turns out to be a bad decision when viewed from a longer or wider perspective. We go through life making decisions about which path to take based largely on "road signs"—the short-run, local reinforcements that we perceive most directly. These types of reinforcements can include monetary incentives, social acceptance or admonishment, and physical pleasure or pain. In general, this strategy of following road signs is quite effective, unless the road signs are inaccurate or misleading. In these cases, we can become trapped into following a path that is ultimately detrimental because of our reliance on the road signs. For example, cigarette smoking has long been a social trap. By following the short-run road signs of the pleasure and social status associated with smoking, we embark on the road to an increased risk of earlier death from smoking-induced cancer. In the last few years, the formerly positive reinforcement associated with smoking has begun to turn into a negative one. As smoking becomes less socially acceptable, we would expect the number of new smokers to decline and many old smokers to escape the trap. The process of escape, however, is much more difficult than the process of avoidance (as most people who have tried to stop smoking can attest).

The elimination of social traps requires intervention—that is, the modification of the reinforcement system. Indeed, it can be argued that the proper role of a democratic government is to eliminate social traps while maintaining as much individual freedom as possible. Cross and Guyer list four broad methods by which we can avoid or escape from traps: education (about the long-term, distributed impacts), insurance, superordinate authority (that is, legal systems, government, and religion), and converting the trap to a trade-off (that is, correcting the road signs).

Education can be used to warn people of long-term impacts that cannot be seen from the road. Examples of this technique include the warning labels now required on cigarette packages and the warnings issued by environmentalists about future hazardous waste problems. People can ignore warnings, however, particularly if the path seems otherwise enticing. For example, warning labels on cigarette packages have had little effect on the number of smokers. The main problem with education as a general method of avoiding and escaping from traps is that it requires that individuals make a significant time commitment to learn the details of each situation. Our current society is so large and complex that we cannot expect professionals—much less the general public—to know the details of all extant traps. In addition, for education to be effective in avoiding traps involving many individuals, *all* of the participants must receive the proper education.

Governments can, of course, forbid or regulate certain actions that have been deemed socially inappropriate. Unfortunately, this approach must be monitored and enforced rigidly, and a strong short-term incentive exists for individuals to ignore or avoid the regulations. A police force and legal system are very expensive to maintain, and increasing their chances of catching violators exponentially increases their costs (in terms of both the cost of maintaining a larger, better-equipped force and the cost of the loss of individual privacy and freedom).

Religion and social customs offer much less expensive ways to avoid certain social traps. If a moral code of action and belief in an ultimate payment for transgressions can be deeply instilled in a person, the probability of that individual's falling into the "sins" (traps) covered by the code will be greatly reduced, with very little enforcement cost. On the other hand, the moral code must also be relatively static to allow beliefs learned early in life to remain in force later. In addition, this approach requires a relatively homogeneous community of like-minded individuals to be truly effective. This system works well in culturally homogeneous societies that are changing very slowly. In modern, heterogeneous, rapidly changing societies, religion and social customs cannot handle all of the newly evolving situations, nor can they deal with the conflict between radically different cultures and belief systems.

Many trap theorists suggest that the most effective method for avoiding and escaping from social traps is to turn the trap into a trade-off. This method does not run counter to our normal tendency to follow the road signs; instead, it merely corrects the signs' inaccuracies by adding compensatory positive or negative reinforcements. A simple example illustrates how this method can prove effective. Playing slot machines is a social trap because the long-term costs and benefits are inconsistent with the short-term costs and benefits. People play the machines because they expect a large short-term jackpot; in fact, the machines are programmed to pay off, for example, 80 cents on the dollar in the long term. People may "win" hundreds of dollars playing the slots (in the short run). If they play long enough, however, they will certainly lose $0.20 for every dollar played. To change this trap into a trade-off, one could simply reprogram the machines so that every time $1 was input, $0.80 would come out. In this way, the short-term reinforcements ($0.80 on a $1 bet) become consistent with the long-term reinforcements ($0.80 on $1 played), and only the dedicated aficionados of spinning wheels with fruit painted on them would continue to play.

The Chesapeake Bay managed to avoid many of the social traps that have afflicted other ecosystems described in this book. From the outset, it achieved a healthy dialog between the interest groups involved and developed a broad consensus on the long-term goals of ecosystem restoration. This strategy guaranteed that efforts focused on long-term, common goals, and short-term, narrow interest group goals did not drive the process. One way to achieve this result in other systems is to employ the methods of policy dialog and consensus building described earlier in this chapter and in chapter 1. Even after this consensus is achieved, however, the system will have trouble inplementing its long-term goals unless they are effectively communicated to the narrow interest groups. Implementation of the Bay management strategy is now stalled on this problem, especially for non–point-sources. In the next section, we describe some new suggestions for solving this problem by modifying the short-term incentives to make them more consistent with long-term goals.

Building Bridges with Incentives

In the context of the theory of social traps, the most effective way to make global and long-term goals consistent with local, private, short-term goals is to somehow change the latter incentives to be consistent with the global, long-term goals (Costanza 1987). These incentives consist of any combination of the reinforcements that are important at the local level, including economic, social, and cultural incentives. The bridges comprise the social and economic instruments and institutions that we must design and build to link the present and the future, the private and the social, and the local and the global. As we saw in chapter 1, one important inconsistency relates to the underdevelopment of the "creative renewal" pathway that can lead to system collapse in the long run. To avoid this trap, we must reinforce creativity to the appropriate degree by creating the proper set of incentives.

An innovative instrument currently being researched to manage the environment under uncertainty is a *flexible environmental assurance bonding system* (Costanza and Perrings 1990). This variation of the deposit-refund system is designed to incorporate environmental criteria and uncertainty into the market, and to induce positive environmental technological innovation. It works in the following way.

In addition to charging a firm directly for known environmental damages, an assurance bond equal to the current best estimate of the largest potential future environmental damages would be levied and kept in an interest-bearing escrow account for a predetermined length of time. In keeping with the precautionary principle, this system requires the commitment of current resources to offset the potentially catastrophic future effects of current activity. Portions of the bond (plus interest) would be returned *if and only if* the firm could demonstrate that the suspected worst case damages had not occurred or would be less than originally assessed. If damages did occur, then portions of the bond would be used to rehabilitate or repair the environment, and possibly to compensate injured parties. By requiring the users of environmental resources to post a bond adequate to cover uncertain future environmental damages (with the possibility for refunds), the burden of proof (and the cost of the uncertainty) shifts from the public to the resource user. At the same time, firms are not charged a final amount for uncertain future damages and can recover portions of their bond depending on how their performance surpasses the worst case.

Deposit-refund systems are not a new concept. In the past, they have been applied to consumer policy, conservation policy, environmental policy, and other efficiency objectives. Such systems can be market-generated or government-initiated, and many are performance-based. For example, deposit-refund systems are currently used with great effectiveness to encourage the proper management of beverage containers and used lubricating oils (Bohm 1981).

Environmental assurance bonds would resemble the producer-paid performance bonds often required for federal, state, or local government work. For example, the Miller Act (40 U.S.C. 270), a 1935 U.S. federal statute, requires contractors who undertake construction for the federal government to secure performance bonds. Such bonds provide a contractual guarantee that the principal (the entity that does the work or provides the service) will perform in a designated way. Likewise, bonds are also required for work done in the private sector.

Performance bonds are frequently posted in the form of corporate surety bonds that are licensed under various insurance laws and, under their charter, have legal authority to act as financial guarantee for others. The unrecoverable cost of this service usually consists of 1% to 5% of the bond amount. Under the Miller Act (FAR 28.203-1 and 28.203-2), however, any contract above a designated amount ($25,000 in the case of construction) can be backed by other types of securities, such as U.S. bonds or notes, in lieu of a bond guaranteed by a surety company. In this case, the contractor provides a duly executed power of attorney and an agreement authorizing collection on the bond or notes if it defaults on the contract (PRC Environmental Management 1986). If the contractor performs all obligations specified in the contract, the securities are returned and the usual cost of the surety is avoided.

Environmental assurance bonds would work in a similar manner, providing a contractual guarantee that the principal would perform in an environmentally benign manner. They would, however, be levied for the current best estimate of the *largest* potential future environmental damages. In most cases, these bonds could be administered by the regulatory authority that currently manages the operation or procedure. For example, in the United States, the EPA could act as the primary authority. In some cases or countries, a completely independent agency might best administer the bonds.

Protocol for "worst case" analysis already exists within the EPA. In 1977, the U.S. Council on Environmental Quality required such an analysis for implementing the National Environmental Policy Act of 1969 (NEPA). This step mandated that the regulatory agency consider the worst environmental consequences of an action when scientific uncertainty was involved (Fogleman 1987).

The bond provides strong economic incentives to reduce pollution, research the true costs of environmentally damaging activities, and develop innovative, cost-effective pollution control technologies. The bonding system extends the "polluter pays" principle into the "polluter pays for uncertainty as well" principle, also known as the "precautionary polluter pays" principle (4P) (Costanza and Cornwell 1992). The new strategy would allow a much more proactive (rather than reactive) approach to environmental problems because the bond is paid upfront, before damage occurs. It would tend to foster prevention rather than cleanup by unleashing the creative resources of firms toward finding more environmentally benign technologies, as these technologies would also be economically attractive. Competition in the marketplace would lead to environmental improvement rather than degradation. In addition, the bonding system would deal more appropriately with scientific uncertainty.

The 4P approach has several potential applications. The remainder of this section describes three very diverse possibilities: growth management, management of toxic chemicals, and global warming. All of these issues involve high-stakes, high-uncertainty problems for which effective management mechanisms do not currently exist.

The traditional approaches to growth management have centered on zoning and other forms of land-use restrictions. While planning and zoning represent better choices than totally uncontrolled growth, they leave much to be desired. In fact, one can certainly argue that they have not improved environmental conditions. The 4P approach suggests the addition of a flexible impact bond system to regional planning. In this approach, the developer would post an initial impact bond that is large enough to cover the worst-case environmental and economic impacts of the proposed development. The developer would be refunded portions of the bond to the extent that the possible impacts did not occur. Innovative developers who could design projects with lower environmental impacts (for example, by using porous surfaces or by avoiding ecologically sensitive areas) or lower economic impacts (for example, by locating their projects near mass transit or by extending mass transit systems) would be directly rewarded through refunds of their impact bonds. In contrast, developers who defaulted on their bonds would suffer through economic competition with their more innovative competitors, and their bonds could still be used to pay for the negative impacts of their projects.

Although impact fees have been proposed before, they have generally been flat, inflexible, one-time fees that offered no incentive to developers to produce anything but the standard fare. In most cases, they covered only a small fraction of the real impacts of the development. A flexible impact bonding system would solve these problems and help to manage growth in a rational yet flexible way, without taking the right to develop away. Rather, it would merely impose the true costs of that growth on the parties who stand to gain from it, while providing strong economic incentives for the developers to minimize the negative impacts.

Another particularly difficult environmental management problem involves the control of toxic chemicals from both point-sources (such as factories) and non–point-sources (such as agriculture and urban areas). These chemicals can damage ecosystems and human health in even extremely low concentrations; literally thousands of these substances are in common use, and enormous uncertainty surrounds their cumulative and individual impacts. The traditional approach to their management has been to develop lists of toxic chemicals and standards for their allowable concentrations in the environment. Because the list is so long and safety standards are so vague, and because we cannot be certain who is producing and releasing what quantities of which chemicals and how they interact once they enter the environment, this approach has not proved very effective.

The 4P approach suggests a flexible toxic chemical bonding system. In this strategy, the size of the bond would depend on the current estimates of the worst-case damages from the release of the chemicals. Refunds would be based on the extent to which each potential polluter peforms better than the worst case. This system would give polluters strong incentives to reduce their releases by recycling and more efficient use. For example, farmers could no longer afford to "overuse" agricultural chemicals just to ensure that they killed all pests. Industries could no longer afford to release new chemicals with poorly known impacts into the environment. Individual homeowners would pay a high price for applying potentially dangerous chemicals to their lawns and would find less expensive alternatives that, under the bonding system, would be more environmentally benign. This system would be designed to complement other regulatory schemes. It would be self-policing and self-funding, and would provide strong economic incentives to correct environmental problems for which few good management alternatives exist.

Finally, the problem of global warming is probably our most severe current example of a high-stakes, high-uncertainty problem. A CO_2 tax has been proposed to deal with this problem using economic incentives, but current thinking on the tax does not incorporate uncertainty. In contrast, the 4P approach suggests a CO_2 bond (with the size of the bond based on the worst-case estimates of the magnitude of future damages) rather than a tax (with the size of the tax based on much more uncertain estimates of actual future damages). In this way, the efficiency advantages of economic incentives can be reaped without suffering the costs of unknown, future damages.

The 4P system has several other potential applications. Any situation with a large, true uncertainty is a likely candidate for its use—and these situations abound in the modern world. To deal with them, we must broaden our understanding of science and change our approach to environmental management accordingly. We can expect science to provide only the limits for our ignorance. Given those limits, we must encourage policy makers to plan for the worst while providing incentives to produce the best.

Admittedly, the administrative details of the functioning of the 4P system must be worked out, and these details would need to differ according to the application. These details, which are critical to the success of the proposal, must be given careful consideration. Several examples exist of well-intentioned performance bonding systems that failed because the details were not adequately considered. For example, mine reclamation bonds have sometimes been set so low that it was less expensive for the mining companies to default than to reclaim the site. Clearly, the bond must be large enough to cover the worst-case damages so that such malfunctions do not occur.

Nevertheless, the assurance bonding system is not a general panacea that can completely replace our existing regulatory system. The bonding system is best envisioned as a complement to management schemes of regulations and incentives, being better suited to handle those situations where uncertainty is high.

We think that working out which situations are most appropriate for the 4P system, and the administrative details of the system for particular applications, can be most effectively addressed via policy dialogs involving all stakeholders affected by the system under consideration, including the environmental, business, regulatory, and scientific communities. Policy dialog techniques involve the use of impartial third parties in structured meeting environments to bring together people with different views and find areas of consensus or agreement. Such dialogs are currently in the planning stages, and we hope to use this process to develop detailed protocols and administrative procedures for pilot studies of the 4P system. Experimental work is also under way (Costanza and Cornwell 1991) to illuminate the underlying incentive structures in the 4P system and to ensure their effective design and use.

SUMMARY AND SYNTHESIS

What can we learn from what happened in the Chesapeake Bay? In many ways, this ecosystem was—and still is—a "best case scenario" for ecosystem management, but it still has a long way to go. What follows is one possible synthesis with an eye toward determining what may be extrapolated to ecosystem management problems in other areas.

A necessary first step for effective action is the *creation of a broad consensus* about both the essentials of the problem and common goals shared by all interest groups. In the Chesapeake Bay, this step proved relatively easy because a large percentage of the population had direct experience with the Bay and could directly perceive its decline. In addition, a broad preexisting consensus had developed about the common goal of protecting the Bay ecosystem and reversing its decline. In most ecosystem management cases, however, these features are missing. Some small groups of people may be directly affected by the decline of a particular ecosystem, but not a broad cross section; alternatively, the impacts may be so subtle and distributed that they are difficult to perceive without the use of sophisticated tools. Achieving a consensus on a common goal across interest groups in ecosystem management is even more challenging, as interest groups are often in direct conflict over management goals—consider, for example, the spotted owl controversy in the Pacific Northwest. These general characteristics often lead to severe "social traps." In most cases, taking special steps is necessary to achieve a broad consensus. One of the most effective tools in this regard is the adaptive environmental management (AEM) technique developed by Holling and Walters (Walters 1987) or, using different terminology, the methods of environmental dispute resolution developed by Bingham (1988). This first step often represents the most formidable barrier to success and, if not overcome, can lead to the cycles of building and destroying bureaucratic structure described in chapter 1. The Chesapeake Bay system was able to avoid most of the lurking social traps by maintaining a relatively broad consensus and effective dialog between interest groups.

The second step relates to achieving broad consensus on the *details of the problem and the methods of solution*. In the Chesapeake Bay, this step involved a large, relatively coordinated EPA-funded study. In general, funding for detailed studies and the development of action plans is difficult to assemble. If the first step can be achieved, however, the second step has a chance.

The third step, *implementation of the remedial action plans,* follows directly from the first two steps. Key concerns at this point include holding the coalition together long enough to reach this stage and finding the resources to put the plans into effect. Implementation is proceeding in the Chesapeake Bay (a best-case scenario), but many rough spots remain. Now that the real extent of the problem and the magnitude of resources necessary to fix it are known, the coalition is showing signs of strain. Indeed, things may eventually fall apart at this stage. To implement a stable solution, the short-term, local incentives that drive the system away from its long-term goals must be corrected. Pollution taxes, environmental assurance bonds, and other incentive-based instruments represent effective ways of accomplishing this task, as do continuing education and dialog on the problems of the Bay ecosystem and the goals of sustainable ecosystem management. If the Chesapeake can successfully implement this final stage, it will truly be a model of sustainable ecosystem management worthy of being copied.

ACKNOWLEDGMENTS

We would like to thank the thousands of scientists, environmental managers, politicians, and citizens who have worked to restore the Chesapeake Bay. We hope that the final steps can be taken to ensure a sustainable ecosystem. In particular, we would like to thank W. Boynton, M. Kemp, J. Bartholomew, and J. Barnes for helpful comments on earlier drafts of this chapter.

REFERENCES

Axelrod R. The evolution of cooperation. New York: Basic Books, 1984.

Blackistone M. Sunup to sundown: watermen of the Chesapeake. Washington, DC: Acropolis Books, 1988.

Bohm P. Deposit-refund systems. Baltimore and London: Johns Hopkins University Press, 1981.

Boynton W, Garber J, Summers R, Kemp M. Inputs, transformations, and transports of nitrogen and phosphorus in the Chesapeake Bay and selected tributaries. Estuaries 1995;18:285–314.

Braat LC, Steetskamp I. Ecological economic analysis for regional sustainable development. In: Costanza R, ed. Ecological economics: the science and management of sustainability. New York: Columbia University Press, 1991: 269–288.

Brockner J, Rubin JZ. Entrapment in escalating conflicts: a social psychological analysis. New York: Springer-Verlag, 1985.

Bureau of the Census. U.S. statistical abstract [various years]. Washington, DC: U.S. Government Printing Office.

———. County and city data book 1988. Washington, DC: U.S. Government Printing Office.

———. County and city data book 1972. Washington, DC: U.S. Government Printing Office.

———. County and city data book 1952. Washington, DC: U.S. Government Printing Office.

Costanza R, ed. Ecological economics: the science and management of sustainability. New York: Columbia University Press, 1991.

———. Toward an operational definition of ecosystem health. In: Costanza R, Norton B, Haskell BJ, eds. Ecosystem health: new goals for environmental management. Washington, DC: Island Press, 1992.

———. Social traps and environmental policy. BioScience 1987;37:407–412.

Costanza R, Cornwell L. The 4P approach to dealing with scientific uncertainty. Environment 1992;34:13–20.

———. 1991. An experimental analysis of the effectiveness of an environmental assurance bonding system on player behavior in a simulated firm. Final report for U.S. EPA, contract #CR-815393-01-0. Washington, DC: Office of Policy, Planning and Evaluation.

Costanza R, Norton B, Haskell BJ, eds. Ecosystem health: new goals for environmental management. Washington, DC: Island Press, 1992.

Costanza R, Perrings C. A flexible assurance bonding system for improved environmental management. Ecol Econ. 1990;2:57–76.

Costanza R, Sklar FH, White ML. Modeling coastal landscape dynamics. BioScience 1990;40:91–107.

Cross JG, Guyer MJ. Social traps. Ann Arbor: University of Michigan Press, 1980.

DeBellevue E, Costanza R. Unified generic ecosystem and land use model (UGEALUM): a modeling approach for simulation and evaluation of landscapes with an application of the Patuxent River watershed. In: Mihursky JA, Chaney A, eds. New perspectives in the Chesapeake system: a research and management partnership. CRC publication no. 137. Solomons, MD: Chesapeake Research Consortium, 1991:265–275.

Dortch MS, Cerco CF, Robey DL. Work plan for three-dimensional time-varying, hydrodynamic and water quality model of Chesapeake Bay. Miscellaneous Paper EL-88-9. Vicksburg, MS: Department of the Army, Waterways Experiment Station, Corps of Engineers, 1988.

Environmental Protection Agency (EPA). Chesapeake Bay: an introduction to an ecosystem. Washington, DC: EPA, 1982.

Fogleman VM. Worst case analysis: a continued requirement under the National Environmental Policy Act? Columbia J Environ Law 1987;13:53.

Funtowicz SO, Ravetz JR. A new scientific methodology for global environmental issues. In: Costanza R, ed. Ecological economics: the science and management of sustainability. New York: Columbia University Press, 1991:137–152.

Hardin G. The tragedy of the commons. Science 1968;162:1243–1248.

Horton T. Bay country. New York: Tickner and Fields, 1989.

Horton T, Eichbaum WM. Turning the tide: Saving the Chesapeake Bay. Washington, DC: Island Press, 1991.

Hwang C-C. Coupling esturine hydrodynamic and water quality models. Ph.D. dissertation. School of Engineering and Applied Sciences, University of Virginia, 1990.

HydroQual. Development and calibration of a coupled hydrodynamic/water quality/sediment flux model of Chesapeake Bay. Final report, task order 2. Contract No. DACW39-88-D0035. Vicksburg, MS: U.S. Army Corps of Engineers, Waterways Experiment Station, 1989.

———. A steady state coupled hydrodynamic/water quality model of the eutrophication and anoxia process in Chesapeake Bay. Report for USEPA contract No. 68-03-3319 to Battelle Ocean Sciences, for USEPA Chesapeake Bay Program, Annapolis, MD, 1987.

Jacoby ME. Working the Chesapeake: watermen on the Bay. College Park, MD: Maryland Sea Grant, 1990.

Perrings C. Reserved rationality and the precautionary principle: technological change, time and uncertainty in environmental decision making. In: Costanza R, ed. Ecological economics: the science and management of sustainability. New York: Columbia University Press, 1991:153–166.

Platt J. Social traps. Am Psychol 1973;28:642–651.

PRC Environmental Management. Performance bonding. A final report prepared for the U.S. Environmental Protection Agency, Office of Waste Programs and Enforcement. Washington, DC: 1986.

Schubel JR. The life and death of the Chesapeake Bay. College Park, MD: Maryland Sea Grant, 1986.

————. The living Chesapeake. Baltimore: Johns Hopkins University Press, 1981.

Summers R. Description of the Patuxent watershed nonpoint source water quality monitoring and modeling program data base and data management system. Technical report #56. Baltimore: Maryland Department of Health and Mental Hygiene, Office of Environmental Programs, Water Management Administration, Division of Modeling and Analysis, 1990.

Teger AI. Too much invested to quit. New York: Pergamon, 1980.

Tippie VK. An environmental characterization of Chesapeake Bay and a framework for action. In: Kennedy VS, ed. The estuary as a filter. Orlando, FL: Academic Press, 1984:467–488.

U.S. Department of Agriculture (USDA). Census of agriculture. Washington, DC: U.S. Government Printing Office, 1987.

————. Census of agriculture. Washington, DC: U.S. Government Printing Office, 1974.

————. Census of agriculture. Washington, DC: U.S. Government Printing Office, 1954.

VanDyne D, Gilbertson C. Estimating U.S. livestock and poultry manure and nutrient production. Washington, DC: U.S. Department of Agriculture, 1978.

Warner W. Beautiful swimmers. New York: Little, Brown, 1976.

White CP. Chesapeake Bay: a field guide. Centreville, MD: Tidewater, 1989.

CHAPTER 19

Paleolimnological Assessments of Ecosystem Health: Lake Acidification in Adirondack Park

John P. Smol

Brian F. Cumming

As with most environmental problems, historical pH data of sufficient quantity and quality to reconstruct past acidification and recovery scenarios are often lacking. In lieu of these missing data sets, indirect proxy methods must be used (Smol 1992, 1995). Fortunately, the distributions of many aquatic organisms (such as diatom and chrysophyte species assemblages) can be quantitatively related to lake water pH and pH-related variables (for a review of these calibration techniques, see Charles and Smol 1994; Birks 1995). Because many of these organisms are well preserved in dated lake sediment cores, past trajectories of lake water pH can be inferred with a high degree of reliability (see chapter 13). The paleolimnological approach has been used extensively throughout North America and Europe to track past changes in lake water acidification and recovery (Charles et al 1989; Battarbee et al 1990; Dixit SS et al 1992; Dixit AS 1992a).

The Adirondack Park in northern New York is a large forested area containing many lakes. A number of these lakes are presently acidic, and this region is known to receive a large amount of acidic precipitation. The link between acid rain and acidic lakes was difficult to prove, however, as long-term pH data were not available. Discussions about the existence of this relationship rapidly developed into a serious and often polarized scientific and political debate, which escalated during the 1980s. The critical questions were as follows:

■ Why did these lakes become acidic?

■ Were they naturally acidic?

■ If not, then how much acidic precipitation could they be subjected to before they began to deteriorate?

■ Were they naturally fishless, or did they lose their fish as a result of acidification?

Paleolimnological approaches were used to answer these pressing questions.

THE ACIDIFICATION HISTORY OF DEEP LAKE, NEW YORK

Because of space limitations, only one paleolimnological example will be described here in detail. Deep Lake (43°36′54″ N, 74°39′52″ W) is a headwater lake in New York's Adirondack Mountains that is known to be presently acidic (pH ≈ 4.8) and fishless. Was this lake naturally acidic, or did it become more acidic as a result of acidic precipitation? A detailed paleolimnological analysis of this site provided these answers (Charles et al 1990). A sediment core was collected, sectioned at fine intervals (1 cm), and dated using ^{210}Pb chronology. It indicated that the short sediment core dated back to the early 1700s. For the purpose of this example, only the post-1800 sediments will be discussed. The major stratigraphic changes are summarized here (Figures 19.1–19.3).

The primary indicators to track past changes in lake water acidity were the siliceous algal remains of diatoms and chrysophytes because, as mentioned in chapter 13, statistically robust transfer functions were developed that could relate lake water pH and other pH-related variables to the distributions of diatom (Dixit et al 1993) and chrysophyte (Cumming et al 1992a) taxa. Even a cursory glance of the paleoindicators recorded in the core (see Fig. 19.1) reveals that a major change in the biological community began to occur in Deep Lake around 1920, with unprecedented changes in the biological communities occurring in the post-1950 sediments. For example, acid-indicating diatoms (such as *Fragilaria acidobiontica* and *Navicula tenuicephala*) and chrysophytes (such as *Synura echinulata*) and then the more acidic taxa *Mallomonas hindonii* and *M. hamata* increase, replacing circumneutral taxa (such as *M. crassisquama*). Changes in fossil midge (chironomid) larvae and cladoceran invertebrates also occur, with the near extirpation of the planktonic Cladocera taxa.

By using the quantitative transfer functions described earlier (see, for example, Dixit et al 1993), the species changes in diatoms could be reconstructed (Charles et al 1990). These pH reconstructions show that Deep Lake had a slightly acidic background pH, but then acidified further during this

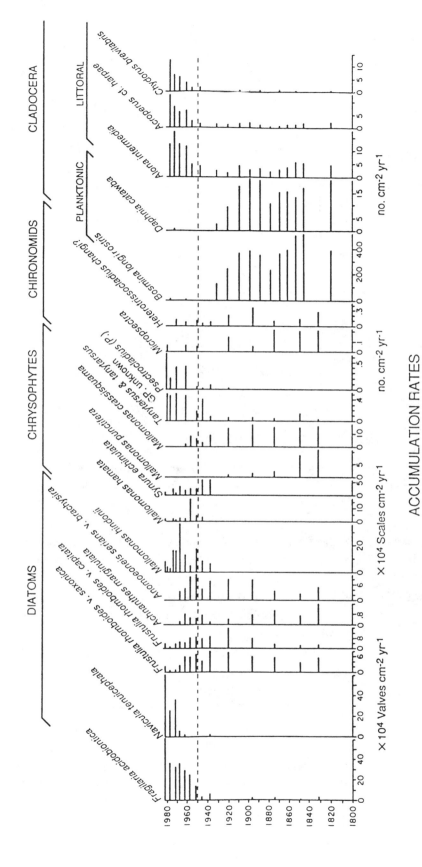

FIGURE 19.1. *Stratigraphic profile of the dominant diatom, chrysophyte, chironomid, and cladocera taxa from a ^{210}Pb-dated sediment core from Deep Lake. All taxa are expressed in terms of accumulation rates (from Charles et al 1990).*

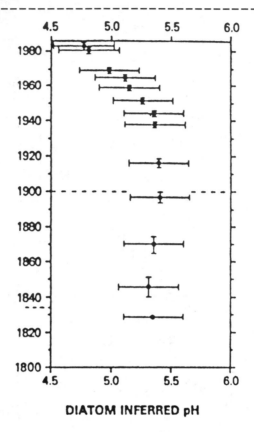

FIGURE 19.2. *Changes in diatom-inferred pH in a* ^{210}Pb*-dated sediment core from Deep Lake (from Charles et al 1990).*

century, coincident with increases in the deposition of strong acids (see Fig. 19.2). Other paleoindicators preserved in the Deep Lake core, such as coal and oil soot particles, polycyclic aromatic hydrocarbons (PAHs), lead (Pb), and vanadium (V), provide a clear record of increased atmospheric input of materials associated with the combustion of fossil fuels beginning in the late 1800s and early 1900s (see Fig. 19.3).

In addition to using these diatom- and chrysophyte-based studies, past trends in fisheries can be assessed using paleolimnological techniques. Although fish do not usually leave reliable fossils in most lake sediments, the mandibles of *Chaoborus* (phantom midge) larvae are generally well preserved and can be identified to the species level (Uutala 1990). The key indicator in these studies for eastern North America is *Chaoborus americanus*, a taxon that can coexist with fish only rarely; thus, its presence indicates periods in a lake's history when the lake has become fishless. Uutala (1990) and Kingston et al (1992) used *Chaoborus* indicators to demonstrate that fishery losses in acidified Adirondack lakes were coincident with major increases in lake water

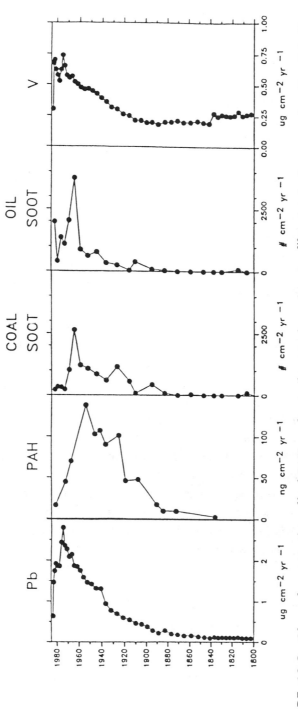

FIGURE 19.3. *Observed concentrations of lead, PAHs, coal soot, oil soot, and vanadium in a ^{210}Pb-dated sediment core from Deep Lake (from Charles et al 1990).*

---

acidity and monomeric aluminum concentrations. In Deep Lake, Uutala (1990) showed that the fish were extirpated by the 1940s, when lake water pH levels had fallen to low levels.

REGIONAL CHANGES IN LAKE ACIDIFICATION

The previous example shows how an individual lake can be studied to determine the effects of past stress (in this case, acidic precipitation) on one lake. Many management questions, however, often require answers to more regional questions, such as: What proportion of Adirondack lakes have acidified since the 1800s, and to what extent? What proportion of lakes were naturally acidic? Analyzing one core in detail is time-consuming, making it difficult to undertake these detailed analyses on a large number of sites. Instead, paleolimnologists can estimate past overall limnological changes by using the "top/bottom" paleolimnological approach. A considerable research effort was recently completed to answer such regional questions in the Adirondack Park (New York), as part of the Paleoecological Investigation of Recent Lake Acidification-II (PIRLA-II) project (Charles and Smol 1990).

The "top/bottom" approach is fairly straightforward. Instead of analyzing 15 to 20 sediment increments over the last century or so of the lake's history, as was done in the Deep Lake example, the paleolimnologist simply examines two samples per core for diatoms, or whatever indicator is being used to infer past limnological conditions. The "top" sample consists of the top 1 cm of sediment, which represents the last few years (approximately present-day conditions) of lake history. The "bottom" sample is a 1-cm interval of sediment taken from the core that predates the period of cultural disturbance. In the PIRLA-II project, this sediment was taken from sediments deposited in the early part of the nineteenth century (which could be dated using ^{210}Pb chronology; see chapter 13). This "bottom" sample represents lake conditions before the advent of strong acidic precipitation. Once the diatoms are enumerated from these two samples, and the environmental inferences are made (for example, diatom-inferred pH), then preimpact conditions can be compared with present-day conditions. Because only two sediment samples per core must be analyzed with the "top/bottom" approach, a larger number of lakes can be used in the subsequent regional assessments.

The results of the top/bottom analyses conducted on diatoms as part of the PIRLA-II project for 37 randomly selected lakes in the Adirondacks are summarized in Figure 19.4. The lakes are listed in order of increasing pH, with the lakes that are presently most acidic appearing at the top. The histograms indicate the inferred changes (since preindustrial times, in the early 1800s) in lake water pH, acid neutralizing capacity (ANC), monomeric alu-

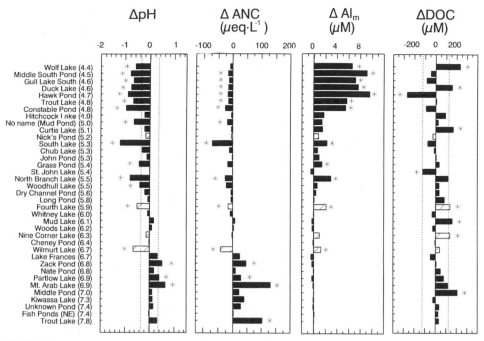

FIGURE 19.4. *Diatom-based estimates of historical change (preindustrial to the present) in pH, acid neutralizing capacity (ANC), total monomeric aluminum (Al_m), and dissolved organic carbon (DOC) from 37 randomly selected Adirondack lakes (listed in order of present-day pH). The estimates are presented as the differences in inferred water chemistry between the top (0–1 cm) and bottom (>20 cm, and usually >25 cm) sediment core intervals. The lakes are arranged according to increasing present-day measured pH. Lakes indicated by hatched bars were limed within five years before coring or assessment of water chemistry. The dotted lines represent ± bootstrap root mean squared error ($RMSE_{boot}$) of prediction for pH and DOC. Estimates of change in ANC and Al_m are presented as differences between back-transformed estimates as these models were developed based on log_e data. Asterisks denote changes that are greater than $RMSE_{boot}$ for the various inference models (from Cumming et al 1992a).*

minum (Al_m, which may be toxic to fish), and dissolved organic carbon (DOC, a source of natural acidity). From this diagram, it is evident that all presently acidic lakes have acidified further since the last century, and that approximately 80% of the target population lakes with present pH between 5.2 and 6.0 have undergone large declines in pH and ANC, and concomitant increases in monomeric aluminum, since preindustrial times (Cumming et al 1992).

Interestingly, these analyses did show that a *few* lakes were naturally acidic, and some lakes were naturally fishless (see, for example, Uutala 1990). This finding has important management considerations. Although some of these lakes may appear to be "damaged" or "unhealthy," others may not be.

Moreover, these lakes are not good candidates for lake liming and fish stocking programs, as these ecosystems have "healthy" biological communities that have developed over thousands of years in, for example, naturally acidic environments and/or naturally fishless conditions. Knowledge of the background (that is, preimpact) conditions is vital for prescribing the correct mitigation programs.

In summary, paleolimnological data have contributed greatly to our knowledge of lake acidification in the Adirondacks by providing missing data on background conditions and natural variability and by allowing us to track changes in lake water pH through time. Examination of the timing and rates of changes from many lakes has provided a weight of evidence that supports acidic deposition as the most likely mechanism for the observed patterns; it has also discounted natural long-term acidification, forest harvesting, and/or forest blowdown as possible mechanisms that could produce the large number of acidic lakes in the Adirondacks (see, for example, Charles et al 1990; Cumming et al 1992, 1994). Furthermore, paleolimnological techniques have been used to estimate loads of sulfate deposition under which lakes have acidified (Cumming et al 1994). The development of paleo-estimates of the amount of acidification in the Adirondacks (see, for example, Sullivan et al 1990, 1992) has prompted a reevaluation and subsequent improvement of computer models that are used to evaluate changes to lakes under different loads of acidic deposition (Sullivan et al 1996).

Acidification constitutes only one example of applying paleolimnological approaches to the study of ecosystem health. Similar techniques are being used to study, for example, lake eutrophication, climatic change, salinification, and the levels of metals and other contaminants.

REFERENCES

Battarbee RW, Mason J, Renberg I, Talling JF. Palaeolimnology and lake acidification. London: Royal Society, 1990.

Birks HJB. Quantitative palaeoenvironmental reconstructions. In: Maddy D, Brew JS, eds. Statistical modelling of Quaternary science data. Technical guide 5. Cambridge: Quaternary Research Association, 1995:161–254.

Charles DF. Relationships between surface sediment diatom assemblages and lake water characteristics in Adirondack lakes. Ecology 1985;66:994–1011.

Charles DF, Battarbee RW, Renberg I, et al. Paleoecological analysis of lake acidification trends in North America and Europe using diatoms and chrysophytes. In: Norton SA, Lindberg SE, Page AL, eds. Acid precipitation. Vol. 2. Stuttgart: Springer-Verlag, 1989:207–276.

Charles DF, Binford MW, Furlong ET, Hites RA, Mitchell MJ, Norton SA, Oldfield F, Paterson MJ, Smol JP, Uutala AJ, White JR, Whitehead DR, Wise J. Paleoecological investigation of recent lake acidification in the Adirondack Mountains, N.Y. J Paleolimnol 1990;3:195–241.

Charles DF, Smol JP. Long-term chemical changes in lakes: quantitative inferences using biotic remains in the sediment record. In: Baker L, ed. Environmental chemistry of lakes and reservoirs. Advances in chemistry series 237. Washington, DC: American Chemical Society, 1994:3–31.

———. The PIRLA II project: regional assessment of lake acidification trends. Verh Internat Verein Limnol 1990;24:474–480.

Cumming BF, Davey K, Smol JP, Birks HJB. When did Adirondack Mountain lakes begin to acidify and are they still acidifying? Can J Fish Aquat Sci 1994;51: 1550–1568.

Cumming BF, Smol JP, Birks HJB. Scaled chrysophytes (Chrysophyceae and Syunurophyceae) from Adirondack (N.Y., USA) drainage lakes and their relationship to measured environmental variables, with special reference to lakewater pH and labile monomeric aluminum. J Phycol 1992a;28:162–178.

Cumming BF, Smol JP, Kingston JC, Charles DF, Birks HJB, Camburn KE, Dixit SS, Uutala AJ, Selle AR. How much acidification has occurred in Adirondack region (New York, USA) lakes since pre-industrial times? Can J Fish Aquat Sci 1992b;49:128–141.

Dixit AS, Dixit SS, Smol JP. Algal microfossils provide high temporal resolution of environmental trends. Water Air Soil Poll 1992a;62:75–87.

———. Long-term trends in lake water pH and metal concentrations inferred from diatoms and chrysophytes in three lakes near Sudbury, Ontario. Can J Fish Aquat Sci 1992b;49(suppl 1):17–24.

Dixit SS, Cumming BF, Kingston JC, Smol JP, Birks HJB, Uutala AJ, Charles DF, Camburn K. Diatom assemblages from Adirondack lakes (N.Y., USA) and the development of inference models for retrospective environmental assessment. J Paleolimnol 1993;8:27–47.

Dixit SS, Smol JP. Diatom evidence of past water quality changes in Adirondack seepage lakes (New York, U.S.A.). Diatom Res 1995;10:113–129.

Dixit SS, Smol JP, Kingston JC, Charles DF. Diatoms: powerful indicators of environmental change. Environ Sci Tech 1992;26:22–33.

Kingston JC, Birks HJB, Uutala AJ, Cumming BF, Smol JP. Assessing trends in fishery resources and lake water aluminum for paleolimnological analyses of siliceous algae. Can J Fish Aquat Sci 1992;49:116–127.

Smol JP. Paleolimnological approaches to the evaluation and monitoring of ecosystem health: providing a history for environmental damage and recovery. In: Rapport D, Gaudet D, Calow P, eds. Evaluating and monitoring the health of large-scale ecosystems. NATO ASI Series. Vol. 128. Stuttgart: Springer-Verlag, 1995:301–318.

———. Paleolimnology: an important tool for effective ecosystem management. J Aquat Ecosys Health 1992;1:49–58.

Sullivan TJ, Charles DF, Smol JP, Cumming BF, Selle AR, Thomas D, Dixit S. Quantification of changes in lakewater chemistry in response to acidic deposition. Nature 1990;345:54–58.

Sullivan TJ, Cosby BJ, Driscoll CT, Charles DF, Hemond HF. Influence of organic acids on model projections of lake acidification. Water Air Soil Poll 1996;91: 271–282.

Sullivan TJ, Turner RS, Charles DF, Cumming BF, Smol JP, Schofield CL, Driscoll CT, Birks HJB, Uutala AJ, Kingston JC, Dixit SS, Bernert JA, Ryan PF, Marmorek DR. Use of historical assessment for evaluation of process-based model projections of future environmental change: lake acidification in the Adirondack Mountains, New York, U.S.A. Environ Poll 1992;77:253–262.

Uutala AJ. *Chaoborus* (Diptera: Chaoboridae) mandibles—paleolimnological indicators of the historical status of fish populations in acid-sensitive lakes. J Paleolimnology 1990;4:139–151.

CHAPTER 20

The Desert Grasslands

Walter G. Whitford

CHARACTERISTICS OF THE DESERT GRASSLAND ECOSYSTEM

The desert grasslands are widely distributed in North America, occupying much of southeastern Arizona, the southern half of New Mexico, and west Texas in the United States and extending southward through 13 states in Mexico, from Sonora to Puebla. They represent more than 500,000 km² of basin and valley lands that skirt the hills and mountain ranges of southwestern North America (McClaran 1995). The entire geographic range of the desert grasslands has seen a rapid decline in grass cover and an increase in cover by woody shrubs, especially mesquite (*Prosopis* spp.) and creosotebush (*Larrea tridentata*). Traditionally, the desert grasslands have supported extensive stands of perennial bunch grasses, known colloquially as the grama grasses, which reproduce primarily by stolons. Other, smaller bunch grasses occupied habitat patches such as infrequently inundated swales that support a cover of tabosa grass, *Hilaria mutica*.

The grasses of the desert grasslands are all C4 (referring to the type of photosynthetic pathway) grasses that thrive in the warm season. Warm-season grasses do not produce leaves until the night-time temperatures remain higher than 10°C. As a result, rainfall during the winter and spring is ineffective for primary production of these grasses.

The woody shrubs that have invaded the desert grasslands are C3 (a photosynthetic pathway) plants that are considered cool-season vegetation. These shrubs have deep root systems and redistribute intercepted moisture efficiently via stem flow to the root channels and to deep storage. Water that is redistributed by shrubs lies largely out of the root zone of grasses (Whitford et al 1995).

At the present time, the desert grasslands exist as small patches of perennial grasses in a sea of woody shrubs or shrub-grass mosaics.

HISTORY OF DEGRADATION

Much of the degradation of the world's arid and semi-arid grasslands has occurred within the last century. In North America, reports of degradation of the desert grasslands surfaced as early as the 1880s (Bahre and Shelton 1993). These early reports appeared during the same decades as similar reports from Australia and southern Africa. Thus, the changes experienced by the desert grasslands do not represent a situation that is unique to North America. The coincidence of timing and the dominant land-use of the desert grasslands strongly suggests that the causes of degradation may be similar across the world's desert grasslands.

In the southwestern United States, human land-use prior to settlement by Europeans consisted of light harvesting of plants and animals by indigenous hunter-gatherers. The first settlement by Europeans began in the late sixteenth century, when small groups of Mexican-Spanish settlers entered the region. Initially they lived in the perennial river valleys, where the Europeans learned about irrigated agriculture from the pueblo tribes. At that time, only limited livestock grazing occurred in the desert grasslands adjacent to the farm settlements. By the 1820s, small Mexican settlements were scattered throughout the region. These settlements' size remained limited by the small amount of rangeland within reach of perennial water and by the depredations by bands of Apaches, a situation that persisted until the 1870s. After the American Civil War, large herds of cattle moved into the unfenced grasslands of the southwestern United States.

By 1900, ranchers in the area had clearly recognized the detrimental effects of grazing by large herds of cattle. The U.S. government sent an agricultural scientist, David Griffiths, to the area in 1900. Griffiths collected data on the environmental degradation by direct observations and by questionnaires sent to local ranchers. He also set up experiments on forage grasses. The following are excerpts from his report to the U.S. Department of Agriculture.

H. C. Hooker, one of the earliest and most successful ranchers in southeastern Arizona, wrote:

> The San Pedro Valley in 1870 had an abundance of willow, cottonwood, sycamore, and mesquite timber, also large beds of sacaton and grama grasses, sagebrush and underbrush of many kinds. The river bed was shallow and grassy and its banks were beautiful with a luxuriant growth of vegetation. Now the river is deep and its banks are washed out, the trees and

underbrush are gone, the sacaton has been cut out by the plow and grub hoe, the mesa has been grazed by thousands of horses and cattle and the valley has been farmed. Cattle and horses going to and from feed and water have made many trails or paths to the mountains. Browse on the hillsides has been eaten off. Fire has destroyed much of the shrubbery as well as the grass giving the winds and rains full sweep to carry away the earth loosened by the feet of the animals. . . . (Griffiths 1901)

In response to a question about maximum numbers of grazing animals and current carrying capacity, Hooker wrote, "There were fully 50,000 head of stock at the head of Sulphur Spring Valley and the valley of the Aravipa in 1890. In 1900 there were not more than one-half that number and they were doing poorly."

In his survey work, Griffiths (1901) learned that the grazing potential of much of southeastern Arizona had been diminished to less than half the stock that had grazed in the same area in 1880. The pattern of degradation of southeastern Arizona desert grasslands was documented in a letter from C. H. Bayless of Oracle, Arizona, to Griffiths:

. . . the San Pedro Valley consisted of a narrow strip of subirrigated and very fertile lands. Beaver dams checked the flow of water and prevented the cutting of a channel. Trappers exterminated the beavers and less grass on the hillsides permitted greater erosion, so that within four or five years a channel varying in depth from 3 to 20 feet was cut almost the whole length of the river. . . . Of the rich grama grasses that originally covered the country so little now remains that no account can be taken of them. . . . Where stock water is far removed, some remnants of perennial grassses can be found. Grasses that grow only from seed sprouted by summer rains are of small and transitory value. The foliage of the mesquite and catsclaw bushes is eaten by most animals and even the various cacti are attempted by starving cattle. . . . No better pasture was ever found in any country than that furnished by our native grama grasses, now almost extinct. . . . The present unproductive conditions are due entirely to overstocking. . . . Twelve years ago, 40,000 cattle grew fat along a certain portion of the San Pedro where now 3,000 can not find sufficient forage for proper growth and development. If instead of 40,000 head, 10,000 had been kept on this range, it would in all probability be furnishing good pasture for the same number today. Very few of these cattle were sold or removed from the range. They were simply left there until the pasture was destroyed and the stock then perished by starvation.

The words of both Bayless and Hooker paint a graphic picture of rapid conversion of a grassland Eden to a desert hell. The degradation processes that Griffiths (1901) described in the San Pedro drainage in southeastern

Arizona did not cease in the early 1900s, however. Data from Landsat (multispectral scanner) imagery of the San Pedro watershed show that the desert grassland fragments continued to diminish in size and become more isolated in the increasingly degraded rangeland (Kepner et al 1995, personal communication). In a comparison of images from 1974 and 1987, the evaluation showed that extensive grassland areas with high connectivity were the ecosystems most vulnerable to fragmentation caused by the expansion of woody shrubs and cacti. In their preliminary analysis, Kepner and Ritters found that the number of grassland patches increased 61% and the average grassland patch size decreased 60% over this period (Figure 20.1). The continuing degradation involved the loss of vegetative cover, which holds soil in place and retains water on-site, and a replacement of grasses with shrubs that are not palatable to livestock. The consequences included increased run-off and sediment input into the major drainage system (the Gila River), reduced groundwater recharge, loss of economic production potential, loss of habitat for grassland specialists like pronghorn antelope, and conflicts between ranchers and other users of the land (including hikers, wildlife viewers, and recreational users from the nearby cities such as Tucson).

Detailed data on changes in vegetation and soils, carrying capacity for livestock, and efficacy of various restoration treatments are available from several research stations that were established in the early 1900s. The Jornada Experimental Range in southern New Mexico, for example, provides data for

 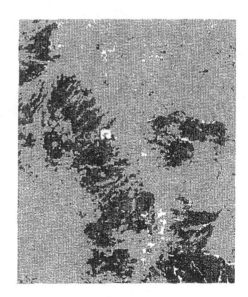

1974   1987

FIGURE 20.1. *Landsat multispectral scanner imagery of the grasslands (black), woodland–shrublands (gray), and irrigated agriculture (white) in the San Pedro river basin.*

the period described by Griffiths (1901) in southeastern Arizona as well as the period from those decades to the present. It encompasses most of the 58,492 hectares in the San Andreas Mountains basin named by the early Spanish settlers as *Jornada del Muerto* (journey of death). In the 1880s, ranches were established around springs in the San Andreas Mountains. The Jornada Basin was not developed until 1901, when ranchers drilled deep wells to provide year-round water for their livestock. Soon after ranches were established in the area, the stockmen began to become concerned about the loss of livestock-carrying capacity and petitioned the government for assistance. The Jornada Experimental Range (originally the Jornada Range Reserve) was established in 1912 with the stated mission of carrying out research designed to improve grazing in the region.

Despite the conservative management of the Jornada Experimental Range, vegetation change continued, as documented by Buffington and Herbel (1965). As in the San Pedro watershed and all other valleys in the desert grassland rangelands of the United States and Mexico, the trajectory of reduction in vegetative cover, replacement of C4 grasses with C3 shrubs, and soil loss has occurred at varying rates over the decades since the Jornada Experimental Range was established.

In the Jornada Basin, desert grasslands (Figure 20.2) have been replaced by creosotebush shrublands on the piedmont slopes of the mountains (Figure

FIGURE 20.2. *Desert grasslands in southern New Mexico. The dominant grasses are black-grama (*Bouteloua eriopoda*) and dropseeds (*Sporobolus spp.*).*

FIGURE 20.3. *Creosotebush* (Larrea tridentata) *shrubland, a vegetation type that has replaced black-grama grassland on piedmont slopes in much of the desert grassland region of the southwestern United States and Mexico.*

20.3) and by mesquite coppice dunes in the badly degraded sites in the basin (Figure 20.4). This pattern has been repeated in Mexican rangelands as far south as the desert grasslands extend and throughout west Texas and southeastern Arizona. Most of the public grazing land administered by the U.S. Bureau of Land Management, for example, has suffered some degree of degradation. In the worst cases (those involving grazing allotments in mesquite coppice dunes), the ranchers have not renewed their allotments because insufficient forage exists to support livestock.

Consequences of Degradation of Desert Grasslands

The economic consequences of the continuing degradation of desert grasslands are the root cause of social conflicts in the rangelands of the West.

FIGURE 20.4. *Mesquite (*Prosopis glandulosa*) coppice dunes, the most degraded and resistant ecosystem that has replaced desert grasslands in the grassland regions of the southwestern United States and Mexico.*

Ranchers in the desert grassland region of the southwestern United States typically realize a return on investment of 1% to 3% (Holechek 1992). "From an investment standpoint Chihuahuan Desert cattle ranching would be considered unprofitable and risky by any Wall Street analyst" (Holechek 1992).

The desert grasslands of New Mexico are in much better condition than their Mexican counterparts. According to assessments by the U.S. Department of the Interior, 15% to 18% of New Mexico desert grassland range is in poor condition, 45% is in fair condition, 35% is in good condition, and only 5% to 10% is in excellent condition. Rangeland in poor to good condition is at risk of degrading to a lower condition. Changes in wholesale cattle prices, government policies such as drought subsidies, and climatic fluctuations all combine to influence ranch management decisions that frequently produce further degradation. Ranches in the poor to fair condition category are frequently "hobby ranches" where the desirable "cowboy lifestyle" is subsidized by employment in another profession (Holechek 1992). Because much of the rangeland exists in an unhealthy state, the present-day cattle grower is forced by economic necessity (the need to repay loans, feed a family, and so on) to maximize herd size. The economic situation compels ranchers to make decisions based on short-term financial returns rather than those that could improve the long-term health of the rangeland resource.

Economic peril is not the only consequence of unhealthy rangelands. Environmental organizations frequently come into conflict with ranchers over management of the most fragile and most biologically diverse parts of the rangeland ecosystems (the riparian zones along creeks and streams). Degradation of the riparian systems followed immediately upon the introduction of the livestock industry into the desert grasslands and continues virtually unabated today. Healthy riparian systems are characterized by grassy banks, no evidence of channel cutting, and reproduction of dominant riparian species such as willows (*Salix* spp.), cottonwoods (*Populus* spp.), and seep willow (*Baccharis* spp.). When they are given access to these areas, domestic livestock tend to concentrate their activity in riparian zones. In their grazing, these animals consume both seedling trees and grass.

In one study, a healthy 1-km stretch of riparian woodland along a dry stream bed was found to support 25 species of breeding birds; in contrast, a 1-km stretch of an adjacent heavily grazed riparian system supported only 6 species of breeding birds (W. G. Whitford, unpublished data). The unhealthy stretch of stream bed had a deep (30–50 cm) incised channel, long stretches of 20–50 m with no fringing vegetation, and no grass or tree shrub seedlings along the channel margins. Because of such findings, environmental activists and other groups of concerned citizens are pressuring the ranching industry to change their management practices so as to protect and perhaps heal damaged riparian ecosystems.

While degradation of desert grassland riparian systems has undoubtedly reduced biodiversity, a growing body of evidence suggests that degradation of the desert grasslands has actually increased biodiversity. That development probably reflects the greater structural diversity of the rangeland ecosystems, which has been transformed from the uniform height and density grassland to shrublands and shrub-grasslands with varying heights and densities. These changes, however, have also produced adverse effects for grassland specialists such as the banner-tailed kangaroo rat and the pronghorn antelope. Other activities that have probably affected biodiversity on desert grassland include the extirpation of the prairie dogs, wolves, and mountain lions by U.S. government animal damage control specialists over most of the U.S. desert grasslands. As a consequence of past management decisions and the continuing loss of habitat, it is clear that many species of desert grassland specialists have disappeared or are at risk of becoming extinct.

RANGELAND REHABILITATION

Efforts to rehabilitate the degraded rangelands have focused on a variety of methods of shrub removal, including herbicide application, root plowing, and burning. These efforts have frequently been combined with seeding either a mixture of seeds of native grasses, a mixture of seeds of native and exotic grasses, or seeds of exotic grasses (Call and Roundy 1991). The reasoning behind these efforts was based on interpretations of the Clementsian conceptual model of succession (Clements 1916, 1928). This model concluded that woody shrubs were invasive and out-competed grasses, thereby creating conditions promoting a decline in grass production and eventual soil loss. More recently constructed conceptual models of vegetation change in semi-arid and arid ecosystems focus on discontinuous and irreversible transitions and alternative stable states (Westoby et al. 1989).

Attempts at rehabilitation of degraded rangelands have been both extensive and expensive. For example, between 1940 and 1981, an average of 600,000 hectares of brush infested rangeland in Texas was treated annually by herbicides, fire, and mechanical means such as root plowing, chaining, and discing (Rappole et al 1986). Despite the intensity of treatment, land managers realized only a transitory and short-term benefit. Regrowth of shrubs occurred in a short period of time, and retreatment became necessary within 15 years of root plowing, 2 years after chaining, and 8 years after treatment with herbicides (Rappole et al 1986).

Similar results were recorded in other parts of the desert grassland rangelands (Herbel et al 1983). Even when combined with livestock exclusion and root plowing, herbicide treatment has not resulted in the reestablishment of grasslands (Roundy and Jordan 1988; Chew 1982). On the Jornada Experimental Range, the shrub vegetation in three 290-hectare grazing exclosures that were built in the early 1930s has remained the same as that in the surrounding grazed areas. Thus, the desert shrublands that now dominate most of the land area that consisted of desert grasslands in the 1820s appear to be both resistant and resilient to efforts to reduce shrub cover and restore grasslands.

Desert shrublands differ from desert grasslands in many key features, such as water infiltration and storage, period of photosynthesis, predictability of growth period, spatial redistribution of water and nutrients, and spatial distribution of the seed bank. In their own unique way, the shrublands are actually healthy ecosystems. At present, the shrublands that replaced the desert grasslands have limited usefulness to humans—thus, they are often considered unhealthy and largely incurable.

Although the attempts to restore native desert grasslands to areas now dominated by woody shrubs have failed both ecologically and economically, partial success in restoring some of the functions of desert grassland has been achieved by seeding and spread of non-native grasses. These grasses have revegetated watersheds and reduced damage by flooding. They also provide forage for livestock at certain times of the year and on some microsites.

After initial screening by the U.S. Department of Agriculture, Lehmann's lovegrass (*Eragrostis lehmanniana*) was seeded on selected sites in southeastern Arizona and New Mexico. Between 1940 and 1980, land managers in Arizona established Lehmann's lovegrass on 70,000 hectares; the grass has since spread to an additional 130,000 hectares (Cox and Ruyle 1986; Anable et al 1992). In New Mexico, *E. lehmanniana* appears to be replacing native grama grasses and dropseed grasses. Despite the benefits of having this perennial grass provide soil cover and forage for livestock in shrub dominated ecosystems, the diversity of the fauna remains severely reduced (Anable et al 1992). Lehmann's lovegrass continues to spread into both shrub-dominated areas and areas of native grassland. The spread of this exotic plant has spurred considerable debate about the costs and benefits of a partial cure.

REFERENCES

Anable ME, McClaran MP, Ruyle GB. Spread of introduced Lehmann lovegrass, *Eragrostis lehmanniana* Nees. in southern Arizona, USA. Biol Conserv 1992; 61:181–188.

Bahre CJ, Shelton ML. Historic vegetation change, mesquite increases and climate in southeastern Arizona. J Biogeography 1993;20:489–504.

Buffington LC, Herbel CH. Vegetational changes on a semi-desert grassland range from 1858 to 1963. Ecol Monographs 1965;35:139–164.

Call CA, Roundy BA. Perspectives and processes in revegetation of arid and semiarid rangelands. J Range Mgt 1991;44:543–549.

Chew RM. Changes in herbaceous and suffrutescent perennials in grazed and ungrazed desertified grassland in southeastern Arizona 1958–1978. Am Midland Naturalist 1982;108:159–169.

Clements FE. Plant succession and indicators. New York: Hafner Publishing, 1928.

————. Plant succession: an analysis of the development of vegetation. Carnegie Institute Publication 242. Washington, DC: 1916.

Cox JR, Ruyle GB. Influence of climatic and edaphic factors on the distribution of *Eragrostis lehmanniana* Nees. in Arizona, USA. J Grassland Soc S Africa 1986; 3:25–29.

Griffiths D. Range improvement in Arizona. USDA, Bureau of Plant Industry Bull 1901;4:1–31.

Herbel CH, Gould WL, Leifeste WF, Gibbens RP. Herbicide treatment and vegetation in response to treatment of mesquites in southern New Mexico. J Range Mgt 1983;36:149–151.

Holecheck JL. Financial aspects of cattle production in the Chihuahuan Desert. Rangelands 1992;14:145–149.

Kepner WG, Ritters RH, Wickham JD. A landscape approach to monitoring and assessing ecological condition of rangelands—San Pedro case study. Proceedings of the Symposium on EMAP Science. Chapel Hill, NC, March 7–9, 1995.

Research Triangle Park, NC: U.S. Environmental Protection Agency, Environmental Monitoring and Assessment Program, 1995.

McClaran MP. Desert grasslands and grasses. In: McClaran MP, Van Devender TR, eds. The desert grassland. Tucson, AZ: University of Arizona Press, 1995:1–30.

Rappole JH, Russell CE, Norwine JR, Fullbright TE. Anthropogenic pressure and impacts on marginal, neotropical, semiarid ecosystems: the case of south Texas. Sci Total Environ 1986;55:91–99.

Roundy BA, Jordan GL. Vegetation changes in relation to livestock exclusion and root plowing in southeastern Arizona. SW Naturalist 1988;33:425–436.

Westoby M, Walker B, Noy-Meir I. Opportunistic management for rangelands not at equilibrium. J Range Mgt 1989;42:266–274.

WhitfordWG, Martinez-Turanzas G, Martinez-Meza E. Persistence of desertified ecosystems: explanations and implications. Environ Monitor Assess 1995;37: 319–332.

CHAPTER 21

The Health of Some Forest Ecosystems of Cuba

Leda Menéndez

Deysi Vilamajó

René Capote

THE notion of ecosystem health is closely related to the concept of sustainability. Although indicators of rapid change are often considered to imply unsustainability, the two concepts are not identical. Some changes predict the further degradation of a habitat; others may merely be part of the normal "physiology" of the system and perhaps even necessary for its preservation.

In this chapter, we will compare two Cuban ecosystems—a montane rainforest and the coastal mangroves. Both of these systems underwent major changes during the period of observation. As we will show, the significance of the changes differs dramatically for the two sites.

The wide range of ecosystems and landscapes, corresponding to the high level of geological and geomorphological diversity, has allowed for the recognition of 45 ecoregions and 24 subregions of the Cuban Archipelago. Of these areas, eight retain a high level of integrity, representing 10% of the national territory (IGACC-ICGC 1989; COMARNA 1993) (Figure 21.1). The Cuban forests range from tropical humid forests, located mainly in the oriental region, to evergreen microphyll coastal forests, to semideciduous forests.

The conditions in these areas have been visibly affected by the modification of the natural habitat caused by intense deforestation, which has occurred throughout Cuba's sociohistorical development (COMARNA 1993). The original vegetation cover in the country has been estimated at between 70% and 80% (95%) (IGACC-ICGC 1989). By 1812, 90% of the original forests still existed. In 1900, a drastic decrease of 54% of the vegeta-

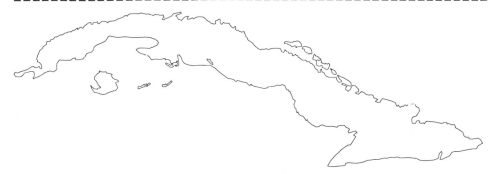

FIGURE 21.1. *Cuban Archipelago.*

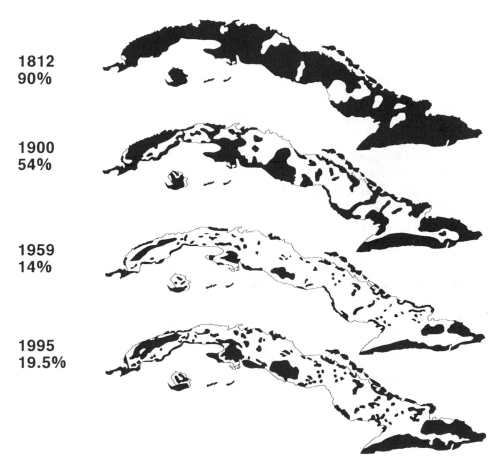

1812
90%

1900
54%

1959
14%

1995
19.5%

FIGURE 21.2. *Changes in forest cover in the Cuban Archipelago (based on data from CIF 1985; IES-CenBio 1995).*

tion cover was observed, due primarily to the intense development of cattle ranching and sugar cane. This dramatic decrease reached its peak in 1959, when only 14% of the area was populated by forests (ICGC 1979). Today the forest cover in the country is estimated to be 19.5% (Figure 21.2).

The areas that conserve the main natural biological resources, with ecosystems and landscapes of high integrity and representativeness, represent 10% of the national territory. They are typically sites that have been maintained at a lower degree of transformation because they are not easily accessible, being located mainly in mountain ranges, swamps, and floodplains. For example, these territories include Cienaga de Zapata (swamp), Sagua-Baracoa and Sierra Maestra (mountain ranges) in the central region, Cordillera de Guamuhaya (mountain range) in the center of the country, and the Cordillera de Guaniguanico (mountain range), as well as coastal areas and island groups that form the Cuban Archipelago (IGACC-ICGC 1989; COMARNA 1993), where the main forests are located.

NATURAL AND ANTHROPOGENIC PERTURBATIONS AFFECTING PRESENT CUBAN FOREST ECOSYSTEMS

One way to accurately understand the structure and function of the tropical forests is through the establishment of medium- and long-term representative or critical areas. This demarcation enables one to identify changes over long cycles or events of low frequency, both autogenic or allogenic in nature, as well as the potential of recovery of these ecosystems facing different types of disturbances.

Sociohistorical Stresses

Perturbations affecting the remaining Cuban ecosystems may be of anthropogenic or natural origin. Among the former are the assimilated effects of the socioeconomic history of the area. Owing to the development of the sugar industry and ranching, the semideciduous forests—once common in the plains of Cuba—have practically disappeared. In their writings, Gomex-Pompa et al (1964) and Gomez-Pompa and Ludlow-Wiechers (1976) point out that an era of secondary vegetation now exists because of the intense transformations of the environment; they consider the value of the secondary plant species that penetrate the jungle, contributing to its heterogeneity, to represent the result of regeneration following the opening of clearings in the canopy.

Natural Meteorological Events

Another major factor in the natural transformation consists of meteorological conditions, especially cyclones (Figure 21.3). In the Caribbean (and especially the Antilles), cyclones frequently perturb ecosystems. The structure, composition, and function of the humid tropical forest are related to the impacts created by these events. According to the classification adopted by the

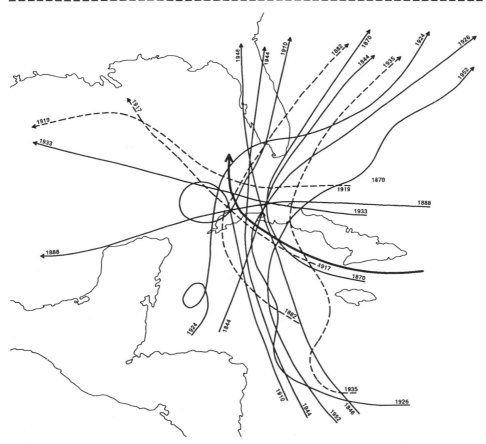

FIGURE 21.3. *Trajectory of hurricanes of great intensity in the western part of Cuba during the months of highest frequency (September and October) from 1844 to 1985. The dashed lines indicate hurricanes that occurred in September, while the solid lines indicate those that occurred in October. The thick arrow in the center of the figure shows the trajectory of cyclone Frederick in September 1979.*

Institute of Meteorology of the Cuban Academy of Sciences (1973), a tropical depression has winds below 35 km/h; a cyclonic disturbance has winds between 35 km/h and 115 km/h, with a well-defined, structured wind circulation; hurricanes occur when the maximum wind velocity exceeds 114 km/h. The hurricanes may be of low intensity (winds as high as 150 km/h), moderate intensity (winds as high as 200 km/h), or high intensity (winds exceeding 200 km/h). Cyclones, hurricanes, and tropical storms are frequent events in the Caribbean region (Tomblin 1981; Scatena and Larsen 1991). Each island is likely to be affected by a hurricane every 15 to 20 years (Scatena and Larsen 1991; Sugden 1992), and the passage of such a storm may exert profound effects on the plants, animals, humans, and landscape in the region (Walker et al 1991).

Vandermeer et al (1990) indicated this type of disturbance had a major influence on the organization of tropical forests. Although the effects of the cyclones on natural ecosystems are recognized, few studies have focused on these effects (Foster 1988).

ECOSYSTEM RESEARCH AND MONITORING

As described briefly above, many natural and anthropogenic perturbations may affect the Cuban forest ecosystems. Understanding and managing the impact of these perturbations are complicated by the fact that the Cuban forest ecosystems are very diverse, ranging from tropical humid rainforests in the east to microphyllous evergreen and semideciduous formations on the coast, corresponding to the general hydroclimatological and ecological conditions. To understand the structure and function of the tropical forests, one can establish medium- and long-term representative or critical areas, permitting the detection of long cyclical changes or rare events of autogenous or allogenous origin. This strategy also enables the researcher to investigate the recuperative capacity of these different ecosystems after different types of perturbations.

Currently, such research must take into account the dynamics, adaptive capacity, fragility, and the spatio-temporal variation of Cuba's ecosystems. Lane (1995) indicates that the health of an ecosystem follows from the relationships between the indicators or direct small-scale measurements, the fundamental or important components such as species or sectors, and estimates of the ecosystem as a whole. The critical properties that define the health of an ecosystem refer to the diversity of its habitats and the reproduction of the organisms, the genotypic and phenotypic diversity, the stocks and cycling of nutrients, energy flow, and the system's capacity to tolerate and attenuate oscillations and harmful effect through mechanisms that restore its equilibrium. The interrelations of the environmental factors with the floristic potential take place on different spatio-temporal scales and conditions than the dynamics and heterogeneity of the forest ecosystem; they also occur at different times than the outcomes.

Using this orientation, studies were conducted to clarify aspects of the health of forest ecosystems in the Sierra del Rosario and the mangroves of the north cape of Matanzas.

The Sierra del Rosario Biosphere Reserve

The establishment of permanent study plots in the Sierra del Rosario Biosphere Reserve is fundamental to assessments undertaken at a national and regional level. It also enables examinations of the changes wrought by

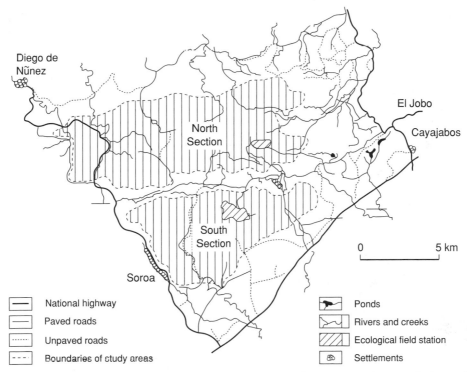

FIGURE 21.4. *Location of the Biosphere Reserve Sierra del Rosario, indicating the research sites, and a climatic diagram of the area.*

cyclones and hurricanes and other perturbations in the forests and contributes to the understanding of functional and adaptive aspects of great importance for the management and conservation of these systems (Figure 21.4).

Methodology. For the study of the diversity and structure of the medium-statured evergreen forest, three plots were established of 400 m² each (20 m × 20 m) and mapped with the location of each tree (Menéndez et al 1988). Two analyses of reciprocal averaging were carried out, taking into account the floristic composition, exposure, and wind protection of each subplot. From 1979 to 1983, researchers recorded the trunk diameters at chest height (DBH) and crowns of the trees 3 m or greater in height and 1 cm diameter (UNESCO 1980). The diameters were correlated with tree height and with annual increase in trunk diameter. Basal areas were calculated from the diameters, and the volume of phytomass was estimated per species (m³/ha) using the formula $r^2h/2$ (according to Brünig et al 1979). The relation of growth to mortality in the forest was calculated from the increments of trunk diameters and the mortality of the recorded trees in each sample.

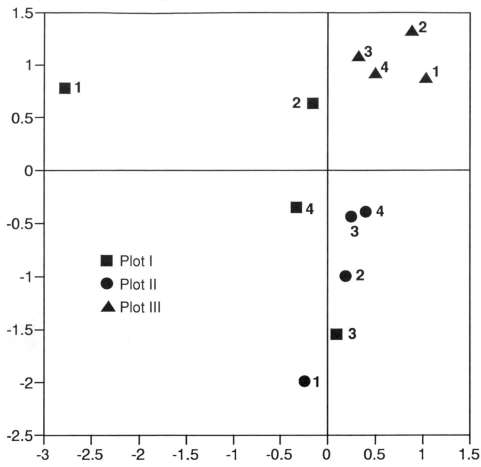

FIGURE 21.5. *Reciprocal averaging analysis according to floristic composition and number of trees per species in the 12 subplots.*

Other aspects investigated included annual production of litter biomass, which was calculated from monthly collection traps located in the plots, and forest regeneration, which was measured through the monitoring of the seedling bank (Menéndez and Vilamajó 1985). Phenological studies were conducted on the tree species to define behavioral patterns so as to understand the functional diversity across the phenotypes that constitute the forest (Vilamajó and Menéndez 1988). From these measurements, the structure and regeneration ability of the forest after disturbances were determined as a measure of the health of these ecosystems (Figure 21.5).

Results. A total of 447 trees were registered and 50 species were identified in the three plots. Among the most abundant of these plants were *Pseudolmedia spuria, Trophis racemosa, Oxandra lanceolata,* and *Matayba apetala.* The remainder of the species presented a lower number of individuals with a rela-

TABLE 21.1   Main tree forest species in the evergreen tropical forest of the Sierra del Rosario identified in the permanent plots

| Species | Plots | | | | |
	I	II	III	Total	%
Pseudolmedia spuria	49	52	41	140	29.4
Trophis racemosa	28	25	30	83	17.4
Oxandra lanceolata	17	39	19	75	15.7
Matayba apelata	9	16	15	40	8.4
Dendropanax arboreus	0	2	11	13	2.7
Hibiscus elatus	3	2	11	16	3.4
Other species (44 species)	33	35	42	110	23.0
Total	**139**	**169**	**169**	**477**	

tively low frequency (Table 21.1). The heterogeneity of the forest became evident through the analysis of the subplots, which showed a separation as a function of the tree species composition and as a function of exposure, topography, and wind protection.

The ordination analysis by reciprocal averaging identified two groups. One was formed by the presence of species of low frequency, explained by the humidity conditions and protection of the sampled sites, which were close to a run-off slope. The second group, in which species were found in less protected sites such as peaks and more exposed slopes, represented a transition from the subgroup found in more protected sites. Each group and subgroup consisted of assemblages of differential tree species. As a result of the differential effect of winds, two continuous floristic assemblages can be recognized as constituting the forest vegetation.

Mortality also differed between plots. It was lower in plot I, which had 8 dead trees, and higher in plot III, which had 32 dead trees (Table 21.2). This separation in groups appears to correspond with variations in forest structure. The distribution of volumes of live biomass and density, considering the different species and plots, shows that in all cases the trees higher than 10 cm DBH have a relatively low density compared with the tree total in the forest. This low density represents most of the live biomass, which indicates the existence of two species groups with particular characteristics of column and density. These results show different growth strategies for these two species groups, classified as tolerant and intolerant to competition (Herrera et al 1988).

TABLE 21.2   Mortality of the trees in the evergreen tropical forest of the Sierra del Rosario

Subplot	Plot I				Plot II				Plot III				Total
	1	2	3	4	1	2	3	4	1	2	3	4	
Trees killed by cyclone Frederick	2			1	1	3	3	1	5	3	4	1	24
Dead trees in 1980						1				1		1	3
Dead trees in 1981	1		1			1	1	1	1		1		7
Dead trees in 1982	2				1	2	2		1	2	1	2	13
Dead trees in 1983		1			1	1		1	3	2	4		13
Total dead trees	**5**	**1**	**1**	**1**	**3**	**8**	**6**	**3**	**10**	**8**	**10**	**4**	**60**

The number of trees by diameter classes (in plots and as a total) decreases in an almost geometric fashion from one diameter range to the next; the same decline appears from the lower size classes (smaller diameters) to the larger ones. An inverse relationship exists between the magnitude of the diameter of one size class and the number of individuals included in this class (Menéndez and Smid 1987). Similar data have been reported for different tropical forests (Golley et al 1969; Mayo 1965; Puig 1979; Proctor et al 1983).

In the Sierra del Rosario, only a few individuals reach diameters larger than 20 cm. Perhaps the topography of this area and the shallow soils influence these characteristics of the forest. It has been observed that the steepest slopes correspond with a larger number of trees having a small diameter.

The vertical pattern of the forest is formed by trees of the most diverse sizes, with the strata often overlapping because of the topography (peaks, slopes, or valleys between mountains). The canopy height does not surpass 20 m. In terms of size classes, tree mortality appears to show the existence of an important selection period, which begins when the trees become part of the class that includes individuals between 5 and 9.9 m in height. The lower mortality among the individuals higher than 10 m indicates the tendency of the forest toward stabilization.

The mean values of the basal areas obtained for the three plots lie in the range reported for manifold tropical forests, although differences have been found between plots. Plot III showed the lowest basal area and its lowest increase (31.3 m²/ha and 0.24 m²/ha, respectively). Plot II, even when it had a greater basal area (41.77 m²/ha), showed a lower increase in basal area (2.17 m²/ha) than plot I (3.1 m²/ha). Plot I showed the highest increase during the study period (0.775 m²/ha/year), and the balance corresponding to the net increase was also the highest, providing evidence of its stability and better health conditions. In plot III, the decreases were higher. The net increases were close to zero in plot II and negative in plot III, showing a high tree mor-

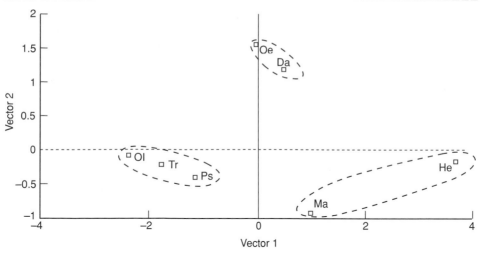

FIGURE 21.6. *Species grouped according to their forest regeneration characteristics. Ps =* Pseudolmedia spuria; *Tr* = Trophis racemosa; *Ol* = Oxandra lanceolata; *Ma* = Matayba apetala; *He* = Hibiscus elatus; *Da* = Dendropanax arboreus; *Oe* = *other species.*

tality in this area, which had greater exposure to wind effects.

An outstanding characteristic of this forest is the absence of emergent trees with crowns completely above the canopy, which have been reported in numerous other tropical forests. The plots found in more protected sites did support a few tree individuals higher than 20 m.

The mean values of crown diameter (CD) divided by diameter at chest height (DBH) for all species in the three plots show that the species tend to both adapt to and resist the effects of the winds. Thus, this forest can be classified as a hurricane or insular type (Odum 1970). In addition to the presence of this type of forest in the Biosphere Reserve Sierra del Rosario, evergreen forests of varying structure can also be found that show a gradation in canopy homogeneity. They range from the homogeneous (in the insular type) to the heterogeneous (with a greater number of the emergents of the continental type). According to the results obtained in the principal component analysis (PCA), at least three species groups that can be distinguished by their structural characteristics and competitive abilities have developed (Figure 21.6).

After cyclone Frederick, 24 trees in the study area died, representing 40% of the dead trees during the study period and 5% of the total of the trees in the three plots in July 1979. This mortality percentage is similar to the one reported for other forests of the Antilles after being struck by hurricanes (Walker 1991; Frangi and Lugo 1991; Bellinghan 1991). The general analysis of the behavior of the forest showed that it remained in a recovery stage in 1993. The distribution of the diameters in the forests and the tree increases by species appeared to be related to the competitive abilities of the species.

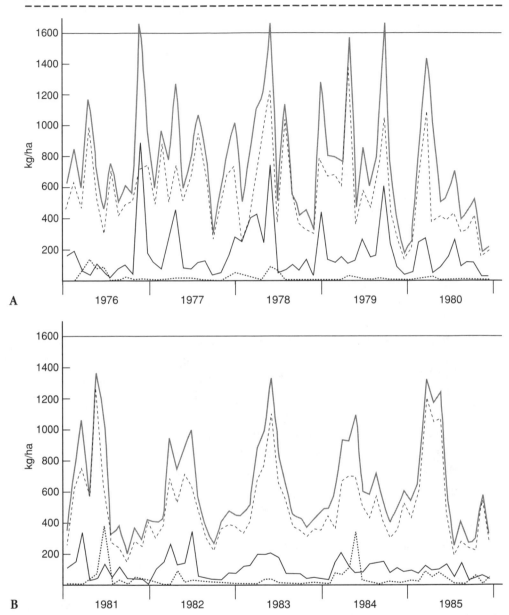

FIGURE 21.7. *Dynamics of the litter biomass of the evergreen tropical forest of the Sierra del Rosario. Note the difference in biomass values before and after cyclone Frederick. The gray line indicates totals; the dashed line indicates leaves; the solid black line indicates stems; the dotted line indicates fruits and flowers.*

After cyclone Frederick struck the island, litter production decreased notably and the annual production rhythms changed. In the reporting period (1976–1985), litter fall occurred in two separate stages: the first spanned the period 1976–1979, being delimited by Frederick, and the second from 1980 to the end of the study (Figure 21.7). The first stage was characterized by high litter fall biomass values. After Frederick hit Cuba, the trend of litter fall (as a

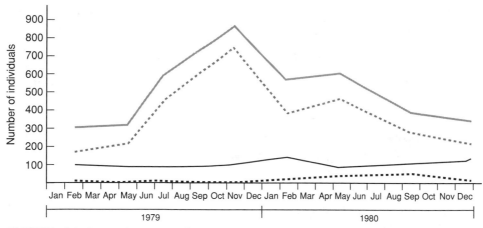

FIGURE 21.8. *Understory seedling dynamics. The gray line indicates totals from all species; the gray dashed line, totals from tree species; the black line, shrubs and herbaceous species; the black dashed line, vine species.*

total or as its components) appeared to be more regular in both areas. In the periods of lowest rainfall, the litter fall biomass increased its production in a stable way. This new behavior of the annual dynamics of litter fall production reflected two trends: one less dependent on rain and more characteristic of a forest not under functional stress (the period before Frederick), and a second characterized by its seasonality, corresponding to a forest subject to functional stress after the loss of green leaves and buds.

The phenological studies corroborated the conditions of the forest of Sierra del Rosario as evergreen, despite the fact that 13% of its tree species are deciduous. The phenomenon of defoliation occurred during the months of February, March, and April—a period of lower rainfall. This phase of higher defoliation occurred simultaneously with the production of young leaves, so that the fall of the leaves during the year was correlated with their aging. A similar behavior was reported by Carabias and Guevara (1985) for an evergreen high-statured forest in Los Tuxtlas, Veracruz. Flowering and fructification were more abundant in the least rainy months, coinciding with the highest absence of leaves.

The study reaffirmed that the ecosystem developed mechanisms to ensure its recovery. It was observed that, after the damage caused by cyclone Frederick, the deciduous tree species did not lose their leaves during the dry period in 1980 and the percentages of flowering and fructification diminished notably. These mechanisms illustrate the ecosystem's vigor and its capacity to maintain its health.

The results obtained in the sampling of density and seedling dynamics in the understory show a considerable change after cyclone Frederick (Figure 21.8). For example, researchers observed a noticeable increase in the seedling populations of the tree species in the forest, especially in *Dendropanax arboreus* (a species that exists but is not abundant in the forest), which accounted for 76% of all seedlings in November 1979.

Conclusions. Therefore, we can say that the evergreen medium-statured forest of the Sierra del Rosario shows floristic groups differentiated as a result of interactions between the topography, exposure, and protection from the winds according to the different regeneration stages. These groups of tree species have developed different structural patterns with respect to the appearance frequency of the species in the plots and subplots, the height and size of the crown of the trees, the distribution in diametric classes, and the relationships of growth and mortality; these patterns provide evidence of the different strategies and competitive abilities of the tree species. The sites least protected from wind action appeared to be more vulnerable to the effects of cyclone Frederick, with a higher proportion of these trees later dying.

The changes occurred in the functioning and dynamics of the forest as a consequence of the cyclonic disturbance. They demonstrate the adaptive capacity of this ecosystem and provide an index of its health, as shown in the following scheme:

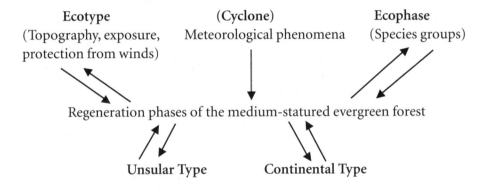

These results undoubtedly point to the fact that natural events such as cyclones are part of the equilibrium of these tropical forests. These events generally do not produce catastrophic effects. Quite the contrary—their occurrence must be understood as a stimulus in the maintenance of heterogeneity and biodiversity.

Mangroves of the Island Group Sabana-Camaguey

The second study case involves the mangroves of the island group Sabana-Camaguey (Alcolado and Menéndez 1993). In numerous capes on the western side of this island group (Figure 21.9), observations on excessive mortality of the red mangrove (*Rhizophora mangle*) began in 1978, raising the attention of government offices and scientific institutions involved with the study of ecology and environmental management.

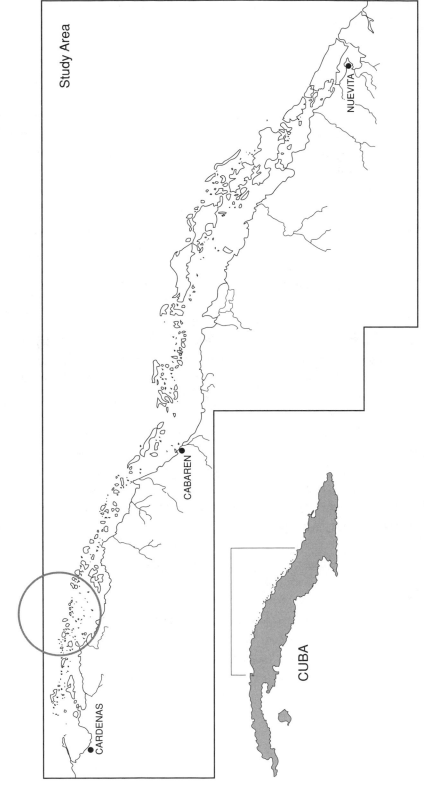

FIGURE 21.9. *Area of higher mortality in the mangroves (in circle) in the island group Sabana-Camaguey.*

Description and Observations. The simultaneous and homogeneous occurrences of large patches of dead and dying trees were noted during inspections that took place between March 1978 and March 1989. The sick trees had small, scarce leaves that tended to be grouped at the apex of the branches. The trunks and branches were a distinct light ash color. Leaves were absent and roots were rotten, fragile, and bad-smelling in the dead mangroves.

Without exception the dead and dying trees were of the permanently flooded type (overwash forest) subject to great tide amplitudes (as high as 1.2 m, according to Rodriguez-Portal and Rodriguez-Ramirez 1983). Mangroves of apparently good health were found immediately next to or in the middle of the dead and dying mangroves. These trees, however, were of the emerged type (on firm land); as such, they were flooded by the tides only periodically.

In all cases, the soil of the dead mangroves was sandy-muddy, having a whitish color but little organic matter. In contrast, the soils of the living mangroves were muddy, compact, and dark, reflecting the presence of humus, abundant litter, and feces of *Capromys* spp. and birds.

Dead mangroves did not appear on the coasts of the main island or on the large capes. These coasts receive nutrients, organic matter, and freshwater from inland even under severe drought conditions, thereby mitigating any stress. On the coasts of the main island, the mangroves are subject to tides of very small amplitude (not higher than 25 cm, according to Rodriguez-Portal and Rodriguez-Ramirez 1983).

In areas of affected mangroves, some of the inferior branches of dead trees remained alive. These branches had aerial roots in the sea, allowing the branches to act independently of the original tree.

The dead and dying mangroves had dense root systems, with the number of roots increasing as nutrients became scarcer (Odum et al 1982). The red mangrove along the river also tended to have scarce short aerial roots, perhaps reflecting the strong competition for nutrients from the soil and water. The decomposition of the masses of roots produced a strong sulfhydric acid smell.

Conclusions. The good health of the marine flora and fauna of the dead mangrove roots ruled out pollution as the causal agent in this phenomenon. Communities of sponges of high species richness (up to 18 species) and of good development were found in the area, as well as anenome, ascidia, gastropods, algae, and other organisms among the roots of the dead mangroves. Research has shown that these sponges can tolerate high salinity.

On the border of an interior lake of very high salinity (99%), located in Cape Corojal Chiquito (the western extreme of the archipelago), red mangroves were observed that had reached a height of 3 m, with slender trunks and few branches but good health. The floor was composed of a thick layer of organic Cyanophyta, humus, and decomposing organic matter that provided a good nutrient source. The roots of the dead or dying trees were infested at variable degrees by piercing larvae of a Lepidoptera. The roots of the living mangroves were barely infected.

We suggest that the massive mortality of the mangroves in the Sabana Archipelago was caused by the unusual and chronic increases in salinity due to drought. The overflooded mangroves could not adapt because the nutrient supply was highly limited. The effect of the Lepidoptera cannot be ignored, however, as the infestation density was higher in the dead and dying mangroves than in the healthy trees.

The main cause of the mangrove mortality cannot be definitively stated, however. The environmental changes weakening the mangroves may have rendered them more susceptible to the butterfly infestation, worsening and accelerating the process; alternatively, this infestation may have occurred independently of the environmental changes. At present, the increasing salinity is considered to be the triggering factor in the process. No massive deaths of trees occurred in wide sections of the Sabana Archipelago where a high exchange of water with the sea occurred, but salinity did not exceed 40%.

Summary

A comparison of the two forest studies shows that the Sierra del Rosario has been characterized by regular restorative processes related to floristic diversity, site differentiation, shifts in biomass production, and phenology. All of these variants are part of the homeostasis of the forest in response to a recurrent perturbation, and a temporary predominance of one or the other phases does not indicate diminished health. On the other hand, the mangrove forest confronted environmental change that was not part of its regular development. It underwent not a brief episode such as a cyclone, but rather a long-term environmental alteration. Species diversity was not available to compensate for the increased mortality, and the insect activity may have behaved as a part of a positive feedback, increasing the perturbation. Only greater root formation may have been an adaptive response. The lack or weakness of homeostatic mechanisms—rather than the magnitude of the changes—serves as an indicator of impaired ecosystem health in this case.

REFERENCES

Academia de Ciencias de Cuba (ACC). Programa científico técnico de turismo. Fundamentación (mecanuscrito). 1993.

Acolado, Menéndez L. Mortalidad masiva de los manglares: un caso en el norte de Cuba. Sinopsis de los manglares y su manejo en los países de Suramérica y la Cuenca del Caribe. Universidad de Miami, FL, 1983:119–129.

los bosques siempreverdes de la Sierra del Rosario. Proyecto MAB no. 1, 1974–1987. 1988:243–260.

Walker LR. Tree damage and recovery from hurricane Hugo in Luquillo Experimental Forest, Puerto Rico. Biotropica 1991;23:379–385.

Walker LR, Lodge, Brokew NVL, Waide RB. An introduction to hurricanes in the Caribbean. Biotropica 1991;23:313–316.

WCMC. Global biodiversity status of Earth's living resources. London: Chapman Hall, 1992.

CHAPTER 22

Environmental Change and Human Health in Honduras

Juan Almendares

Paul R. Epstein

Manuel Sierra

Pamela K. Anderson

HONDURAS is a mountainous nation located in Central America that supports 5.5 million inhabitants. Changes in the environment and human health conditions of this nation over the past two decades have been related to accelerated deforestation, natural disasters, overgrazing, monoculture agriculture, shrimp production, pesticide abuse, military expenditures, burdens brought by external debt, and declines in the country's health and education budgets. In the southern regions, changes in rain patterns and soil fertility have been accompanied by an increase in median ambient temperatures, and storms and natural disasters over the last decade have produced extensive landslides and floods (Almendares et al 1993).

These conditions have exerted a profound impact on biodiversity and exposed flora and fauna to new predators (Almendares et al 1993). Deforestation in central and northern Honduras has affected water basins and water availability, for example. Likewise, climatic and other interacting changes have influenced the distribution of vector-borne diseases. This combination of factors has rendered Honduras extremely vulnerable to new threats, to the point that the country is considered a "critical endangered region"—one where basic life-support systems are threatened (Almendares et al 1993). Other regions similarly designated as critically endangered include Haiti and the Philippines, the basin of Mexico, southeastern Kenya (Ukambani), eastern Borneo, Nepal's middle mountains, China's Ordos Plateau, Amazonia, and the Aral Sea (former USSR) (Kasperson and Kasperson 1995).

DEFORESTATION AND NATURAL DISASTERS

Between 1964 and 1990, Honduras lost 34% (25,899 km^2) of its pine and deciduous forest through logging, military construction, and use of wood as fuel (which represents 60% of the nation's total energy expenditure). Deforestation continues at the rate of 80,000 hectares annually, or 25% of the total forest each decade. This trend has already affected the condition of the water basins in Honduras (Table 22.1) as well as water availability. Currently, 35% of all Honduran housing does not have a water supply and 40% is not equipped with sewage disposal.

Water availability has emerged as a major concern as the major river basins have been cleared (to date, more than 50% of their lengths have suffered this fate). Deforestation reduces land stability and, as in Bangladesh, heavy rains bring first flooding and then landslides.

In 1993, four major natural disasters took place in Honduras. In June, tropical depression Bret struck, followed by hurricane Gert in September. Both of these storms affected the central, southern, northern, eastern, and western areas of the country, damaging banana and sugar cane plantations and harming cereal production. Subsequently, flooding in the Aguan Valley (October 31 to November 1) damaged the city of Tocoa and the palm oil plantations. The cold front (affecting the northern areas of Cortés and Yoro) occurred in November (COPECO 1994). Together, these four disasters resulted in 90,607 injured persons, 245 deaths, 255 missing persons, and an economic loss of approximately $81 million (Table 22.2).

At the end of 1993, an energy crisis developed in Honduras—the result of deforestation, changes in climate, and a failure of a major hydroelectric plant. Between January and August 1994, the available supply of electricity declined by 27.5% following a drop of 44% in hydrological production. Electricity for

TABLE 22.1 Deforestation of the water basins in Honduras

Water River Basins	Deforestation (%)
Choluteca	68
Nacaome	85
Chamelecón	73
Ulua	64
Aguan	76
Sico	69

Source: PNMC, COHDEFOR, BID 1992.

TABLE 22.2   Natural disasters in Honduras, 1993

Loss	Disasters			
	Bret and Gert	Flooding	Cold Front	Total
Injured families	13,639	1967	1863	17,469
Injured persons	67,440	11,990	11,177	90,607
Deaths	27	103*	115	245
Disappeared	12	221*	22	255
Cost (US$ million)	77,917,750	0.101616	3,260,934	81,280,300

* Ella Visser, Flooding in Aguan Valley, Honduras, November 5, 1993. Program for Disaster Costa Rica. PAHO/WHO.
Source: Report of the Permanent Commission for Contingency (COPECO).

domestic and industrial use became limited to 12 hours per day. To solve this problem, thermic generators have been installed; these generators require $29.5 million to finance their diesel consumption (Economic Advisory Team of the Honduras Government, 1994). The higher oil consumption and greater noise pollution by the thermic plants are contributing to an already highly contaminated environment (Advisory and Technical Group of the Honduras Government, 1994).

INCREASES IN INFECTIOUS DISEASE

Water-Borne Infections

Cases of *Vibrio cholerae* have increased significantly in Honduras. This increase could be related to the changes in climate and biodiversity, natural disasters (as discussed earlier), and an increase in severe poverty (PAHO/WHO 1994) (Table 22.3).

TABLE 22.3   Incidence of *Vibrio cholerae* in Honduras

Year	Number of Cases
1991	17
1992	407
1993	4013
1994	5049

Source: Minister of Public Health and PAHO. Health Situation and Their Services, 1994.

Vector-Borne Disease

Malaria. In southern Honduras (Choluteca), the desiccation and soil erosion caused by cattle grazing and intensive sugar cane and cotton cultivation have altered the regional hydrological cycle. The sustained increase in ambient temperature renders this region too hot for anopheline mosquitoes, causing incidence of malaria to decline. On the other hand, semi-desertification has forced people to move to cities, plantations, and assembly plants further north. In other areas, the indiscriminate use of pesticides in the cultivation of bananas, pineapples, and melons has helped anopheline mosquitoes develop widespread resistance to these chemical agents. In 1987, 20,000 cases of malaria were registered. In 1992, 70,000 new cases were reported. By 1993, however, incidence dropped to 53,502 cases.

Chagas Disease. American tripanosomes cause intestinal "megasyndromes" and a cardiomyopathy with arrhythmia that is the major cause of sudden death in rural Latin America. Kissing bugs (*Rhodnius prolixis* and *Triatoma dimitata*) are the chief vectors, and reservoirs of disease include opossums (*Didelphis marsupialis*), armadillos, cats, and dogs. Environmental change has altered all components of the life cycle. Deprived of their habitats, sustenance, and blood meals, reservoirs and insects alike tend to move to the peri-urban "misery belts." In all areas, higher temperatures shorten the generation time, stimulate the organisms to seek blood meals, and increase the number of active metacyclic parasites in insect feces, thus amplifying vector abundance and transmission.

Of 10,601 residents in endemic areas studied in 1992, the Honduran Minister of Health found 2494 (23.5%) were seropositive. Of triatomes sampled in that year in the capital (Tegucigalpa), 45% were infected with *Trypanosoma cruzi*. Blood transfusions have now become a significant means of Chagas and malaria transmission in Honduras (Almendares et al 1993).

Leishmaniasis. The life cycle of the pathogenic *Leishmania* spp. involves sandflies, rodents, and dogs. Intrusion into Honduran forests by road builders and by refugees and troops from Nicaragua, El Salvador, and Guatemala has brought an increase in cutaneous and visceral forms of leishmaniasis.

La Mosca Blanca. The sweet potato whitefly, *Bemicia tabaci*, has been present in Honduras for many years. Until the end of 1980s, it was considered a secondary pest. Since 1989, however, it has been recognized as a serious vector of disease, transmitting geminiviruses in cotton, tomatoes, and beans. In 1991, 80% to 100% of the total cotton produced was infected, with a complete yield loss in some fields. During 1990–1992, yields of tomatoes in the Comayagua Valley declined 70%. Damage from bean golden mosaic geminivirus (BGMV) was also esimated at 70%, with the most serious damage occurring in 1989–1990. Expansion of traditional crops and introduction of "nontraditional" exports have also led to increased whitefly population. Outbreaks of this pest have been further exacerbated by the occurrence of more frequent and severe droughts. At the same time, intensive and accelerating use of pesticides has consistently reduced populations of whitefly predators.

CLINICAL IMPACTS

The common bean (*Phaseolus vulgaris*) is the principal element in the diet of poor residents of Latin America. More than 60% of Honduran children younger than 10 years have at least grade 1 malnutrition (according to the Gomez classification). The loss of staple crops to vector-borne plant pathogens further stunts these individuals' growth and development, reduces their immunity, and increases the burden of communicable illness.

Dengue fever is increasing as *Aedes* sp. breeding sites swell in peri-urban areas. The advance of *Aedes aegypti* (another vector whose maturation and generation are accelerated by warm temperatures) and the spread of the cold-hardy *Aedes albopictus* pose an increasing concern throughout Latin America, and yellow fever is already in resurgence.

CONCLUSIONS

Honduras has undergone a number of sustained stresses simultaneously, making it extremely fragile and rendering it vulnerable to further stress. The tropical ecosystem in the country is suffering an accelerated destruction because of economic and social policies that favor environmental changes that contribute to loss of biodiversity. The same changes also promote increases in diseases transmitted by insect vectors and water.

REFERENCES

Advisory and Technical Group of the Honduras Government. The impact of electrical energy crisis in the economy. Honduras: October 1994.

Almendares J, Sierra M, Anderson PK, Epstein P. Critical regions, a profile of Honduras. Lancet 1993;342:1400–1402.

COPECO. Statistical information of the natural disasters of Honduras 1990–1994. Report of the Permanent Commission for Contingency. Honduras: 1994.

Kasperson JX, Kasperson RE. Global environmental risk. Tokyo: United Nations University, 1995.

PAHO/WHO. Final report of the Joint Evaluation of the Cooperation. Minister of Health of Honduras. November 1994.

CHAPTER 23

Who Framed the Kyrönjoki?

Mikael Hildén

SOME ecosystems obviously suffer from an ecosystem distress syndrome. The Kyrönjoki and its estuary on the Ostrobothnian coast in the catchment of the Gulf of Bothnia, the Baltic Sea, has been and partly remains one of these systems. It takes no expertise to discover that something is wrong in a river in which mass kills of fish have occurred, in which the water occasionally turns sterile blue instead of its normal dark brown, and whose valued resources of sea trout, whitefish (Coregonids), lamprey, crayfish, and burbot have diminished (Hildén and Rapport 1993).

We can describe the distress of ecosystems in ever more sophisticated ways using additional variables and details. A description of the obvious is, however, not particularly important. Instead we seek to understand the systems and the causes of their present condition so as to plan and introduce remedies. This goal forces us to shift our attention from describing the condition to searching for culprits, by using available data and knowledge and occasionally by conducting experiments to collect additional evidence or rule out options. "Culprits" here are meant in a broad sense, rather than a strictly legal sense.

THE FACTS

The Kyrönjoki is a medium-size Finnish river with a catchment of approximately 5000 km² and a mean annual discharge of 43 m³s⁻¹. The scarcity of lakes leads to large seasonal fluctuations in its level, with mean annual minimal and maximal discharges of 4.8 m³s⁻¹ and 329 m³s⁻¹, respectively. The catchment has largely risen and still rises from the sea at a rate of 0.7 to 0.8 m per century. From about 7000 to 12,000 B.P., sulfate-rich sediments were deposited in Finland's coastal waters. In today's Kyrönjoki catchment, these sediments are found mostly in areas less than 50 m above sea level and have become sulfide-bearing clay soils that cover 7% to 8% of the catchment. Because of the low gradient, these soils are found at distances of more than 100 km from the present river mouth. When they are oxidized, their drainage water can be highly acidic (pH values of 3–4). The drainage water also contains high levels of toxic metals, such as aluminum.

TABLE 23.1   The subareas of the Kyrönjoki catchment

Subarea	Height Above Sea Level (m) and Area	Dominant Land-Uses	Characteristics of the River	Present Water Quality
Archipelago	0; the influence of the river water is felt for 5–25 km	Forestry, small-scale agriculture	Dominance of river water in surface waters	Occasional humic or acidic pulses
Estuary	0; ⁻10,000 ha	Extensive agricultural lands	Slow flow, dykes along river, extensive bays with river water	Low buffering capacity, permanently acidified areas, acidic pulses, high levels of nutrients and organic substances
Lower reach	5–20 m; 68,744 ha	Extensive agricultural lands, forestry, villages along river	Several rapids, long reaches of slow flow	Low buffering capacity, some tributaries very acidic, high levels of nutrients and organic substances
Middle reach	20–45 m; 116,553 ha	Extensive agricultural lands, forestry, the town of Seinäjoki	Long reaches of slow flow, dykes along parts of the river	Steadily decreasing buffering capacity, high levels of nutrients and organic substances
Upper reach	>45 m; 298,031 ha	Forestry, agriculture in river valleys, peat extraction	Many diverse streams, brooks and rivers with flows that vary from slow to rapid; several small reservoirs	Varying, humic, not seriously acidic; in certain areas high nutrient levels and suspended solids

The river and its estuary can be classified into subareas with characteristic environmental conditions, water quality, and land-use (Table 23.1). The upper reaches consist of many tributaries, partly with rapid flow. Some tributaries originate in peat areas, whereas others are fed by springs in sandy moraine areas. Although forests and forestry dominate the land around the smaller tributaries, agriculture remains important in the larger river valleys.

TABLE 23.2   The observed symptoms

Subarea	Symptom	Time Period
Archipelago	Decline of salmonid fishery	19th century
	"Dirty" water	1960s–1970s
	Abnormal deaths of fish in passive gear	1970s–1980s
	Loss of fishing places for passive gear	1960s–1970s
	Small catches of some valued fish	1960s–1990s
Estuary	Decline of salmonid and coregonid fishery	19th–early 20th century
	Decline of lamprey fishery	
	Fish kills	1960s–1970s
	Decline of fish catches	1970s–1990s
	"Dirty" water	1960s–1970s
	Extensive algal growth	1960s–1990s
	Disappearance of mollusks	1960s(?)–1970s
	Areas of transparent greenish water	1960s–
Lower reach	Decline of salmonid and coregonid fishery	19th–early 20th
	Decline of lamprey fishery	century
	Fish kills	Earliest recorded in 19th century, 20th century: 1960s–
	Loss of crayfish	1960s(?)–1970s
	Decline of fish catches	1960s–
	Loss of recreational fishing opportunities	1960s–1980s
	Quality problems in drinking water supply	1960s–
Middle reach	Decline of salmonid fishery	19th–early 20th century
	Loss of crayfish	1970s–
	Decline of fish catches	1960s(?)–
	Loss of recreational fishing opportunities	1960s–1980s
	Poor water quality	1960s–
Upper reach	Loss of crayfish	1960s–1980s
	Loss of brook trout and grayling	1960s(?)–1980s
	Loss of recreational fishing opportunities	1960s–1980s
	Poor water quality in certain tributaries (high level of suspended solids, organics, nutrients) and in reservoirs (low oxygen during winter)	1960s–1990s
	Mercury in reservoir fish	1970s–1990s

The middle reach is characterized by a mostly slow-flowing river that moves through vast agricultural areas. Forests are found in the catchments of smaller tributaries. The lower reach is also dominated by the slow-flowing river and extensive agricultural lands. In the estuary and archipelago, forestry and small-scale agriculture are important land-use practices.

From an environmental point of view, the Kyrönjoki reached its worst state in the 1960s and 1970s. Although the record of symptoms of an ecosystem distress syndrome is not complete, even a partial list of woes convincingly shows that many functions of river systems were impaired in the Kyrönjoki (Table 23.2). In some cases, the immediate cause of an adverse change is known. We know, for example, that the acidification of the river water—possibly in combination with high concentrations of toxic metals—caused the fish kills. We also know that high levels of nutrients lead to intense primary production. This knowledge is only a starting point, however, in our search for culprits that caused the degradation of the Kyrönjoki system.

THE SUSPECTS

A simple solution would be to blame all adverse changes on human activities. Communities have developed along the river since the earliest settlements; today, more than 200,000 people live in the catchment or its immediate vicinity. Assigning a general blame does not help in rehabilitation, however, as it does not clarify the processes involved in the degradation. It also gives us no basis for altering existing activities in a direction that would prove less harmful for the river.

We can find the suspects by examining activities and processes in the catchment (Table 23.3). For each of these features, some circumstantial evidence ties the activity to adverse changes in the river system. Because uncertain circumstantial evidence is not always sufficient to justify action for the improvement of the situation, an important question is therefore the standard of proof that we require to make conclusions about causes and effects.

An examination of the available evidence does not completely exclude any of the suspects from further examination. The observations that adverse changes can be observed throughout the system and that, for example, nutrient levels have not decreased dramatically despite investments in waste-water treatment suggest, however, that further inquiries should focus on non–point-sources and on activities and processes with a wide geographical distribution. In addition, the temporal information suggests that the degradation did not proceed smoothly, but rather coincided with intensified use of the Kyrönjoki catchment. This finding does not completely rule out natural

TABLE 23.3   Possible culprits and the circumstantial evidence

Activity	Evidence
Natural processes	8% of the soils in the catchment are sulfate soils with high levels of sulfuric compounds; changing groundwater level oxidizes and leaches acidifying substances. Historical documents show that the rising of land in combination with siltation forces fishing places further offshore.
Agriculture	Altogether one-third of the catchment is agricultural land, but agricultural land covers as much as 60% of the catchment in the areas of accentuated degradation. The sulfate soils with very acidic drainage water constitute 22% of the agricultural land. Intense drainage activities in 1960s–1980s. Use of fertilizers has increased in agriculture, especially since the 1960s. In the nineteenth century, lakes were lowered and 15 lakes were drained completely to increase agricultural land. Agriculture has demanded flood protection, which has led to the clearing of rapids, building of reservoirs, and construction of dykes and drainage pumps, many of which discharge acidic water (1960s–1970s).
Forestry	Half the catchment is forest land, and most of it is used for timber production. Forestry has introduced draining of forest land. Meandering brooks have been converted to straight drains and rapids have been cleared, especially in the 1960s and 1970s. Drainage of forest land in sandy areas has increased suspended solids and siltation. Drainage of peatlands has increased loads of nutrients and organics, especially since the 1960s. Clear-cuts have increased nutrient loads. From the eighteenth to mid-twentieth century, extensive timber rafting and clearing of rapids have been carried out.
Peat production	Peatlands represent approximately 26% of the catchment. In some subareas, large portions of the peatland are used for peat production: the discharge water is humic and has elevated nutrient levels. Peat production has increased since the 1970s. Humic discharge water can be acidic, but not as acidic as the sulfate soil waters.
Industries and municipalities (point-source polluters)	Poor wastewater treatment along the river until 1970s. Discharges of nutrients and organics. The ratio of wastewater to total flow is high in summer, especially in upper parts of the middle reach.
Water development	Although its origins lie in the sixteenth century, water development has intensified in the twentieth century with the building of large dams and building of reservoirs in the 1960s and 1970s. Large-scale development and regulation of flow for flood protection and power production have occurred since the 1960s.

processes, but makes them an unlikely single explanation for the observed changes.

We are left with forestry, peat production, agriculture, and water development, possibly in combination with natural processes, as major candidates for the cause of the adverse changes. One way to approach identification of the true culprit is to combine the information of Tables 23.1 and 23.3. A pooling of these data could specify the role played by each activity and process in degrading the Kyrönjoki for different subareas.

THE OPTIONS

When we have narrowed down the problem areas, we can consider options for rehabilitation. In the Kyrönjoki, this work was initiated in 1995 when environmental authorities, municipalities, and other authorities and organizations formed a commission intended to manage the entire catchment.

A basic choice must be made between alleviating symptoms and attacking causes of the problems. For the Kyrönjoki, the actions that aim at reducing symptoms of degradation include the following: liming of the land or water, stocking of fish and crayfish, aeration of reservoirs, and improved purification of water supply. To be effective, these measures must be continuous. They offer an advantage in that they are easy to adjust and can be abandoned easily if monitoring suggests that they are inefficient.

Measures intended to solve the problems of the Kyrönjoki can be classified into two groups. The first group of options would impose (reversible) conditions on activities in the catchment. Examples of these tactics include reduction of fertilization, changes in water-level regulation at power stations, and reduction in the size of clear-cuts in forestry. The second group consists of more permanent measures, which are meaningful only in a longer time perspective and are more difficult to reverse. Examples include the rehabilitation of rapids, the establishment of protected banks and areas around fields to reduce nutrient leakage, the protection of particular biotopes (notably brooks and small streams) in forestry, and the rehabilitation of exploited peatlands.

THE DECISION PROBLEMS

Anyone attempting to rehabilitate a distressed system for which less than perfect knowledge exists faces several decision problems. First, what are the objectives of the rehabilitation? It is unlikely that a "pristine" state can be recovered. In fact, for systems with a long history of human intervention, the whole concept is fuzzy. For example, in the Kyrönjoki the vast fields with their barns in the river valleys are considered a valuable cultural heritage, but they represent only the last few hundred years of the river's history. Objectives must therefore be specific to provide a meaningful reference for the rehabilitation.

Second, what are the possible consequences of the rehabilitation measures? It is not sufficient to specify the type of measure (removing symptoms versus solving underlying problems; continuous reversible measures versus permanent changes in the system). To facilitate the choice between alternatives, possible consequences are analyzed in different dimensions. For example, one might examine the following issues:

- *Ecological:* Under what circumstances is it possible to recreate a self-reproducing population of crayfish in the river?

- *Economic:* What are the costs of liming?

- *Sociocultural:* How will the change of status of the river affect local communities?

These dimensions reflect the subjective nature of the objectives for rehabilitation.

Third, what are the probabilities of the different consequences occurring? Some consequences are straightforward and their likelihood can be specified. For example, the cost of liming the river water is a function of the discharge, the acidity of the water, and the target pH level. Available monitoring data provide probability distributions for the occurrence of different combinations of acidity and discharge. For other consequences, the probabilities are more difficult to specify. The likelihood that a restored rapid will harbor a self-reproducing, exploitable stock of crayfish cannot be specified with a high degree of accuracy in the Kyrönjoki; this series of events depends on the technical structure, the fluctuations of the water quality, the occurrence of crayfish disease, and the development of exploitation practices.

In ecosystem rehabilitation, the decision problems are clearly multidimensional. Furthermore, the decisions are made by heterogeneous groups of stakeholders who perceive the system differently. The catchment of the Kyrönjoki stretches over 21 municipalities, for example, and many other authorities and private actors are involved in decisions concerning its future. To facilitate discussions among these diverse groups, the problems must be presented in a clear structure that highlights the possible consequences. A suggestion is given in Table 23.4.

THE FUTURE

The Commission for the Kyrönjoki, which was founded in 1995, has already begun working to rehabilitate the river. In 1996, plans for special fund-raising were initiated. In addition, some voluntary changes in agriculture and

TABLE 23.4 A possible structure for the rehabilitation problems in the Kyrönjoki using liming as an example

Ecosystem Problem	Possible Measures	Ecological Consequences	Sociocultural Consequences	Economic Consequences
Low pH values of river water in the lower reach, especially in spring and autumn	Liming the river water Liming the underground drainage system for fields	Improved reproduction of fish in river and estuary Rehabilitation of lamprey stock Possibilities for return of mollusk and other pH-sensitive invertebrates Increased primary production due to interaction with high nutrient levels?	New possibilities for river fishing Possible competition between coastal and river fishing for salmonids Possibilities for fishing tourism Change of agricultural practices	*Costs:* River water liming Renewal of underground drains *Benefits:* Development of river fishing and associated activities Savings in water treatment for water works
		Risks: Occasional extreme acidification cannot be managed and kills organisms	*Risks:* Intense coastal fishery does not give river fishery a chance to recover Restrictions reduce positive consequences	*Risks:* Financing of continuous liming uncertain Change of underground drains requires long-term commitment Measures may be insufficient to solve problem because of the large amount of acidifying substances

forestry practices that can reduce the load on the Kyrönjoki have been implemented. The future, however, is open-ended.

REFERENCE

Hildén M, Rapport D. Four centuries of cumulative impacts on a Finnish river and its estuary: an ecosystem health-approach. J Aquat Ecosys Health 1993;2:261–275.

INDEX

Note: Page numbers followed by an *f* indicate figures; those followed by a *t* indicate tables.